# Audel™
# Pumps and Hydraulics

**All New 6th Edition**

## Rex Miller
## Mark Richard Miller
## Harry Stewart

WILEY

Wiley Publishing, Inc.

Vice President and Executive Group Publisher: Richard Swadley
Vice President and Publisher: Joseph B. Wikert
Executive Editor: Carol A. Long
Editorial Manager: Kathryn A. Malm
Development Editor: Kevin Shafer
Production Editor: Vincent Kunkemueller
Text Design & Composition: TechBooks

*Library of Congress Control Number:*

ISBN: 0-7645-7116-8

Printed in the United States of America

SKY10070411_032124

# Contents

# Acknowledgments

No book can be written without the aid of many people. It takes a great number of individuals to put together the information available about any particular technical field into a book. The field of pumps and hydraulics is no exception. Many firms have contributed information, illustrations, and analysis of the book.

The authors would like to thank every person involved for his or her contributions. Following are some of the firms that supplied technical information and illustrations.

Abex Corp., Denison Division
ABS
The Aldrich Pump Company, Standard Pump Div.
Becker Pumps
Brown and Sharpe Mfg. Co.
Buffalo Forge Company
Buffalo Pumps
Caterpillar Tractor Co.
Commercial Shearing Inc.
Continental Hydraulics
Deming Division, Crane Co.
Double A Products Co.
Gold Pumps
Gould Pumps
Hydreco, Div. of General Signal
Imperial-Eastern Corp.
Logansport Machine Co., Inc.
Lynair, Inc.
Marvel Engineering Co.
Mobile Aerial Towers, Inc.
Oilgear Company
Parker-Hannifin Corp.
Pathon Manufacturing Company, Div. of Parker-Hannifin Corp.
Pleuger Submersible Pumps, Inc.
Rexnard, Inc., Hydraulic Component Div.
Roper Pump Compan
Schrader Div., Scovil Mfg. Co.
Sherwood

Snap-Tite, Inc.
Sperry Vickers, Division of Sperry Rand Corp.
Sunstrand Hydro-Transmission, Div. of Sundstrand Corp.
Superior Hydraulics, Div. of Superior Pipe Specialties
TAT Engineering
Viking Pump Division
The Weatherhead Co.

# About the Authors

**Rex Miller** was a Professor of Industrial Technology at The State University of New York, College at Buffalo for more than 35 years. He has taught at the technical school, high school, and college level for more than 40 years. He is the author or co-author of more than 100 textbooks ranging from electronics through carpentry and sheet metal work. He has contributed more than 50 magazine articles over the years to technical publications. He is also the author of seven civil war regimental histories.

**Mark Richard Miller** finished his BS in New York and moved on to Ball State University, where he earned a master's degree, then went to work in San Antonio. He taught high school and finished his doctorate in College Station, Texas. He took a position at Texas A&M University in Kingsville, Texas, where he now teaches in the Industrial Technology Department as a Professor and Department Chairman. He has co-authored 11 books and contributed many articles to technical magazines. His hobbies include refinishing a 1970 Plymouth Super Bird and a 1971 Road-runner.

**Harry L. Stewart** was a professional engineer and is the author of numerous books for the trades covering pumps, hydraulics, pneumatics, and fluid power.

# Introduction

The purpose of this book is to provide a better understanding of the fundamentals and operating principles of pumps, pump controls, and hydraulics. A thorough knowledge of pumps has become more important, due to the large number of applications of pump equipment in industry.

The applied principles and practical features of pumps and hydraulics are discussed in detail. Various installations, operations, and maintenance procedures are also covered. The information contained will be of help to engineering students, junior engineers and designers, installation and maintenance technicians, shop mechanics, and others who are interested in technical education and self-advancement.

The correct servicing methods are of the utmost importance to the service technician, since time and money can be lost when repeated repairs are required. With the aid of this book, you should be able to install and service pumps for nearly any application.

The authors would like to thank those manufacturers that provided illustrations, technical information, and constructive criticism. Special thanks to TAT Engineering and Sherwood Pumps.

# Part I

# Introduction to Basic Principles of Pumps and Hydraulics

# Chapter 1

## Basic Fluid Principles

Pumps are devices that expend energy to raise, transport, or compress fluids. The earliest pumps were made for raising water. These are known today as *Persian* and *Roman waterwheels* and the more sophisticated *Archimedes screw*.

Mining operations of the Middle Ages led to development of the *suction* or *piston pump*. There are many types of suction pumps. They were described by Georgius Agricola in his *De re Metallica* written in 1556 A.D. A suction pump works by atmospheric pressure. That means when the piston is raised, it creates a partial vacuum. The outside atmospheric pressure then forces water into the cylinder. From there, it is permitted to escape by way of an outlet valve. Atmospheric pressure alone can force water to a maximum height of about 34 feet (10 meters). So, the force pump was developed to drain deeper mines. The downward stroke of the force pump forces water out through a side valve. The height raised depends on the force applied to the piston.

Fluid is employed in a closed system as a medium to cause motion, either linear or rotary. Because of improvements in seals, materials, and machining techniques, the use of fluids to control motions has greatly increased in the recent past.

Fluid can be either in a liquid or gaseous state. Air, oil, water, oxygen, and nitrogen are examples of fluids. They can all be pumped by today's highly improved devices.

### Physics

A branch of science that deals with matter and energy and their interactions in the field of mechanics, electricity, nuclear phenomena, and others is called *physics*. Some of the basic principles of fluids must be studied before subsequent chapters in this book can be understood properly.

### Matter

Matter can be defined as anything that occupies space, and all matter has inertia. Inertia is that property of matter by which it will remain at rest or in uniform motion in the same straight line or direction unless acted upon by some external force. *Matter* is any substance that can be weighed or measured. Matter may exist in one of three states:

- Solid (coal, iron, ice)
- Liquid (oil, alcohol, water)
- Gas (air, hydrogen, helium)

Water is the familiar example of a substance that exists in each of the three states of matter (see Figure 1-1) as ice (solid), water (liquid), and steam (gas).

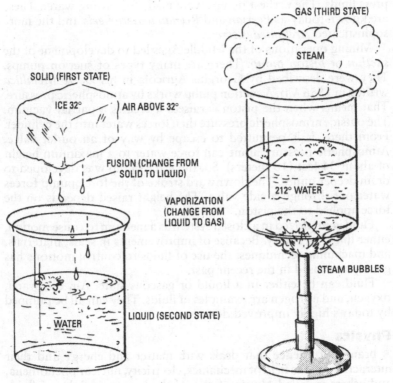

**Figure 1-1** The three states of matter: solid, liquid, and gas. Note that the change of state from a solid to a liquid is called fusion, and the change of state from liquid to a gas is called vaporization.

## Body
A body is a mass of matter that has a definite quantity. For example, a mass of iron 3 inches × 3 inches × 3 inches has a definite quantity of 27 cubic inches. It also has a definite weight. This weight can be determined by placing the body on a scale (either a lever or platform scale or a spring scale). If an accurate weight is required, a lever or platform scale should be employed. Since weight depends on gravity, and since gravity decreases with elevation, the reading on a spring scale varies, as shown in Figure 1-2.

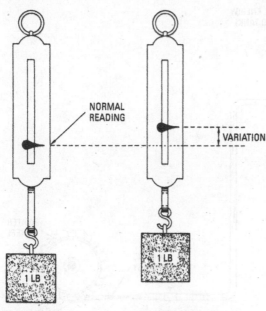

(A) At sea level.          (B) At higher elevation.

**Figure 1-2** Variation in readings of a spring scale for different elevations.

## Energy

Energy is the capacity for doing work and overcoming resistance. Two types of energy are *potential* and *kinetic* (see Figure 1-3).

Potential energy is the energy that a body has because of its relative position. For example, if a ball of steel is suspended by a chain, the position of the ball is such that if the chain is cut, work can be done by the ball.

Kinetic energy is energy that a body has when it is moving with some velocity. An example would be a steel ball rolling down an incline. Energy is expressed in the same units as work (foot-pounds).

As shown in Figure 1-3, water stored in an elevated reservoir or tank represents potential energy, because it may be used to do work as it is liberated to a lower elevation.

### Conservation of Energy

It is a principle of physics that energy can be transmitted from one body to another (or transformed) in its manifestations, but energy may be neither created nor destroyed. Energy may be dissipated.

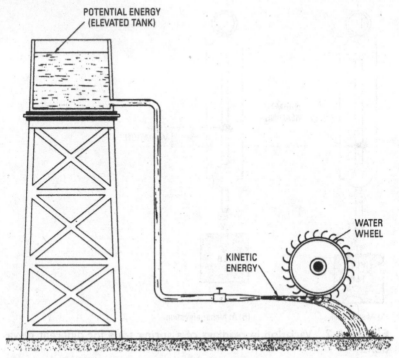

**Figure 1-3** Potential energy and kinetic energy.

That is, it may be converted into a form from which it cannot be recovered (the heat that escapes with the exhaust from a locomotive, for example, or the condensed water from a steamship). However, the total amount of energy in the universe remains constant, but variable in form.

### Joule's Experiment

This experiment is a classic illustration (see Figure 1-4) of the conservation of energy principle. In 1843, Dr. Joule of Manchester, England, performed his classic experiment that demonstrated to the world the mechanical equivalent of heat. It was discovered that the work performed by the descending weight ($W$ in Figure 1-4) was not lost, but appeared as heat in the water—the agitation of the paddles having increased the water temperature by an amount that can be measured by a thermometer. According to Joule's experiment, when 772 foot-pounds of work energy had been expended on the 1 pound of water, the temperature of the water had increased 1°F.

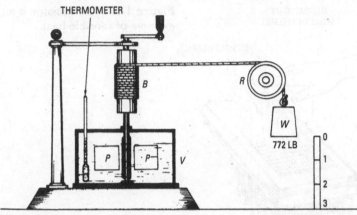

**Figure 1-4** Joule's experiment revealed the mechanical equivalent of heat.

This is known as *Joule's equivalent*: That is, 1 unit of heat equals 772 foot-pounds (ft-lb) of work. (It is generally accepted today that ft-lb. be changed to lb.ft. in the meantime or transistion period you will find it as ft-lb. or lb.ft.)

Experiments by Prof. Rowland (1880) and others provide higher values. A value of 778 ft-lb is generally accepted, but 777.5 ft-lb is probably more nearly correct, the value 777.52 ft-lb being used by Marks and Davis in their steam tables. The value 778 ft-lb is sufficiently accurate for most calculations.

## Heat

Heat is a form of energy that is known by its effects. The effect of heat is produced by the accelerated vibration of molecules. Theoretically, all molecular vibration stops at –273°C (known as absolute zero), and there is no heat formed. The two types of heat are *sensible* heat and *latent* heat.

### Sensible Heat

The effect of this form of heat is indicated by the sense of touch or feeling (see Figure 1-5).

Sensible heat is measured by a thermometer. A thermometer is an instrument used to measure the temperature of gases, solids, and liquids. The three most common types of thermometers are *liquid-in-glass*, *electrical*, and *deformation*.

The liquid-in-glass generally employs mercury as the liquid unless the temperature should drop below the freezing point of mercury,

INDICATED BY
SENSE OF FEELING

**Figure 1-5** The radiator is an example of sensible heat.

SENSIBLE HEAT

RADIATOR

in which case alcohol is used. The liquid-in-glass is relatively inexpensive, easy to read, reliable, and requires no maintenance. The thermometer consists of a glass tube with a small uniform bore that has a bulb at the bottom and a sealed end at the top. The bulb and part of the tube are filled with liquid. As the temperature rises, the liquid in the bulb and tube expand and the liquid rises in the tube. When the liquid in the thermometer reaches the same temperature as the temperature outside of the thermometer, the liquid ceases to rise.

In 1714, Gabriel Daniel Fahrenheit built a mercury thermometer of the type now commonly in use.

Electrical thermometers are of the more sophisticated type. A *thermocouple* is a good example. This thermometer measures temperatures by measuring the small voltage that exists at the junction of two dissimilar metals. Electrical thermometers are made that can measure temperatures up to 1500°C.

*Deformation thermometers* use the principle that liquids increase in volume and solids increase in length as temperatures rise. The *Bourdon tube thermometer* is a deformation thermometer.

Extremely high temperatures are measured by a *pyrometer*. One type of pyrometer matches the color (such as that of the inside of a furnace) against known temperatures of red-hot wires.

Figure 1-6 shows the Fahrenheit, Celsius, and Reaumur thermometer scales. Figure 1-7 illustrates the basic principle of a thermocouple pyrometer.

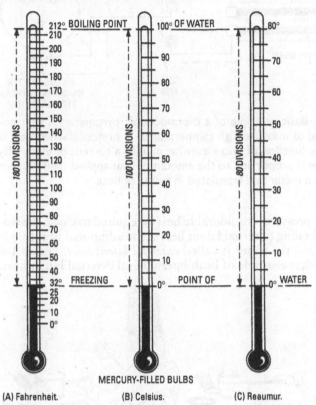

MERCURY-FILLED BULBS

(A) Fahrenheit.  (B) Celsius.  (C) Reaumur.

**Figure 1-6** Three types of thermometer scales.

**Latent Heat**

This form of heat is the quantity of heat that becomes concealed or hidden inside a body while producing some change in the body other than an increase in temperature.

When water at atmospheric pressure is heated to 212°F, a further increase in temperature does not occur, even though the supply of heat is continued. Instead of an increase in temperature, vaporization occurs, and a considerable quantity of heat must be added to the liquid to transform it into steam. The total heat consists of *internal* and *external* latent heats. Thus, in water at 212°F and at

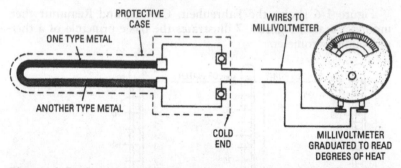

**Figure 1-7** Basic principle of a thermocouple pyrometer. A thermocouple is used to measure high temperatures. In principle, when heat is applied to the junction of two dissimilar metals, a current of electricity begins to flow in proportion to the amount of heat applied. This current is brought to a meter and translated in terms of heat.

atmospheric pressure, considerable heat is required to cause the water to begin boiling (internal latent heat). The additional heat that is required to boil the water is called *external latent heat*. Figure 1-8 shows a familiar example of both internal and external latent heat.

(A) Internal latent heat—waiting for the pot to boil.

(B) External latent heat—when the pot boils.

**Figure 1-8** Domestic setting for illustrating internal (left) and external (right) latent heat.

## Unit of Heat
The *heat unit* is the amount of heat required to raise the temperature of 1 pound of water 1°F at the maximum density of the water. The *British thermal unit* (abbreviated Btu) is the standard for heat measure. A unit of heat (Btu) is equal to 252 calories, which is the

quantity of heat required to raise the temperature of 1 pound of water from 62°F to 63°F.

Assuming no loss of heat, 180 Btu are required to raise the temperature of 1 pound of water from 32°F to 212°F. If the transfer of heat occurs at a uniform rate and if six minutes are required to increase the temperature of the water from 32°F to 212°F, 1 Btu is transferred to the water in (6 × 60) ÷ 180, or 2 seconds.

## Specific Heat

This is the ratio of the number of Btu required to raise the temperature of a substance 1°F to the number of Btu required to raise the temperature of an equal amount of water 1°F. Some substances can be heated more quickly than other substances. Metal, for example, can be heated more quickly than glass, wood, or air. If a given substance requires one-tenth the amount of heat to bring it to a given temperature than is required for an equal weight of water, the number of Btu required is $1/10$ (0.1), and its specific heat is $1/10$ (0.1).

## Example

The quantity of heat required to raise the temperature of 1 pound of water 1°F is equal to the quantity of heat required to raise the temperature of 8.4 pounds of cast iron 1°F. Since the specific heat of water is 1.0, the specific heat of cast iron is 0.1189 (1.0 ÷ 8.4).

Thus, the specific heat is the ratio between the two quantities of heat. Table 1-1 shows the specific heat of some common substances.

## Transfer of Heat

Heat may be transferred from one body to another that is at a lower temperature (see Figure 1-9) by the following:

- Radiation
- Conduction
- Convection

When heat is transmitted by radiation, the hot material (such as burning fuel) sets up waves in the air. In a boiler-type furnace, the heat is given off by *radiation* (the heat rays radiating in straight lines in all directions). The heat is transferred to the crown sheet and the sides of the furnace by means of radiation.

Contrary to popular belief that heat is transferred through solids by radiation; heat is transferred through solids (such as a boiler-plate) by conduction (see Figure 1-10). The temperature of the furnace boilerplate is only slightly higher than the temperature of the water that is in contact with the boilerplate. This is because of the extremely high conductivity of the plate.

*Conduction* of heat is the process of transferring heat from molecule to molecule. If one end of a metal rod is held in a flame and

**Table 1-1   Specific Heat of Common Substances**

| Solids | | Liquids | |
| --- | --- | --- | --- |
| Copper | 0.0951 | Water | 1.0000 |
| Wrought Iron | 0.1138 | Sulfuric acid | 0.3350 |
| Glass | 0.1937 | Mercury | 0.0333 |
| Cast Iron | 0.1298 | Alcohol | 0.7000 |
| Lead | 0.0314 | Benzene | 0.9500 |
| Tin | 0.0562 | Ether | 0.5034 |
| Steel, Hard | 0.1175 | | |
| Steel, Soft | 0.1165 | | |
| Brass | 0.0939 | | |
| Ice | 0.5040 | | |

| Gases | | |
| --- | --- | --- |
| Type | At Constant Pressure | At Constant Volume |
| Air | 0.23751 | 0.16847 |
| Oxygen | 0.21751 | 0.15507 |
| Hydrogen | 3.40900 | 2.41226 |
| Nitrogen | 0.24380 | 0.17273 |
| Ammonia | 0.50800 | 0.29900 |
| Alcohol | 0.45340 | 0.39900 |

the other end in the hand, the end in the hand will become warm or hot. The reason for this is that the molecules in the rod near the flame become hot and move rapidly, striking the molecules next to them. This action is repeated all along the rod until the opposite end is reached. Heat is transferred from one end of the rod to the other by conduction. Conduction depends upon unequal temperatures in the various portions of a given body.

*Convection* of heat is the process of transmitting heat by means of the movement of heated matter from one location to another. Convection is accomplished in gases and liquids.

In a place heated by a radiator, the air next to the radiator becomes warm and expands. The heated air becomes less dense than the surrounding cold air. It is forced up from the radiator by the denser, colder air. Most home heating systems operate on the principle of transmission of heat by convection.

Nearly all substances expand with an increase in temperature, and they contract or shrink with a decrease in temperature. There

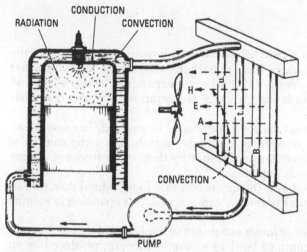

**Figure 1-9** Transfer of heat by radiation, conduction, and convection. It should be noted that the air, not the water, is the cooling agent. The water is only the medium for transferring the heat to the point where it is extracted and dissipated by the air.

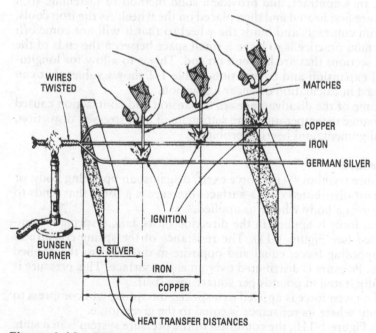

**Figure 1-10** Differences in heat conductivity of various metals.

13

is one exception to this statement for all temperature changes, the exception being water. It is a remarkable characteristic of water that at its maximum density (39.1°F) water expands as heat is added and that it also expands slightly as the temperature decreases from that point.

Increase in heat causes a substance to expand, because of an increase in the velocity of molecular action. Since the molecules become more separated in distance by their more frequent violent collisions, the body expands.

*Linear expansion* is the expansion in a longitudinal direction of solid bodies, while *volumetric expansion* is the expansion in volume of a substance.

The *coefficient of linear expansion* of a solid substance is the ratio of increase in length of body to its original length, produced by an increase in temperature of 1°F.

Expansion and contraction caused by a change in temperature have some advantages, but also pose some disadvantages. For example, on the plus side, rivets are heated red-hot for applying to bridge girders, structural steel, and large boilerplates. As the rivets cool, they contract, and provide a solid method of fastening. Iron rims are first heated and then placed on the wheel. As the iron cools, the rim contracts and binds the wheel so that it will not come off. Common practice is to leave a small space between the ends of the steel sections that are laid end on end. This is to allow for longitudinal expansion and contraction. Table 1-2 shows values that can be used in calculation of linear expansion.

Some of the disadvantages of expansion and contraction caused by change in temperatures are setting up of high stresses, distortion, misalignment, and bearing problems.

## Pressure

Pressure (symbol $P$) is a force exerted against an opposing body, or a thrust distributed over a surface. Pressure is a force that tends to compress a body when it is applied.

If a force is applied in the direction of its axis, a spring is compressed (see Figure 1-11). The resistance of the spring constitutes an opposing force, equal and opposite in direction to the applied force. Pressure is distributed over an entire surface. This pressure is usually stated in pounds per square inch (psi).

If a given force is applied to a spring, the spring will compress to a point where its resistance is equal to the given force.

In Figure 1-11b, the condition of the pressure system is in a state of equilibrium.

### Table 1-2    Linear Expansion of Common Metals
### (between 32°F and 212°F)

| Metal | Linear Expansion per Unit Length per Degree F |
|---|---|
| Aluminum | 0.00001234 |
| Antimony | 0.00000627 |
| Bismuth | 0.00000975 |
| Brass | 0.00000957 |
| Bronze | 0.00000986 |
| Copper | 0.00000887 |
| Gold | 0.00000786 |
| Iron, cast | 0.00000556 |
| Iron, wrought | 0.00000648 |
| Lead | 0.00001571 |
| Nickel | 0.00000695 |
| Steel | 0.00000636 |
| Tin | 0.00001163 |
| Zinc, cast | 0.00001407 |
| Zinc, rolled | 0.00001407 |

*Volumetric expansion = 3 times linear expansion.*

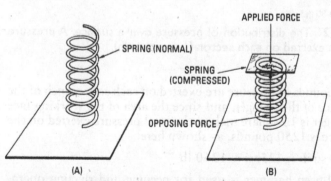

**Figure 1-11** The nature of pressure: (a) spring in its normal state; and (b) pressure system in state of equilibrium.

### Problem

The total working area of the plunger of a pump is 10 square inches. What is the amount of pressure on the plunger when pumping against 125 psi (see Figure 1-12)?

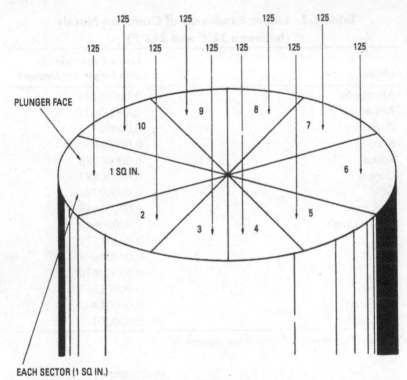

125  125                    125    125
125        125       125  125        125              125

PLUNGER FACE

9          8

10              7

1 SQ IN.                         6

2              5

3    4

EACH SECTOR (1 SQ IN.)

**Figure 1-12** The distribution of pressure over a surface. A pressure of 125 psi is exerted on each sector (1 square inch).

**Solution**

Since 125 pounds of pressure are exerted on each square inch of the working face of the plunger, and since the area of the working face of the plunger is 10 square inches, the total pressure exerted on the plunger face is 1250 pounds, as shown here:

$$10 \text{ sq in} \times 125\text{psi} = 1250 \text{ lb}$$

The ball-peen hammer is used for peening and riveting operations. The peening operation indents or compresses the surface of the metal, expanding or stretching that portion of the metal adjacent to the indentation. As shown in Figure 1-13, the contact area is nearly zero if the flat and special surfaces are perfectly smooth. However, perfectly smooth surfaces do not exist. The most polished surfaces (as seen under a microscope) are similar to emery paper. Therefore, the contact area is very small. As shown in Figure 1-14,

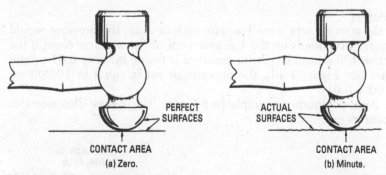

**Figure 1-13** Theoretical contact area (a) and actual contact area (b) of flat and spherical surfaces.

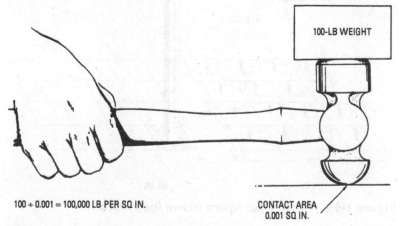

100 ÷ 0.001 = 100,000 LB PER SQ IN.

CONTACT AREA
0.001 SQ IN.

**Figure 1-14** The pressure (psi) is multiplied when it is applied to the flat surface through a spherical contact area.

the pressure, in psi, is multiplied when applied through a spherical contact surface.

### Problem
If the ball-peen of a machinist's hammer is placed in contact with a flat surface (see Figure 1-14) and a weight of 100 lb is placed on the hammer (not including the weight of the hammer), how many pounds of pressure are exerted at the point of contact if the contact area is 0.01 square inch?

## Solution

If the contact area were 1 square inch in area, the pressure would equal 100 pounds on the 1 square inch of flat surface. Now, if the entire 100-pound weight or pressure is borne on only 0.01 square inch (see Figure 1-14), the pressure in psi is equal to 10,000 psi (100 ÷ 0.01).

Perhaps another example (see Figure 1-15) may illustrate this point more clearly.

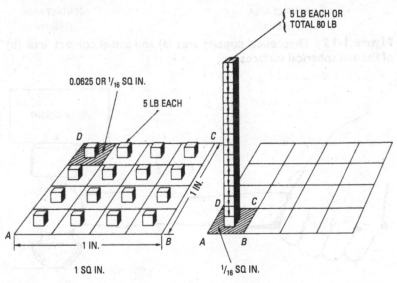

**Figure 1-15** Pressure per square inch of flat surface.

## Problem

Lay out entire surface ABDC equal to 1 square inch, and divide the surface into 16 small squares ($\frac{1}{16}$ square inch), placing a 5-pound weight on each small square. The area of each small square is $(\frac{1}{16})^2$, or 0.0625 square inches. If all the 5-pound weights are placed on one small square (as in the diagram), the total weight or pressure on that small square is 80 pounds (5 × 16), or, on 0.0625 square inches of surface.

In the left-hand diagram (see Figure 1-15), the 5-pound weights are distributed over the entire 1 square inch of area, the pressure totaling 80 psi of surface (5 × 16). In the right-hand diagram, the sixteen 5-pound weights (80 lb) are borne on only 0.0625 square inches of surface. This means the total weight or pressure (if each of the sixteen small squares were to bear 80 pounds) would be 1280 psi of surface (16 × 80).

## Atmospheric Pressure

Unless stated otherwise, the term *pressure* indicates pressure psi. The various qualifications of pressure are *initial pressure, mean effective pressure, terminal pressure, backpressure,* and *total pressure.*

The atmospheric pressure is due to the weight of the Earth's atmosphere. At sea level it is equal to approximately 14.69 psi. The pressure of the atmosphere does not remain constant at a given location, because weather conditions are changing continually.

Figure 1-16 illustrates atmospheric pressure. If a piston having a surface area of 1 square inch is connected to a weight by a string passing over a pulley, then, a weight of 14.69 pounds is required to raise the weight from the bottom of the cylinder (assuming air tightness and no friction) against the atmosphere that distributes a pressure of 14.69 pounds over the entire face area of the piston

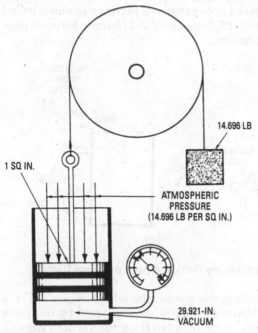

14.696 LB

1 SQ IN.

ATMOSPHERIC
PRESSURE
(14.696 LB PER SQ IN.)

29.921-IN.
VACUUM

**Figure 1-16**   Atmospheric pressure.

(area = 1 square inch). Then the system is in a *state of equilibrium*, the weight balancing the resistance or weight of the atmosphere. A slight excess pressure is then required to move the piston.

Atmospheric pressure decreases approximately 0.5 pounds for each 1000-foot increase in elevation. When an automobile climbs a high mountain, the engine gradually loses power because air expands at higher altitudes. The volume of air taken in by the engine does not weigh as much at the higher altitudes as it weighs at sea level. The mixture becomes too rich at higher altitudes, causing a poor combustion of fuel.

A *perfect vacuum* is a space that has no matter in it. This is unattainable even with the present pumps and chemical processes. Space in which the air pressure is about one-thousandth of that of the atmosphere is generally called a vacuum. *Partial vacuum* has been obtained in which there are only a few billion molecules in each cubic inch. In normal air, there are about four hundred billion times a billion molecules of gas to each cubic inch.

### Gage Pressure

Pressure measured above that of atmospheric pressure is called *gage pressure*. Pressure measured above that of a perfect vacuum is called *absolute pressure*. Figure 1-17 illustrates the difference between gage pressure and absolute pressure.

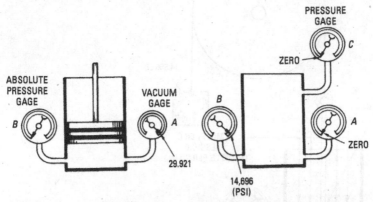

**Figure 1-17**  Absolute pressure (left) and gage pressure (right).

In the cylinder containing the piston (the left of Figure 1-17), a perfect vacuum exists below the piston, as registered by the value 29.921 inches of mercury (explained later) on the vacuum gage *A*. The equivalent reading on the absolute pressure gage *B* is zero psi.

If the piston is removed from the cylinder (the right of Figure 1-17), air rushes into the cylinder. That is, the vacuum is replaced by air at atmospheric pressure, the vacuum gage *A* drops to zero, the absolute pressure gage *B* reads 14.696, and the pressure gage *C* indicates a gage pressure of zero.

## Barometer

A *barometer* is an instrument that is used to measure atmospheric pressure. The instrument can be used to determine height or altitude above sea level, and it can be used in forecasting weather.

The barometer reading is expressed in terms of *inches of mercury* (in. Hg). This can be shown (see Figure 1-18) by filling a 34-inch length of glass tubing with mercury and then inverting the tubing in an open cup of mercury. The mercury inside the glass tubing falls until its height above the level of the mercury in the cup is approximately 30 inches (standard atmosphere). The weight of the 30-inch column of mercury is equivalent to the weight of a similar column of air approximately 50 miles in height.

The barometer reading in inches of mercury can be converted to psi by multiplying the barometer reading by 0.49116. This value corresponds to the weight of a 1-inch column of mercury that has a cross-sectional area of 1 square inch.

The barometer readings (in. Hg) are converted to atmospheric pressure (psi) in Table 1-3. The table calculations are based on the standard atmosphere (29.92 inches of mercury) and pressure (14.696 psi). Thus, 1 inch of mercury is equivalent to 0.49116 psi (14.696 ÷ 29.921).

### Problem

What absolute pressure reading corresponds to a barometer reading of 20 inches of mercury?

### Solution

The absolute pressure reading can be calculated by means of the formula:

$$\text{barometer reading (in. Hg)} \times 0.49116 = \text{psi}$$

Therefore, the absolute pressure reading is (20 × 0.49116), or 9.82 psi.

In an engine room, for example, the expression "28-inch vacuum" signifies an absolute pressure in the condenser of 0.946 psi (14.696 − 13.75). This indicates that the mercury in a column connected to a condenser having a 28-inch vacuum rises to a height of 28 inches, which represents the difference between the atmospheric

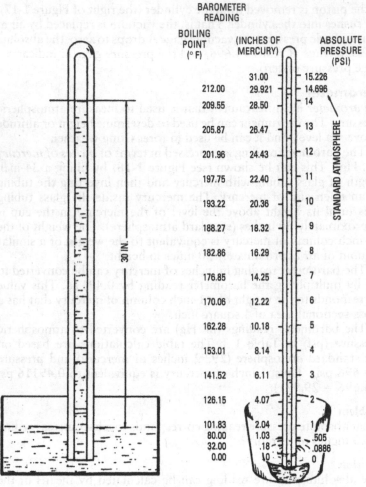

**Figure 1-18** The basic principle of the barometer and the relation of the Fahrenheit scale, barometric pressure reading, and absolute pressure.

pressure and the pressure inside the condenser 13.804 pounds (14.73 − 0.946).

## Gravity
The force that tends to attract all bodies in the Earth's sphere toward the center point of the earth is known as *gravity*. The symbol for

### Table 1-3  Conversion of Barometer Reading to Absolute Pressure

| Barometer (in. Hg) | Pressure (psi) |
|---|---|
| 28.00 | 13.75 |
| 28.25 | 13.88 |
| 28.50 | 14.00 |
| 28.75 | 14.12 |
| 29.00 | 14.24 |
| 29.25 | 14.37 |
| 29.50 | 14.49 |
| 29.75 | 14.61 |
| 29.921 | 14.696 |
| 30.00 | 14.74 |
| 30.25 | 14.86 |
| 30.50 | 14.98 |
| 30.75 | 15.10 |
| 31.00 | 15.23 |

gravity is $g$. The rate of acceleration of gravity is 32.16 feet per second. Starting from a state of rest, a free-falling body falls 32.16 feet during the first second; at the end of the next second, the body is falling at a velocity of 64.32 feet per second (32.16 + 32.16).

### Center of Gravity

That point in a body about which all its weight or parts are evenly distributed or balanced is known as its *center of gravity* (abbreviated *c.g.*). If the body is supported at its center of gravity, the entire body remains at rest, even though it is attracted by gravity. A higher center of gravity and a lower center of gravity are compared in Figure 1-19, as related to the center of gravity in automobiles.

### Centrifugal Force

The force that tends to move rotating bodies away from the center of rotation is called *centrifugal force*. It is caused by inertia. A body moving in a circular path tends to be forced farther from the axis (or center point) of the circle described by its path.

If the centrifugal force balances the attraction of the mass around which it revolves, the body continues to move in a uniform path. The operating principle of the centrifugal pump (see Figure 1-20) is based on centrifugal force.

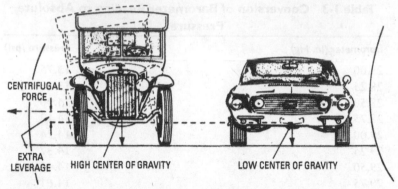

CENTRIFUGAL FORCE

EXTRA LEVERAGE    HIGH CENTER OF GRAVITY    LOW CENTER OF GRAVITY

**Figure 1-19** Comparison of the height of the center of gravity in an earlier model automobile (left) and later model (right).

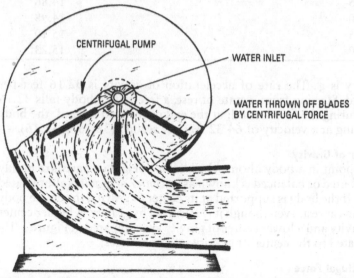

CENTRIFUGAL PUMP

WATER INLET

WATER THROWN OFF BLADES BY CENTRIFUGAL FORCE

**Figure 1-20** The use of centrifugal force in the basic operation of a centrifugal pump.

### Centripetal Force

The force that tends to move rotating bodies toward the center of rotation is called *centripetal force*. Centripetal force resists centrifugal force, and the moving body revolves in a circular path when these opposing forces are equal—that is, the system is in a *state of equilibrium* (see Figure 1-21).

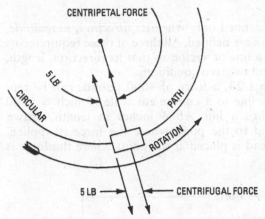

**Figure 1-21** The state of equilibrium between centrifugal and centripetal force.

If a body O (see Figure 1-22) is acted upon by two directly opposed forces OA and OC, those forces are equal. If it is also acted upon by another pair of directly opposed forces OB and OD, the various forces balance and the resultant reaction on the body O is zero (that is, the body remains in a state of rest).

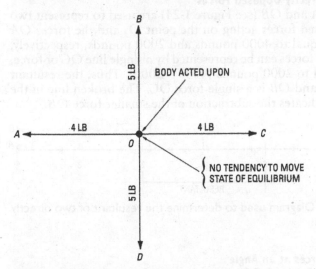

**Figure 1-22** State of equilibrium existing as a resultant of directly opposed forces.

## Force

A force is completely defined only when its *direction, magnitude*, and *point of application* are defined. All three of these requirements can be represented by a line or vector, so that its direction, length, and location correspond to given conditions.

As shown in Figure 1-23, a force of 4000 pounds can be represented by drawing a line to a convenient scale (1 inch = 1000 pounds), which requires a line *AB* 4 inches in length, drawn in the direction of and to the point where the force is applied. Note that the arrowhead is placed at the point where the force is applied.

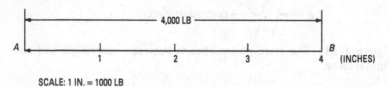

SCALE: 1 IN. = 1000 LB

**Figure 1-23**  A line or vector is used to represent a force and its intervals, its direction, and its point of application. The arrowhead indicates the point of application of the force.

### Resultant of Directly Opposed Forces

If the lines *OA* and *OB* (see Figure 1-24) are used to represent two directly opposed forces acting on the point *O*, and the forces *OA* and *OB* are equal to 4000 pounds and 2000 pounds, respectively, these opposed forces can be represented by a single line *OC* or force, which is equal to 2000 pounds (4000 − 2000). Thus, the resultant of forces *OA* and *OB* is a single force *OC*. The broken line in the illustration indicates the subtraction of the smaller force *OB*.

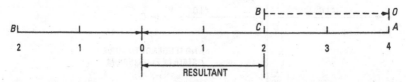

**Figure 1-24**  Diagram used to determine the resultant of two directly opposed forces.

### Resultant of Forces at an Angle

If two forces *OA* and *OB* are acting on a common point *O*, an angle is formed (see Figure 1-25) in which the two forces can be

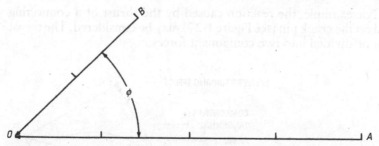

**Figure 1-25**  Diagram of two forces acting on a common point at an angle to each other.

represented by the lines OA and OB whose lengths represent 4000 pounds and 2000 pounds, respectively.

To determine the direction and intensity of the *resultant* force, a parallelogram of force can be constructed (see Figure 1-26). The parallelogram can be constructed from the diagram in Figure 1-25. The broken line BC is constructed parallel to line OA, and the broken line AC is drawn parallel to line OB. The diagonal OC represents the direction and intensity (by measuring its length) of the resultant force that is equivalent to the forces OA and OB.

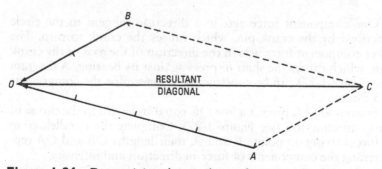

**Figure 1-26**  Determining the resultant of two angular forces by the parallelogram-of-forces method.

## Components of a Force
The *components of a force* can be determined by reversing the process of determining the resultant of two forces. A component of force is a single force that was used to compound the resultant force derived by the parallelogram-of-forces principle.

For example, the reaction caused by the thrust of a connecting rod on the crank pin (see Figure 1-27) may be considered. The thrust can be divided into two component forces.

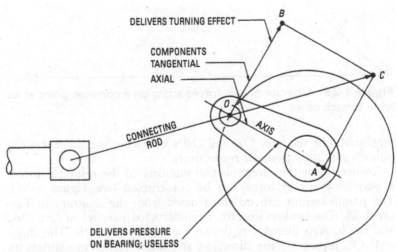

**Figure 1-27** Determining the components of a force by means of the parallelogram-of-forces method.

One component force acts in a direction tangent to the circle described by the crank pin, which causes the crank to turn. The other component force acts in the direction of the axis of the crank arm, which causes the shaft to press against its bearing. A diagram (see Figure 1-27) can be constructed to determine the components of a force.

From point $O$, project a line $OC$ equal in length to the thrust of the connecting rod (see Figure 1-27). Complete the parallelogram of forces to obtain points $B$ and $A$, their lengths $OB$ and $OA$ representing the components of force in direction and intensity.

## Motion

*Motion* is usually described as a change in position in relation to an assumed fixed point. Motion is strictly a relative matter because there can be no motion unless some point or object is regarded as stationary (see Figure 1-28).

As shown in Figure 1-28, the man is rowing the boat at a speed of 4 miles per hour against a current flowing at 2 miles per hour in the opposite direction. The boat is moving at 4 miles per hour with

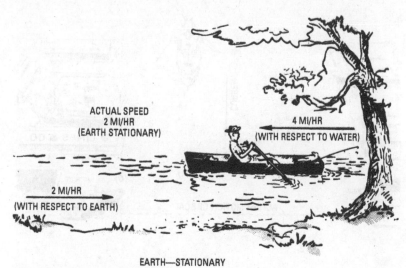

**Figure 1-28** Motion.

respect to the water, and the water is moving at 2 miles per hour with respect to the Earth.

The familiar example of the ferryboat crossing the river, pointing upstream to counteract the motion of the water, is used to illustrate *apparent* and *actual* motion (see Figure 1-29). The line *OA* represents the apparent motion (both distance and direction) of the boat. However, regarding the Earth as stationary, the line *OB* represents the actual motion of the boat. If the water is regarded as stationary, the boat is moving in the direction represented by the line *OA*.

### Newton's Laws of Motion
The noted physicist, Sir Isaac Newton, announced the three laws of motion as follows:

- *First Law of Motion*—If a body is at rest, it tends to remain at rest. If a body is in motion, it tends to remain in motion in a straight line until acted upon by a force.
- *Second Law of Motion*—If a body is acted on by several forces, it tends to obey each force as though the other forces do not exist, whether the body is at rest or in motion.
- *Third Law of Motion*—If a force acts to change the state of a body with respect to rest or motion, the body offers a resistance

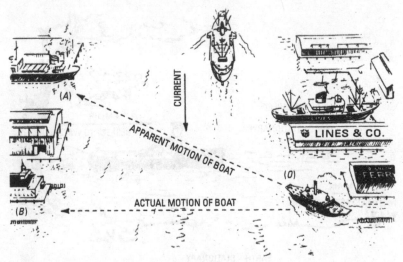

**Figure 1-29** Apparent and actual motion.

that is equal and directly opposed to the force. In other words, to every action there is an equal and opposite reaction.

### Types of Motion
The rate of change of position in relation to time is termed *velocity*. Velocity is also the rate of motion in a given direction (as the rotation of a sphere) in relation to time. The rate of increase in velocity or the average increase of velocity in a given unit of time is called *acceleration*.

A train traveling at a rate of 30 miles per hour is an example of *linear* velocity. A line shaft rotating at a rate of 125 revolutions per minute is an example of *rotary* velocity. Figure 1-30 shows linear motion and rotary motion.

*Tangential* motion is the equivalent of rotary motion, but is regarded as moving in a straight line or tangential direction. Tangential motion or velocity is used in belting calculations. As shown in Figure 1-31, the circumference of a 1-foot diameter pulley is 3.1416 feet. Thus, for each revolution of the pulley, the belt travels the tangential equivalent distance *AB*, or 3.1416 feet.

### Example
If a 4-foot diameter pulley is rotating at 100 rpm, what is the tangential speed of the belt? The calculation is as follows:

$$\text{tangential speed} = \text{circumference} \times \text{rpm}$$

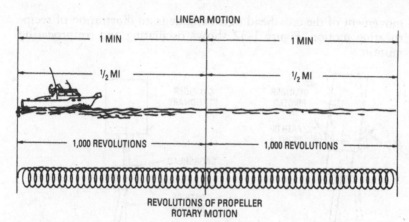

**Figure 1-30** Linear motion and rotary motion.

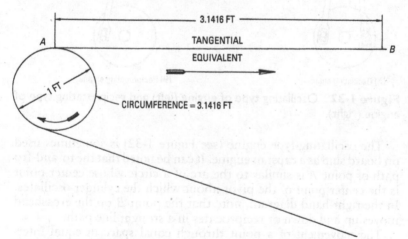

**Figure 1-31** The tangential equivalent distance of rotary motion.

tangential speed = 4 × 3.1416 × 100

tangential speed = 12.566 × 100

tangential speed = 1256.6 feet per minute

Vibrating motion that describes a path similar to the arc of a circle is called *oscillating* motion. A familiar example of oscillating motion is the pendulum of a clock. A vibrating motion that makes a path similar to a straight line is called *reciprocating* motion. The movement of the crosshead of an engine is an illustration of recip-rocating motion. Figure 1-32 shows oscillating and reciprocating motion.

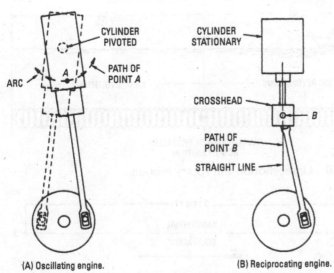

(A) Oscillating engine.                    (B) Reciprocating engine.

**Figure I-32**  Oscillating type of engine (left) and reciprocating type of engine (right).

The oscillating-type engine (see Figure 1-32) is sometimes used on board ship as a capstan engine. It can be noted that the to-and-fro path of point A is similar to the arc of a circle whose center point is the center point of the pivot about which the cylinder oscillates. In the right-hand diagram, note that the point B on the crosshead moves up and down or reciprocates in a straight line path.

The movement of a point through equal space in equal inter-vals of time is called *constant* motion. The movement of a point through unequal spaces in equal intervals of time is called *variable* motion. Figure 1-33 illustrates both constant motion and variable motion.

In the movements of the crank pin and piston of an engine (see Figure 1-33), the rate of motion from the position of the crank pin at the beginning of the stroke (position A) is constant as it rotates to position B. This means that it passes through equal arcs in equal intervals of time. Perpendiculars from points 1, 2, 3, and so on, locate the corresponding positions a, b, c, and so on, of the piston. As shown in the diagram, the traversed spaces (Aa, ab, and so on) are

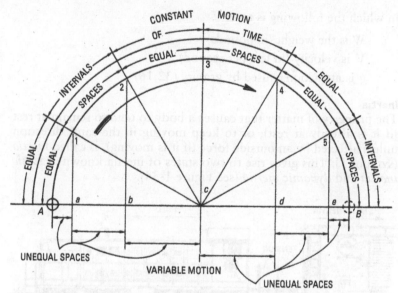

**Figure 1-33**  Constant motion and variable motion.

unequal. Therefore, the motion is variable. The diagram represents the true relation where there is no distortion, as with a Scotch-yoke mechanism. However, when a connecting rod is used, there is distortion caused by the angularity of the connecting rod.

## Momentum

The power of a body to overcome resistance by means of its motion is termed *momentum*. Momentum is the quantity of motion in a moving body.

Momentum is measured by multiplying the quantity of matter in a body by its velocity. A numerical value of momentum can be expressed as the force (in pounds) steadily applied that can stop a moving body in 1 second. Therefore, momentum is equal to the mass of a body multiplied by its velocity in feet per second or:

$$\text{momentum} = \frac{\text{weight}}{32.16} \times \text{velocity (ft/s)}$$

Following is the formula for determining momentum:

$$M = \frac{WV}{g}$$

in which the following is true:

W is the weight (in pounds)

V is velocity (in feet per second)

g is attraction caused by gravity (32.16)

## Inertia

The property of matter that causes a body to tend to remain at rest (if it is already at rest); or to keep moving in the same direction unless affected by an outside force (if it is moving), is called *inertia* (symbol $I$). This gives rise to two states of inertia known as *static inertia* and *dynamic inertia* (see Figure 1-34).

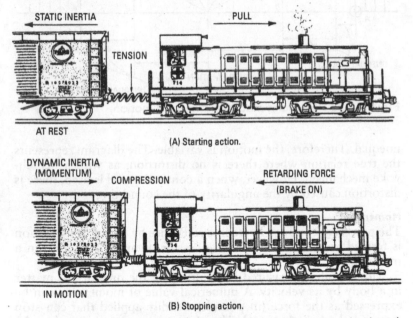

STATIC INERTIA          PULL

TENSION

AT REST

(A) Starting action.

DYNAMIC INERTIA (MOMENTUM)    COMPRESSION    RETARDING FORCE (BRAKE ON)

IN MOTION

(B) Stopping action.

**Figure 1-34** Static inertia with respect to a body at rest (top) and dynamic inertia with respect to a body in motion (bottom).

## Friction

The resistance to motion of two moving objects that touch is called *friction*. Friction is caused partially by the natural adhesion of one body to another. However, its chief cause is the roughness of surfaces that are in contact. Even a glossy, polished surface is not smooth when viewed with a powerful magnifying glass or microscope.

## Coefficient of Friction

The ratio of the force required to slide a body along a horizontal plane surface to the weight of the body is the *coefficient of friction*. The coefficient of friction is equivalent to the tangent of the *angle of repose*.

The angle of repose is the largest angle with the horizontal at which a mass of material (such as an embankment or pile of coal) can remain at rest without sliding. This angle varies with different materials.

## Laws of Friction

The first laws of friction were given by Morin in about 1830, but they have been modified by later experiments. As summarized by Kent, the laws of friction are as follows:

- Friction varies approximately as the normal pressure with which the rubbing surfaces are pressed together.
- Friction is approximately independent of the area of the surfaces, but it is slightly greater for smaller surfaces than for larger surfaces.
- Friction decreases with an increase in velocity, except at an extremely low velocity and with soft surfaces.

As applied to lubricated surfaces, the laws of friction for *perfect lubrication* (surfaces completely separated by a film of lubricant) are as follows:

- The coefficient of friction is independent of the materials making up the surfaces.
- The coefficient of friction varies directly with the viscosity of the lubricant, which varies inversely with the temperature of the lubricant.
- The coefficient of friction varies inversely as the unit pressure and varies directly as the velocity.
- The coefficient of friction varies inversely as the mean film thickness of the lubricating medium.
- Mean film thickness varies directly with velocity and inversely as the temperature and unit pressure.

As applied to *imperfect lubrication*, that is, surfaces partially separated (which may range from nearly complete separation to nearly complete contact) by a film of lubricant, the laws of friction are as follows:

- The coefficient of friction increases with an increase in pressure between the surfaces.
- The coefficient of friction increases with an increase in relative velocity between the surfaces.

Machinery cannot be operated without lubrication (notwithstanding the alleged antifriction metals) because of the tiny irregularities in a smooth metal surface. A lubricant is used to keep the rubbing parts separated by a thin film of oil, thus preventing actual contact so far as possible.

## Work and Power

The expenditure of energy to overcome resistance through a certain distance is *work*. It is difficult to understand horsepower without an understanding of the difference between work and power (see Figure 1-35). *Power* is the rate at which work is done (that is, work divided by the time in which it is done).

The standard unit for measuring work is the *foot-pound* (ft-lb). The foot-pound is the amount of work that is done in raising a

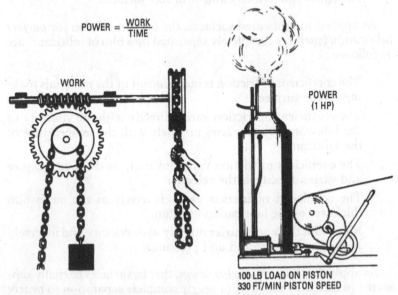

$$POWER = \frac{WORK}{TIME}$$

WORK

POWER
(1 HP)

100 LB LOAD ON PISTON
330 FT/MIN PISTON SPEED

**Figure 1-35** The difference between work (left) and power (right).

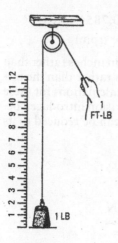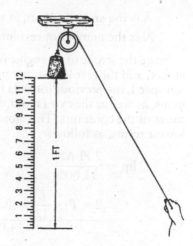

**Figure 1-36** One foot-pound.

weight of 1 pound through a distance of 1 foot, or in overcoming a pressure of 1 pound through a distance of 1 foot (see Figure 1-36).

James Watt is believed to have adopted the term *horsepower*. He used it for his steam engine, to represent the power or capacity of a strong London draft-type horse for doing work during a short time interval. The term was used as a power rating for his steam engines. The present standard unit for measuring power is horsepower (*hp*). It is defined as 33,000 foot-pounds per minute. In other words, one horsepower is required to raise a weight of the following:

- 33,000 lb a height of 1 foot in one minute
- 3300 lb a height of 10 feet in one minute
- 33 lb a height of 1000 feet in one minute
- 3.3 lb a height of 10,000 feet in one minute
- 1 lb a height of 33,000 feet in one minute

A formula that is generally used to calculate engine horsepower is as follows:

$$hp = \frac{2 \text{ PLAN}}{33,000}$$

in which the following is true:

    *P* is the mean effective pressure in psi

    *L* is the length of stroke, in feet

$A$ is the area of piston, in square inches ($0.7854 \times d^2$)

$N$ is the number of revolutions per minute (rpm)

Since the stroke of an engine is usually given in inches rather than in feet, and the revolutions per minute are given rather than the piston speed, the previous formula involves extra calculations for these items, as well as the extra multiplication and division introduced because of the constants. Therefore, the formula can be reduced to its lowest terms, as follows:

$$hp = \frac{2 \, PLAN}{33,000}$$

$$= \frac{2 \times P \times \dfrac{L}{12} \times (0.7854 \times d^2) \times N}{33,000}$$

$$= \frac{0.1309 \times PLD^2N}{33,000}$$

$$= 0.000003967 \, PLD^2N$$

Thus, the constant 0.000004 can be used for most calculations. By changing the order of the factors, the formula is simplified to:

$$hp = 0.000004 \, d^2LNP$$

According to its definitions and the manner in which it is derived, the various types of horsepower are *nominal, indicated, brake, effective, hydraulic, boiler,* and *electrical.* Figure 1-37 shows the various types of horsepower.

## Basic Machines

For many years the basic mechanical contrivances that enter into the composition or formation of machines were referred to as *mechanical powers.* Since these mechanical contrivances are regarded in a more static than dynamic sense (that is, the consideration of opposing forces in equilibrium rather than tending to produce motion), it is more correct to refer to them as *basic machines.* Strictly speaking, the term *power* is a dynamic term relating to the time rate of doing work. When the elements of a machine are in equilibrium, no work is done. Therefore, it is incorrect to refer to the basic machines as mechanical powers.

It should be understood that the action of all the basic machines depends on the principle of work, which is: The applied force,

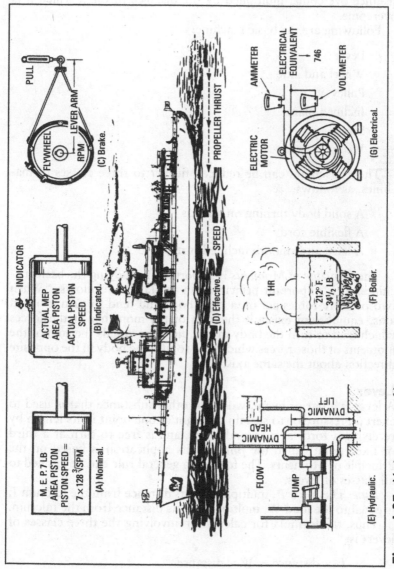

**Figure 1-37** Various types of horsepower.

- (A) Nominal.

  M. E. P. 7 LB
  AREA PISTON
  PISTON SPEED = $\frac{7 \times 128 \sqrt[3]{SPM}}{}$

- (B) Indicated.

  INDICATOR

  ACTUAL MEP
  AREA PISTON
  ACTUAL PISTON SPEED

- (C) Brake.

  PULL
  LEVER ARM
  FLYWHEEL
  RPM

- (D) Effective.

  PROPELLER THRUST
  SPEED

- (E) Hydraulic.

  FLOW
  PUMP
  DYNAMIC HEAD
  DYNAMIC LIFT

- (F) Boiler.

  1 HR
  212° F.
  34½ LB

- (G) Electrical.

  AMMETER
  ELECTRICAL EQUIVALENT
  746
  VOLTMETER
  ELECTRIC MOTOR

multiplied by the distance through which it moves, equals the resistance overcome, multiplied by the distance through which it is overcome.

Following are the basic machine:

- Lever
- Wheel and axle
- Pulley
- Inclined plane
- Screw
- Wedge

These machines can be reduced further to three classes of machines, as follows:

- A solid body turning on an axis
- A flexible cord
- A hard and smooth inclined surface

The *Principle of Moments* is important in studying the basic machines. This important principle can be stated as follows: "When two or more forces act on a rigid body and tend to turn it on an axis, equilibrium exists if the sum of the moments of the forces which tend to turn the body in one direction equals the sum of the moments of those forces which tend to turn the body in the opposite direction about the same axis.

## Lever

A lever is a bar of metal, wood, or other substance that is used to exert a pressure or to sustain a weight at one point in its length by receiving a force at a second point, and is free to turn at a third or fixed point called the *fulcrum*. Its application is based on the Principle of Moments. The following general rule can be applied to all classes of levers.

*Rule:* The force $P$, multiplied by its distance from the fulcrum $F$, is equal to the load $W$, multiplied by its distance from the fulcrum.

Thus, the formula for calculation involving the three classes of levers is:

$$F \times \text{distance} = W \times \text{distance}$$

As shown in Figure 1-38, there are three classes of levers.

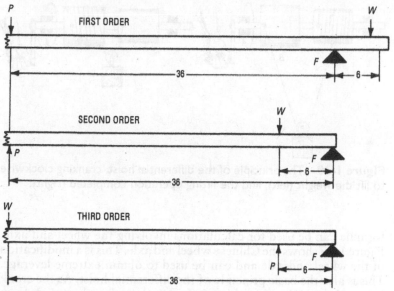

**Figure 1-38** The three classes of levers.

## Problem

What force $P$ is required at a point 3 feet from the fulcrum $F$ to balance a weight of 112 pounds applied at a point 6 inches from the fulcrum?

## Solution

The distances or lengths of the levers are 3 feet and 6 inches, respectively. Since the distances must be of the same denomination, the 3 feet must be reduced to inches (3 × 12), or 36 inches. Then, applying the rule:

$$P \times 36 = 112 \times 6$$
$$P = \frac{112 \times 6}{36}$$
$$= 18.67 \text{ lb}$$

## Wheel and Axle

A comparison of the wheel and axle with the first-order lever (see Figure 1-38) indicates that they are similar in principle. The same

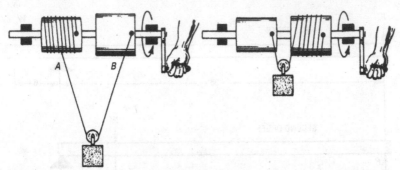

**Figure 1-39** The principle of the differential hoist, cranking clockwise to lift the weight (left), and the lifting operation completed (right).

formula can be used for calculations involving the wheel and axle. Figure 1-39 shows the Chinese wheel and axle. This is a modification of the wheel and axle and can be used to obtain extreme leverage. This is also the basic principle of the differential hoist. As the crank is turned clockwise, the cable winds onto drum B and unwinds from drum A. Since drum B is larger in diameter, the length of cable between the two drums is gradually taken up, lifting the load. Thus, if the difference in the diameters of the two drums is small, an extreme leverage is obtained, enabling heavy weights to be lifted with little effort. Also, the load remains suspended at any point, because the difference in diameters of the two drums is too small to overbalance the friction of the parts.

## Pulley

The two types of pulleys are *fixed* and *movable*. No mechanical advantage is obtained from the fixed pulley. Its use is important in accomplishing work that is appropriate (raising water from a well, for example).

The movable pulley, by distributing its weight into separate portions, is attended by mechanical advantages proportional to the number of points of support. As illustrated in Figure 1-40, the relation between the force applied and the load lifted is changed by the various basic pulley combinations. Of course, an even greater range may be obtained by additional pulleys, but there is a practical limit to which this is mechanically expedient. The following rule states the relation between force and load.

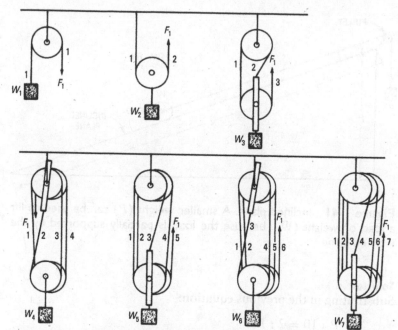

**Figure 1-40**  Relation between force applied and load lifted in the various basic pulley combinations.

*Rule:* The load $W$ that can be lifted by a combination of pulleys is equal to the force $F$ times the number of ropes supporting the lower or movable block.

## Inclined Plane

If a sloping path or incline is substituted for a direct upward line of ascent, a given weight can be raised by a smaller weight. Thus, the inclined plane is a basic machine, because a lesser force can be applied to lift a load (see Figure 1-41).

*Rule:* As the applied force $P$ is to the load $W$, so is the height $H$ to the length of the inclined plane $L$.

Thus, the calculation is:

$$\text{force} : \text{load} = \text{height} : \text{length of plane}$$

### Problem

What force $P$ is required to lift a load of 10 pounds if the height is 2 feet and the plane is 12 feet in length?

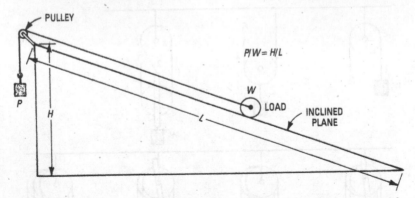

**Figure 1-41** Inclined plane. A smaller weight ($P$) can be used to lift a load or weight ($W$), because the load is partially supported by the inclined plane.

## Solution
Substituting in the previous equation:

$$P : 10 = 2 : 12$$

$$P \times 12 = 2 \times 10$$

$$P = \frac{10 \times 2}{12} = \frac{20}{12} = 1\frac{2}{3}$$

## Screw
This type of basic machine is merely an inclined plane wrapped around a cylinder. The screw is used to exert a severe pressure through a small space. Since it is subjected to a high loss from friction, the screw usually exerts a small amount of power in itself, but a large amount of power may be derived when combined with the lever or wheel.

*Rule:* As the applied force $P$ is to the load $W$, so is the pitch to the length of thread per turn.

Thus, the calculation is:

applied force $P$ : load $W$ = pitch : length of thread per turn

## Problem
If the pitch or distance between threads is $\frac{1}{4}$ inch and a force $P$ of 100 pounds is applied, what load of weight $W$ can be moved by the screw if the length of thread per turn of the screw is 10 inches?

### Solution

Substituting in the above equation:

$$100 : \text{load } W = \frac{1}{4} : 10$$

$$\text{load } W \times \frac{1}{4} = 10 \times 100$$

$$\text{load } W = \frac{10 \times 100}{\frac{1}{4}}$$

$$= 4000 \text{ lb}$$

## Wedge

The wedge is virtually a pair of inclined planes that are placed back-to-back or in contact along their bases (see Figure 1-42).

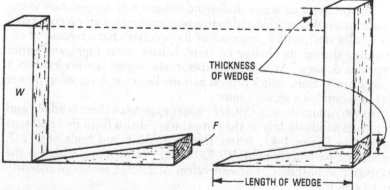

**Figure 1-42** Application of a wedge (P) to raise a heavy load or weight (W).

*Rule:* As the applied force $P$ is to the load $W$, so is the thickness of the wedge to its length.

Thus, the calculation is:

applied force $P$ : load $W$ = thickness of wedge : length of wedge

### Problem

What applied force $P$ is required to raise a load $W$ of 2000 pounds, using a wedge that is 4 inches in thickness and 20 inches in length?

**Solution**

Substituting in the equation:

$$P(\text{applied force}) : 2000 = 4 : 20$$
$$P \times 20 = 4 \times 2000$$
$$P = \frac{4 \times 2000}{20}$$
$$= 400 \text{ lb}$$

## Water

In the study of hydraulics and pumps, it is important that water and its characteristics be understood. Water is a most remarkable substance. By definition, water is a compound of hydrogen and oxygen in the proportion of 2 parts by weight of hydrogen to 16 parts by weight of oxygen.

The behavior of water under the influence of temperature is extraordinary. When subjected to low temperatures, water is converted to a solid (ice), which, because of its peculiar characteristic of expanding during its change of state, causes burst pipes and other types of damage. At higher temperatures, water is converted to a gas (steam). Thus, water is used as a medium for developing power (as in steam for a steam engine).

At maximum density (39.1°F) water expands as heat is added and it also expands slightly as the temperature drops from this point, as shown in Figure 1-43. Water freezes at 32°F and boils at 212°F, when the barometer reads 29.921 inches of mercury, which is the standard atmosphere. The equivalent of 29.921 inches of mercury is 14.696 psi.

Water contains approximately 5 percent air by volume (see Figure 1-44), mechanically mixed with it. This is the reason that steam engines that condense moisture should have air pumps attached to the condenser. Otherwise, the necessary vacuum cannot be maintained.

In the operation of steam heating plants, air is liberated when the water boils and passes into radiators with the steam. Therefore, automatic air valves must be provided to rid the system of the air. If automatic air valves are not provided (or if they become clogged because of corrosion), radiators may become air-bound and ineffective.

The boiling point of water rises as the pressure increases. Thus, the boiling point is 212°F for standard atmospheric pressure at sea level. The boiling point is 327.8°F at 100 pounds (absolute) pressure.

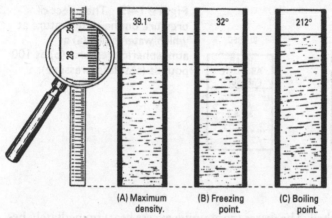

**Figure 1-43** A remarkable property of water—expansion at temperatures both above and below its temperature at point of maximum density (39.1°F). (A) If one pound of water is placed in a cylinder having a cross-sectional area of 1 square inch at 39.1°F, the water rises to a height of 27.68 inches; (B) as the temperature drops to 32°F, the water rises to a height of 27.7 inches; and (C) as the temperature is increased to 212°F, the water rises to a height of 28.88 inches in the tube.

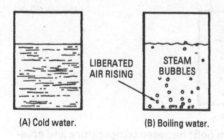

(A) Cold water.    (B) Boiling water.

**Figure 1-44** Boiling of water to liberate air that is mechanically mixed in the water.

Figure 1-45 shows water boiling (as in a tea kettle) with the addition of heat. If the vessel were closed, the water would continue to boil, causing the pressure to rise. If no more heat is added when the pressure reaches 100 pounds, for example, the water would cease to boil and the pressure would remain constant if no heat were lost. The temperature of the water, steam, and pressure are said to be in a *state of equilibrium.* The least variation in temperature (either upward or downward) destroys the state of equilibrium and causes a change. By permitting some of the confined steam to escape (see Figure 1-46), the water can be made to boil again. This is because of the reduction in pressure, which causes the equilibrium of the system

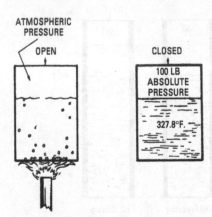

**Figure 1-45** The effect of pressure on the temperature at which water boils: (a) at atmospheric pressure; (b) at 100 pounds (absolute pressure).

to be disturbed. The water (containing excess heat) immediately begins to boil and tends to keep the pressure constant. If the process is continued, a gradual reduction in temperature and pressure results until all of the heat originally introduced into the system is used up.

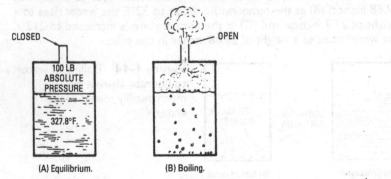

(A) Equilibrium.  (B) Boiling.

**Figure 1-46** State of equilibrium (left) between temperature and pressure. The equilibrium is upset (right) by reducing the pressure, resulting in boiling.

The boiling point of water is lowered as the elevation increases, because of the lower pressure of the atmosphere. At an elevation of 5000 feet, water boils at 202°F. Eggs cannot be boiled at high altitudes. Baking of cakes and bread sometimes presents a problem at high altitudes.

It is not essential that water be hot before it can be brought to a boil. For example, water at 28-inch vacuum pressure boils at 100°F. However, if the vacuum pressure is increased to 29.74 inches, the water boils at 32°F (see Figure 1-47).

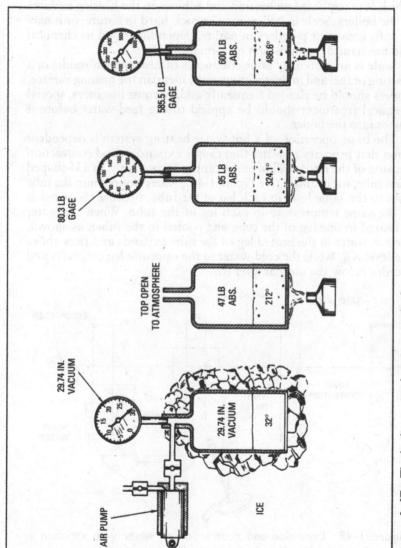

**Figure I-47** The boiling point (temperature) of water varies with a change in pressure. Entirely different values are obtained for other liquids.

29.74 IN. VACUUM

AIR PUMP

29.74 IN. VACUUM

32°

ICE

TOP OPEN TO ATMOSPHERE

47 LB ABS.

212°

80.3 LB GAGE

95 LB ABS.

324.1°

585.5 LB GAGE

600 LB ABS.

486.6°

For efficient operation of boilers, the water used must be as pure as possible. Impure water often contains ingredients that form scale, which is precipitated on heating and adheres to the heating surfaces of the boilers. Scale in boilers may be rock-hard in nature; or it may be soft, greasy, or powdery in nature, depending upon its chemical and mechanical composition or formation.

Scale is an extremely poor conductor of heat, which results in a wasting of fuel and in overheating of the metal in the heating surface. Boilers should be cleaned frequently and, in most instances, special chemical treatment should be applied to the feed-water before it passes into the boiler.

The basic operation of a hot-water heating system is dependent upon that property of water that causes expansion and contraction because of the rise or fall of temperature, respectively. In a U-shaped glass tube, for example (see Figure 1-48), water poured into the tube rises to the same level in each leg of the tube, because the water is at the same temperature in each leg of the tube. When the water is heated in one leg of the tube and cooled in the other, as shown, the hot water in the heated leg of the tube expands and rises above the level AB, while the cold water in the opposite leg contracts and recedes below the normal level AB.

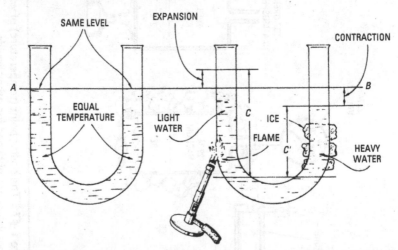

**Figure 1-48** Expansion and contraction of water with variation in temperature, resulting in a change in weight per unit volume.

Equilibrium can occur within the tube even though the water is at different levels, because the longer column C (see Figure 1-48), consisting of expanded and lighter water, weighs the same as the shorter column consisting of contracted and heavier water.

In the hot-water heating system, the weight of the hot and expanded water in the upflow column $C$ (see Figure 1-49), being less than that of the cold and contracted water in the downflow column, upsets the equilibrium of the system and results in a continuous circulation of water, as indicated by the arrows. This type of circulation is known as *thermocirculation*.

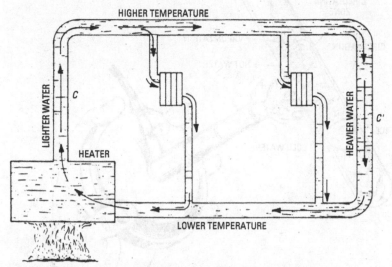

**Figure 1-49** Thermocirculation in a hot-water heating system.

In Figure 1-50, there is no circulation of water even though heat is applied. If this should happen in a boiler, practically no generation of steam would result, except for a film of steam that separates the water from the heating surface. The latter surface becomes red-hot and the boiler may be damaged or destroyed. Such a situation is known as the *spheroidal state* (see Figure 1-51). As in the illustration, a small quantity of water poured on a red-hot plate separates into drops and moves all around the plate, being supported by a thin film of steam. Since the water, after steam has formed, is not in contact with the plate, there is practically no cooling effect on the plate.

Figure 1-52 shows cooling by re-evaporation. Some regard re-evaporation as a loss, which, in fact, is incorrect. Since the area of the indicator diagram from point $L$ (see Figure 1-52) to a point of release is increased, re-evaporation represents a gain.

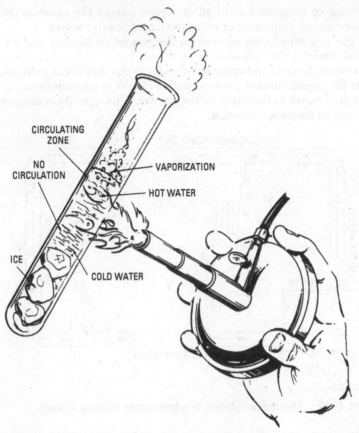

CIRCULATING ZONE

NO CIRCULATION

VAPORIZATION

HOT WATER

ICE

COLD WATER

**Figure 1-50** Experiment illustrating effect of no circulation. If ice is placed in the bottom of the test tube and heat is applied near the surface of the water, the water boils at that point. However, the heat does not melt the ice, because the cold water around the ice is heavier than the hot water at the top, which prevents thermocirculation. If the heat is applied at the bottom of the tube, the ice melts and all the water is vaporized if the heat is applied long enough.

The reason that the loss concept is incorrect is that it is the cost of re-evaporation that is a loss, and not the re-evaporation process itself. That is, re-evaporation robs the cylinder walls of a quantity of heat corresponding to the latent heat of re-evaporation. This additional cooling of the cylinder walls increases condensation during the first portion of the stroke, which is the loss. Since this loss exceeds

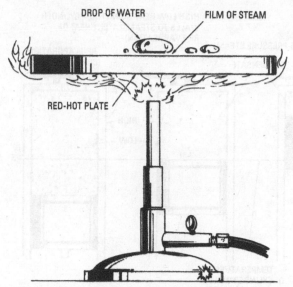

**Figure 1-51**   The spheroidal state in which a drop of water on a red-hot plate changes to steam.

the gain because of evaporation, re-evaporation is erroneously considered a loss.

The U.S. gallon of water weighs 8.33111 pounds only at the standard temperature of 62°F. At any other temperature reading, its weight is different. For calculations at most temperature readings, the weight of a gallon of water is considered to be $8^{1}/_{3}$ pounds, which is near enough in most instances. However, it should be understood that this is an approximate value. For precision calculations, the weight of a gallon of water at the given temperature should be used. Table 1-4 shows the weight of water per cubic foot at various temperatures. Table 1-5 shows the relative volume of water at various temperatures compared with its volume at 4°C.

## Properties of Water with Respect to Pump Design

Experience in the design of pumps has shown that water is nearly an unyielding substance when it is confined in pipes and pump passages, which necessitates substantial construction for withstanding the pressure (especially from periodic shocks or water hammer). Accordingly, in pump design, a liberal factor of safety should be used.

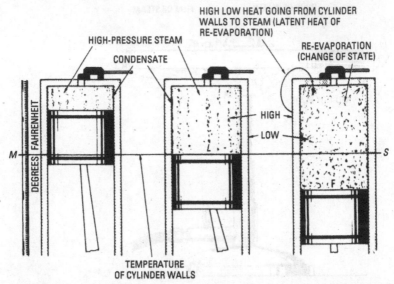

**Figure 1-52** A cylinder in a steam engine, illustrating the cooling action by change of state. The line (MS) represents the average temperature of the cylinder walls. In actual operation, when steam is admitted to the cylinder and during a portion of its stroke, its temperature is higher than that of the cylinder walls (left). If the point (L) is assumed to be the position of the piston at equal temperatures (center), condensation takes place. As the piston advances beyond point (L), the temperature of the steam is lower than that of the cylinder walls. The excess heat in the cylinder walls causes the condensate to boil (that is, re-evaporation occurs), which robs the cylinder walls of some of their heat.

### Pressure at Different Depths

The pressure of water varies with the head. This is equal to 0.43302 psi for each foot of static head. Thus, a head of 2.31 feet exerts a pressure of 1 psi (2.31 × 0.43302), as shown in Table 1-6.

### Compressibility of Water

Water is only slightly compressible. According to Kent, its compressibility ranges from 0.00004 to 0.000051 inch for one atmosphere of pressure, decreasing as temperature increases. For each 1 cubic foot, distilled water diminishes in volume from 0.0000015 to 0.0000013 inches. Water is so incompressible that, even at a depth of 1 mile, 1 cubic foot of water weighs approximately ½ pound more than at the surface.

Table 1-4    Weight of Water per Cubic Foot at Various
Temperatures

| Temp., °F | Lb.per cu. ft. | Temp., °F | Lb.per cu. ft. | Temp., °F | Lb.per cu. ft. | Temp., °F | Lb.per cu. ft. |
|---|---|---|---|---|---|---|---|
| 32 | 60.41 | 66 | 62.32 | 100 | 61.99 | 134 | 61.48 |
| 33 | 62.41 | 67 | 62.32 | 101 | 61.98 | 135 | 61.46 |
| 34 | 62.42 | 68 | 62.31 | 102 | 61.96 | 136 | 61.44 |
| 35 | 62.42 | 69 | 62.30 | 103 | 61.95 | 137 | 61.43 |
| 36 | 62.42 | 70 | 62.30 | 104 | 61.94 | 138 | 61.41 |
| 37 | 62.42 | 71 | 62.29 | 105 | 61.93 | 139 | 61.39 |
| 38 | 62.42 | 72 | 62.28 | 106 | 61.91 | 140 | 61.37 |
| 39 | 62.42 | 73 | 62.27 | 107 | 61.90 | 141 | 61.36 |
| 40 | 62.42 | 74 | 62.26 | 108 | 61.89 | 142 | 61.34 |
| 41 | 62.42 | 75 | 62.25 | 109 | 61.87 | 143 | 61.32 |
| 42 | 62.42 | 76 | 62.25 | 110 | 61.86 | 144 | 61.30 |
| 43 | 62.42 | 77 | 62.24 | 111 | 61.84 | 145 | 61.28 |
| 44 | 62.42 | 78 | 62.23 | 112 | 61.83 | 146 | 61.26 |
| 45 | 62.42 | 79 | 62.22 | 113 | 61.81 | 147 | 61.25 |
| 46 | 62.41 | 80 | 62.21 | 114 | 61.80 | 148 | 61.23 |
| 47 | 62.41 | 81 | 62.20 | 115 | 61.78 | 149 | 61.21 |
| 48 | 62.41 | 82 | 62.19 | 116 | 61.77 | 150 | 61.19 |
| 49 | 62.41 | 83 | 62.18 | 117 | 61.75 | 151 | 61.17 |
| 50 | 62.40 | 84 | 62.17 | 118 | 61.74 | 152 | 61.15 |
| 51 | 62.40 | 85 | 62.16 | 119 | 61.72 | 153 | 61.13 |
| 52 | 62.40 | 86 | 62.15 | 120 | 61.71 | 154 | 61.11 |
| 53 | 62.39 | 87 | 62.14 | 121 | 61.69 | 155 | 61.09 |
| 54 | 62.39 | 88 | 62.13 | 122 | 61.68 | 156 | 61.07 |
| 55 | 62.38 | 89 | 62.12 | 123 | 61.66 | 157 | 61.05 |
| 56 | 62.38 | 90 | 62.11 | 124 | 61.64 | 158 | 61.03 |
| 57 | 62.38 | 91 | 62.10 | 125 | 61.63 | 159 | 61.01 |
| 58 | 62.37 | 92 | 62.08 | 126 | 61.61 | 160 | 60.99 |
| 59 | 62.37 | 93 | 62.07 | 127 | 61.60 | 161 | 60.97 |
| 60 | 62.36 | 94 | 62.06 | 128 | 61.58 | 162 | 60.95 |
| 61 | 62.35 | 95 | 62.05 | 129 | 61.56 | 163 | 60.93 |
| 62 | 62.35 | 96 | 62.04 | 130 | 61.55 | 164 | 60.91 |
| 63 | 62.34 | 97 | 62.02 | 131 | 61.53 | 165 | 60.89 |
| 64 | 62.34 | 98 | 62.01 | 132 | 61.51 | 166 | 60.87 |
| 65 | 62.33 | 99 | 62.00 | 133 | 61.50 | 167 | 60.85 |

(*continued*)

## Table 1-4 (continued)

| Temp., °F | Lb.per cu. ft. | Temp., °F | Lb.per cu. ft. | Temp., °F | Lb.per cu. ft. | Temp., °F | Lb.per cu. ft. |
|---|---|---|---|---|---|---|---|
| 168 | 60.83 | 193 | 60.28 | 230 | 59.42 | 480 | 50.00 |
| 169 | 60.81 | 194 | 60.26 | 240 | 59.17 | 490 | 49.50 |
| 170 | 60.79 | 195 | 60.23 | 250 | 58.89 | 500 | 48.78 |
| 171 | 60.77 | 196 | 60.21 | 260 | 58.62 | 510 | 48.31 |
| 172 | 60.75 | 197 | 60.19 | 270 | 58.34 | 520 | 47.62 |
| 173 | 60.73 | 198 | 60.16 | 280 | 58.04 | 530 | 46.95 |
| 174 | 60.71 | 199 | 60.14 | 290 | 57.74 | 540 | 46.30 |
| 175 | 60.68 | 200 | 60.11 | 300 | 57.41 | 550 | 45.66 |
| 176 | 60.66 | 201 | 60.09 | 310 | 57.08 | 560 | 44.84 |
| 177 | 60.64 | 202 | 60.07 | 320 | 56.75 | 570 | 44.05 |
| 178 | 60.62 | 203 | 60.04 | 330 | 56.40 | 580 | 43.29 |
| 179 | 60.60 | 204 | 60.02 | 340 | 56.02 | 590 | 42.37 |
| 180 | 60.57 | 205 | 59.99 | 350 | 55.65 | 600 | 41.49 |
| 181 | 60.55 | 206 | 59.97 | 360 | 55.25 | 610 | 40.49 |
| 182 | 60.53 | 207 | 59.95 | 370 | 54.85 | 620 | 39.37 |
| 183 | 60.51 | 208 | 59.92 | 380 | 54.47 | 630 | 38.31 |
| 184 | 60.49 | 209 | 59.90 | 390 | 54.05 | 640 | 37.17 |
| 185 | 60.46 | 210 | 59.87 | 400 | 53.62 | 650 | 35.97 |
| 186 | 60.44 | 211 | 59.85 | 410 | 53.19 | 660 | 34.48 |
| 187 | 60.42 | 212 | 59.82 | 420 | 52.74 | 670 | 32.89 |
| 188 | 60.40 | 214 | 59.81 | 430 | 52.33 | 680 | 31.06 |
| 189 | 60.37 | 216 | 59.77 | 440 | 51.87 | 690 | 28.82 |
| 190 | 60.35 | 218 | 59.70 | 450 | 51.28 | 700 | 25.38 |
| 191 | 60.33 | 220 | 59.67 | 460 | 51.02 | 706.1 | 19.16 |
| 192 | 60.30 | | | 470 | 50.51 | | |

## Table 1-5   Expansion of Water

| °C | °F | Volume | °C | °F | Volume | °C | °F | Volume |
|---|---|---|---|---|---|---|---|---|
| 4 | 39 | 1.00000 | 35 | 95 | 1.00586 | 70 | 158 | 1.02241 |
| 5 | 41 | 1.00001 | 40 | 104 | 1.00767 | 75 | 167 | 1.02548 |
| 10 | 50 | 1.00025 | 45 | 113 | 1.00967 | 80 | 176 | 1.02872 |
| 15 | 59 | 1.00083 | 50 | 122 | 1.01186 | 85 | 185 | 1.03213 |
| 20 | 68 | 1.00171 | 55 | 131 | 1.01423 | 90 | 194 | 1.03570 |
| 25 | 77 | 1.00286 | 60 | 140 | 1.01678 | 95 | 203 | 1.03943 |
| 30 | 86 | 1.00425 | 65 | 149 | 1.01951 | 100 | 212 | 1.04332 |

# Air

Air is a gas that is a mixture of 23.2 percent (by weight) oxygen, 75.5 percent nitrogen, and 1.3 percent argon. Other substances present in the air or atmosphere in small amounts are 0.03 to 0.04 percent carbonic acid, or carbon dioxide, 0.01 percent krypton, and small amounts of several other gases. The air or atmosphere is a mixture of the following gases by volume: 21.0 percent oxygen, 78.0 percent nitrogen, and, 0.94 percent argon.

The term *free air* refers to the air at atmospheric pressure. It does not refer to air under identical conditions. Barometer and temperature readings vary with the altitude of a locality and at different times. Thus, free air is not necessarily the air at sea level conditions— or an absolute pressure of 14.7 psi at a temperature of 60°F. It is correct to refer to the air at atmospheric condition at the point where a compressor is installed as free air.

The average condition of the atmosphere in a temperate climate is referred to as *normal air*. This term is used to indicate air with 36 percent relative humidity at 68°F.

# Humidity

Water vapor is always present in the atmosphere. The actual quantity of water present in the air is referred to as *absolute humidity*, and it is usually expressed as grains of moisture per cubic foot of air. A *grain* is $1/7000$ (0.00014285) part of 1 pound. The temperature of the air determines the amount of water that the air is capable of holding—the warmer the air, the more moisture it can hold. For example, the air at 80°F can hold nearly twice the moisture as it can hold at 60°F.

The actual amount of moisture in the air as compared with the maximum amount of moisture that the air is capable of holding at a given temperature, expressed as a percentage, is called the *relative humidity*. Two thermometers (a wet-bulb and a dry-bulb thermometer) are required to determine relative humidity. This is a form of *hydrometer* and it consists of two thermometers mounted side by side; the bulb of one thermometer is kept moist by means of a loose cotton wick tied around its bulb, the lower end of the wick dipping into a vessel that contains water. The wet bulb is cooled by evaporation of water from the bulb. Therefore, the wet-bulb thermometer indicates a lower temperature reading than the dry-bulb thermometer (the difference depending on the rate of evaporation, which, in turn, is determined by the amount of water vapor in the atmosphere). If the air is saturated with moisture, its relative humidity is 100 percent. Air at the same temperature, but holding one-half the saturation amount, has a relative humidity of 50 percent. A table

## Table 1-6  Pound per Square Inch to Feet (Head) of Water (Based on Water at Its Greatest Density)

| Pressure, Pounds Per Square Inch | Feet Head | Pressure, Pounds Per Square Inch | Feet Head | Pressure, Pounds Per Square Inch | Feet Head | Pressure, Pounds Per Square Inch | Feet Head | Pressure, Pounds Per Square Inch | Feet Head | Pressure, Pounds Per Square Inch | Feet Head | Pressure, Pounds Per Square Inch | Feet Head |
|---|---|---|---|---|---|---|---|---|---|---|---|---|---|
| 1 | 2.31 | 53 | 122.43 | 105 | 242.55 | 157 | 382.67 | 209 | 482.79 | 261 | 602.91 | 365 | 843.15 |
| 2 | 4.62 | 54 | 124.74 | 106 | 244.86 | 158 | 364.98 | 210 | 485.10 | 262 | 605.22 | 370 | 854.70 |
| 3 | 6.93 | 55 | 127.05 | 107 | 247.17 | 159 | 367.29 | 211 | 487.41 | 263 | 607.53 | 375 | 856.28 |
| 4 | 9.23 | 56 | 129.36 | 108 | 249.48 | 160 | 369.60 | 212 | 489.72 | 264 | 609.84 | 380 | 877.80 |
| 5 | 11.55 | 57 | 131.67 | 109 | 251.79 | 161 | 371.91 | 213 | 492.03 | 265 | 612.15 | 385 | 889.35 |
| 6 | 13.86 | 58 | 133.98 | 110 | 254.10 | 162 | 374.22 | 214 | 494.04 | 266 | 614.46 | 390 | 900.90 |
| 7 | 16.17 | 59 | 136.29 | 111 | 256.41 | 163 | 375.53 | 215 | 496.65 | 267 | 616.77 | 395 | 912.45 |
| 8 | 18.48 | 60 | 138.60 | 112 | 258.72 | 164 | 378.84 | 216 | 498.96 | 268 | 619.08 | 400 | 924.00 |
| 9 | 20.79 | 61 | 140.91 | 113 | 261.03 | 165 | 381.15 | 217 | 501.27 | 269 | 621.39 | 405 | 931.55 |
| 10 | 23.10 | 62 | 143.22 | 114 | 263.34 | 166 | 383.45 | 218 | 503.58 | 270 | 623.70 | 410 | 947.10 |
| 11 | 25.41 | 63 | 145.53 | 115 | 265.65 | 167 | 385.77 | 219 | 505.89 | 271 | 626.01 | 415 | 958.65 |
| 12 | 27.72 | 64 | 147.84 | 116 | 267.96 | 168 | 388.08 | 220 | 508.20 | 272 | 628.32 | 420 | 970.20 |
| 13 | 30.03 | 65 | 150.15 | 117 | 270.27 | 169 | 390.39 | 221 | 510.51 | 273 | 630.63 | 425 | 981.75 |
| 14 | 32.34 | 66 | 152.46 | 118 | 272.58 | 170 | 392.70 | 222 | 512.82 | 274 | 632.94 | 430 | 993.30 |
| 15 | 34.65 | 67 | 154.77 | 119 | 274.89 | 171 | 395.01 | 223 | 515.13 | 275 | 635.25 | 435 | 1004.85 |
| 16 | 36.96 | 68 | 157.08 | 120 | 277.20 | 172 | 397.32 | 224 | 517.44 | 276 | 637.56 | 440 | 1016.40 |
| 17 | 39.27 | 69 | 159.39 | 121 | 279.51 | 173 | 399.63 | 225 | 519.75 | 277 | 639.87 | 445 | 1027.95 |
| 18 | 41.58 | 70 | 161.70 | 122 | 281.82 | 174 | 401.94 | 226 | 522.06 | 278 | 642.18 | 450 | 1039.50 |
| 19 | 43.89 | 71 | 164.01 | 123 | 284.13 | 175 | 404.25 | 227 | 524.37 | 279 | 644.49 | 455 | 1051.06 |
| 20 | 46.20 | 72 | 166.32 | 124 | 286.44 | 176 | 406.56 | 228 | 526.65 | 280 | 646.80 | 460 | 1062.60 |
| 21 | 48.51 | 73 | 168.83 | 125 | 288.75 | 177 | 408.87 | 229 | 528.99 | 281 | 649.11 | 465 | 1074.15 |
| 22 | 50.82 | 74 | 170.94 | 126 | 291.06 | 178 | 411.18 | 230 | 531.30 | 282 | 851.42 | 470 | 1085.70 |
| 23 | 53.13 | 75 | 173.25 | 127 | 293.37 | 179 | 413.49 | 231 | 533.61 | 283 | 653.73 | 475 | 1097.25 |
| 24 | 55.44 | 76 | 175.56 | 128 | 295.68 | 180 | 415.80 | 232 | 535.92 | 284 | 655.04 | 480 | 1106.80 |

| | | | | | | | | | | | | |
|---|---|---|---|---|---|---|---|---|---|---|---|---|---|
| 25 | 57.75 | 77 | 177.87 | 129 | 297.99 | 181 | 418.11 | 233 | 538.23 | 285 | 668.35 | 485 | 1120.35 |
| 26 | 60.06 | 78 | 180.18 | 130 | 300.30 | 182 | 420.42 | 234 | 540.54 | 286 | 660.66 | 490 | 1131.90 |
| 27 | 62.37 | 79 | 182.49 | 131 | 302.61 | 183 | 422.73 | 235 | 542.85 | 287 | 662.97 | 495 | 1143.45 |
| 28 | 64.68 | 80 | 184.80 | 132 | 304.92 | 184 | 425.04 | 236 | 545.16 | 288 | 665.28 | 500 | 1155.00 |
| 29 | 66.99 | 81 | 187.11 | 133 | 307.23 | 185 | 427.35 | 237 | 547.47 | 289 | 667.59 | 525 | 1212.75 |
| 30 | 69.30 | 82 | 189.42 | 134 | 309.54 | 186 | 429.66 | 238 | 549.78 | 290 | 669.90 | 550 | 1270.50 |
| 31 | 71.61 | 83 | 191.73 | 135 | 311.85 | 187 | 431.97 | 239 | 552.09 | 291 | 672.21 | 575 | 1328.25 |
| 32 | 73.92 | 84 | 194.04 | 136 | 314.16 | 188 | 434.28 | 240 | 554.40 | 292 | 674.52 | 600 | 1386.00 |
| 33 | 76.23 | 85 | 196.35 | 137 | 316.47 | 189 | 436.59 | 241 | 556.71 | 293 | 676.83 | 625 | 1443.75 |
| 34 | 78.54 | 86 | 198.55 | 138 | 318.78 | 190 | 438.90 | 242 | 559.02 | 294 | 679.14 | 650 | 1501.50 |
| 35 | 80.85 | 87 | 200.97 | 139 | 321.09 | 191 | 441.21 | 243 | 561.33 | 295 | 681.45 | 675 | 1559.25 |
| 36 | 83.18 | 88 | 203.28 | 140 | 323.40 | 192 | 443.52 | 244 | 563.64 | 296 | 683.76 | 700 | 1617.00 |
| 37 | 85.47 | 89 | 205.59 | 141 | 325.71 | 193 | 445.83 | 245 | 565.95 | 297 | 686.07 | 725 | 1674.75 |
| 38 | 87.78 | 90 | 207.90 | 142 | 328.02 | 194 | 448.14 | 246 | 568.26 | 298 | 688.38 | 750 | 1732.60 |
| 39 | 90.09 | 91 | 210.21 | 143 | 330.33 | 195 | 450.45 | 247 | 570.57 | 299 | 690.69 | 775 | 1790.25 |
| 40 | 92.40 | 92 | 212.52 | 144 | 332.64 | 196 | 452.76 | 248 | 572.88 | 300 | 693.00 | 800 | 1848.00 |
| 41 | 94.71 | 93 | 214.83 | 145 | 334.95 | 197 | 455.07 | 249 | 575.19 | 305 | 704.55 | 825 | 1905.75 |
| 42 | 97.02 | 94 | 217.14 | 146 | 337.26 | 198 | 457.38 | 250 | 577.50 | 310 | 716.10 | 850 | 1963.60 |
| 43 | 99.33 | 95 | 219.45 | 147 | 339.57 | 199 | 459.69 | 251 | 579.81 | 315 | 727.55 | 875 | 2021.25 |
| 44 | 101.64 | 96 | 221.76 | 148 | 341.88 | 200 | 462.00 | 252 | 582.12 | 320 | 739.20 | 900 | 2079.00 |
| 45 | 103.95 | 97 | 224.07 | 149 | 344.19 | 201 | 464.31 | 253 | 584.43 | 325 | 750.75 | 925 | 2136.75 |
| 46 | 106.26 | 98 | 226.38 | 150 | 345.50 | 202 | 466.62 | 254 | 586.74 | 330 | 762.30 | 950 | 2194.50 |
| 47 | 108.57 | 99 | 228.69 | 151 | 348.81 | 203 | 468.93 | 255 | 589.05 | 335 | 773.85 | 975 | 2252.25 |
| 48 | 110.88 | 100 | 231.00 | 152 | 351.12 | 204 | 471.24 | 256 | 591.36 | 340 | 785.40 | 1000 | 2310.00 |
| 49 | 113.19 | 101 | 233.31 | 153 | 353.43 | 205 | 473.55 | 257 | 593.67 | 345 | 796.95 | 1500 | 3465.00 |
| 50 | 115.50 | 102 | 235.62 | 154 | 355.74 | 206 | 476.86 | 258 | 595.98 | 350 | 808.50 | 2000 | 4620.00 |
| 51 | 117.81 | 103 | 237.93 | 155 | 358.05 | 207 | 478.17 | 259 | 598.29 | 355 | 820.05 | 3000 | 6930.00 |
| 52 | 120.12 | 104 | 240.24 | 156 | 380.36 | 208 | 480.48 | 260 | 600.60 | 360 | 831.60 | | |

can be used to determine the percentage of relative humidity after the wet-bulb and dry-bulb readings have been obtained.

## Weight of Air

Pure air (at 32°F and a barometric pressure of 14.696 psi) weighs 0.08071 pounds per cubic foot. The weight and volume of air change with variations in temperature and pressure (Table 1-7).

**Table 1-7  Volume and Weight of Air at Atmospheric Pressure for Different Temperatures**

| Temperature °F | Volume of 1 Pound of Air (cu.ft.) | Weight per Cubic Foot (lbs.) | Temperature °F | Volume of 1 Pound of Air (cu.ft.) | Weight per Cubic Foot (lbs.) |
|---|---|---|---|---|---|
| 0 | 11.57 | 0.0864 | 325 | 19.76 | 0.0508 |
| 12 | 11.88 | 0.0842 | 350 | 20.41 | 0.0490 |
| 22 | 12.14 | 0.0824 | 375 | 20.96 | 0.0477 |
| 32 | 12.39 | 0.0807 | 400 | 21.69 | 0.0461 |
| 42 | 12.64 | 0.0791 | 450 | 22.94 | 0.0436 |
| 52 | 12.89 | 0.0776 | 500 | 24.21 | 0.0413 |
| 62 | 13.14 | 0.0761 | 600 | 26.60 | 0.0376 |
| 72 | 13.39 | 0.0747 | 700 | 29.59 | 0.0338 |
| 82 | 13.64 | 0.0733 | 800 | 31.75 | 0.0315 |
| 92 | 13.89 | 0.0720 | 900 | 34.25 | 0.0292 |
| 102 | 14.14 | 0.0707 | 1000 | 37.31 | 0.0268 |
| 112 | 14.41 | 0.0694 | 1100 | 39.37 | 2.0254 |
| 122 | 14.66 | 0.0682 | 1200 | 41.84 | 0.0239 |
| 132 | 14.90 | 0.0671 | 1300 | 44.44 | 0.0225 |
| 142 | 15.17 | 0.0659 | 1400 | 46.95 | 0.0213 |
| 152 | 15.41 | 0.0649 | 1500 | 49.51 | 0.0202 |
| 162 | 15.67 | 0.0638 | 1600 | 52.08 | 0.0192 |
| 172 | 15.92 | 0.0628 | 1700 | 54.64 | 0.0183 |
| 182 | 16.18 | 0.0618 | 1800 | 57.14 | 0.0175 |
| 192 | 16.42 | 0.0609 | 2000 | 62.11 | 0.0161 |
| 202 | 16.67 | 0.0600 | 2200 | 67.11 | 0.0149 |
| 212 | 16.92 | 0.0591 | 2400 | 72.46 | 0.0138 |
| 230 | 17.39 | 0.0575 | 2600 | 76.92 | 0.0130 |
| 250 | 17.89 | 0.0559 | 2800 | 82.64 | 0.0121 |
| 275 | 18.52 | 0.0540 | 3000 | 87.72 | 0.0114 |
| 300 | 19.16 | 0.0522 | | | |

*Volumetric expansion = linear expansion.*

Moisture in the air has an adverse effect on an air compressor. The efficiency of the air compressor is reduced because the presence of water vapor in the air being compressed increases the total heating capacity of the air. This is due to the latent heat of the water vapor. The increased temperature increases the pressure and power required for compression.

## Summary

The three states in which matter may exist are known as solid, liquid, and gas. Water is a familiar example of a substance that exists in each of the three states of matter as ice, water, and steam, respectively.

Energy is the capacity for doing work and for overcoming resistance. The two types of energy are potential and kinetic.

The two types of heat are sensible and latent. The effect of heat is produced by the accelerated vibration of molecules. Heat is transferred from one body to another by radiation, conduction, and convection.

Pressure (P) is a force exerted against an opposing body, or a thrust distributed over a surface. Pressure is considered to be distributed over a unit area of the surface.

Atmospheric pressure is caused by the weight of the Earth's atmosphere. At sea level it is equal to about 14.69 psi. The pressure of the atmosphere does not remain constant at a given location, because weather conditions vary continually.

Pressure measured above that of atmospheric pressure is termed gage pressure. Pressure measured above that of a perfect vacuum is termed absolute pressure.

The barometer is used to measure atmospheric pressure. The barometer reading is expressed in inches of mercury. At standard atmospheric pressure, the barometer reads approximately 30 inches of mercury. The barometer reading in inches of mercury can be converted to psi by multiplying the barometer reading by 0.49116.

The force that tends to draw all bodies in the Earth's sphere toward the center of the Earth is known as gravity. The rate of acceleration of gravity is approximately 32 feet per second per second. Centrifugal force tends to move a rotating body away from its center of rotation. Centripetal force tends to move a rotating body toward its center of rotation.

A force is defined completely only when its direction, magnitude, and point of application are defined. All these factors can be represented by a line or vector with an arrowhead.

Motion is described as a change in position in relation to an assumed fixed point. Motion is strictly a relative matter. Velocity is the rate of change of position in relation to time, and acceleration is the rate of increase or average increase in velocity in a given unit of time.

The resistance to motion of two moving objects that touch is called friction. The ratio of the force required to slide a body along a horizontal plane surface to the weight of the body is called the coefficient of friction.

The expenditure of energy to overcome resistance through a distance is work. The standard unit for measuring work is the foot-pound (ft-lb). The foot-pound is the amount of work that is done in raising 1 pound a distance of 1 foot, or in overcoming a pressure of 1 pound through a distance of 1 foot.

Power is the rate at which work is done, or work divided by the time in which it is done. The standard unit for measuring power is the horsepower (hp), which is defined as 33,000 foot-pounds per minute. The formula that can be used to calculate engine horsepower is as follows:

$$hp = \frac{2\ PLAN}{33,000}$$

The basic machines are lever, wheel and axle, pulley, inclined plane, screw, and wedge. The Principle of Moments is important in studying the basic machines.

An important property of water is that it varies in weight (pound per unit volume) with changes in temperature, giving rise to circulation in boilers and heating systems. A U.S. gallon of water (231 cubic inches) weights 8.33111 pounds at 62°F.

Air is a gas that is a mixture of 23.2 percent (by weight) oxygen, 75.5 percent nitrogen, and 1.3 percent argon. Other substances present are 0.03 to 0.04 percent carbonic acid, or carbon dioxide, 0.01 percent krypton, and small quantities of several other gases. Air is a mixture (by volume) of 21.0 percent oxygen, 78.0 percent nitrogen, and 0.94 percent argon.

Humidity is the water vapor that is always present in the atmosphere. The actual quantity of water present is absolute humidity, and it is usually expressed as grains of moisture per cubic foot of air. A grain is $\frac{1}{7000}$ part of one pound. The actual amount of moisture in the air as compared with the maximum amount of moisture that the air is capable of holding, expressed as a percentage, is called relative humidity.

## Review Questions

1. What is the definition of *matter*?
2. What are the three states of matter?

3. What is the definition of *energy*?
4. What are the two forms of energy?
5. State the Law of Conservation of Energy.
6. What are the two forms of heat?
7. What is the unit of heat?
8. What is meant by the *specific heat* of a substance?
9. What are the three methods of transferring heat?
10. What is meant by *pressure*?
11. What is the value for standard atmospheric pressure?
12. What is the difference between *gage pressure* and *absolute pressure*?
13. How is atmospheric pressure measured?
14. What is the definition of *gravity*?
15. What is the difference between *centrifugal force* and *centripetal force*?
16. What is the definition of *motion*?
17. Explain the difference between *momentum* and *inertia*.
18. What is *friction*, and what is its cause?
19. Explain the difference between *work* and *power*.
20. What is the standard unit of work? Of power?
21. State the Principle of Moments.
22. What elements are found in air?
23. What is the definition of *humidity*?
24. What is *inertia*?
25. What did Joule's experiment reveal?
26. What is the difference between *latent heat* and *sensible heat*?
27. What are some of the disadvantages of expansion and contraction caused by changes in temperature?
28. What does a *barometer* measure?
29. What is meant by *state of equilibrium*?
30. What are Newton's three laws of motion?
31. What are the two states of inertia?
32. What does *coefficient of friction* mean?
33. What are the three laws of friction?
34. What does the term *fulcrum* mean?

**35.** What is the *point of maximum density* of water?

**36.** What does a gallon of water weigh?

**37.** How compressible is water?

**38.** What is the weight of pure air at 32°F and a barometric pressure of 14.696 pounds?

**39.** How is relative humidity found?

**40.** What causes the efficiency of an air compressor to decrease?

# Chapter 2

## Principles of Hydraulics

An understanding of hydraulics is necessary before the basic principles of the various types of pumps can be understood. *Hydraulics* is the branch of physics that deals with the mechanical properties of water and other liquids, and with the application of these properties in engineering.

While hydraulics came into prominence in the past half century, it is not new. Blaise Pascal, a French scientist in the seventeenth century, evolved the fundamental law for the science of hydraulics. The inability to produce satisfactory sealing materials, machined finishes, and machined parts restricted the growth of hydraulics prior to the early part of the twentieth century.

### Basic Principles

As stated earlier, water is a truly remarkable substance. At its maximum density (39.1°F), water expands as heat is applied. It also expands slightly as the temperature decreases from that point. Under the influence of temperature and pressure, water can exist as a solid (ice), a liquid (water), and a gas (steam).

*The pressure exerted by a liquid on a surface is proportional to the area of the surface.* Figure 2-1 shows this hydraulic principle. Two cylinders with different diameters are jointed by a tube and filled with water. If the area of the larger piston $M$, for example, is 30 times the area of the smaller piston $S$, and a 2-pound weight is placed on the smaller piston, the pressure is transmitted to the water and to the larger piston. That pressure is equal to 2 pounds for each portion of the larger piston's surface that is equal in area to the total area of the surface of the smaller piston. The larger piston is exposed to an upward pressure equal to 30 times that of the smaller piston, or 60 pounds. In addition, if a 60-pound weight is placed on the larger piston, the two pistons remain in equilibrium. This equilibrium is destroyed if either a greater or a lesser weight is applied.

*A small quantity of water can be made to balance a much larger weight.* As shown in Figure 2-2, a locomotive weighing 101,790 pounds can be balanced on a hydraulic lift against a smaller quantity of water, assuming no leakage or friction and that the vertical pipe leading to the plunger cylinder is exceedingly small in diameter. For example, if the area of the plunger requires 100 psi pressure on the plunger to balance the 101,790-pound locomotive, the load is balanced when the pipe is filled with water to a height of 231 feet (100 × 2.31).

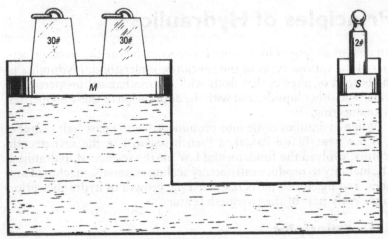

**Figure 2-1** The hydraulic principle that the pressure exerted by a liquid on a surface is proportional to the area of the surface.

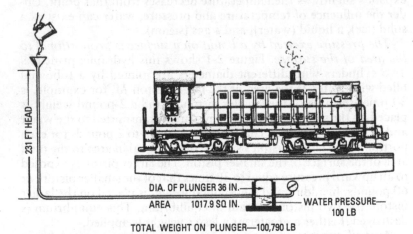

**Figure 2-2** The principle that a smaller quantity of water can be made to balance a much larger weight.

*The pressure on any portion of a fluid having uniform density is proportional to its depth below the surface* (see Figure 2-3). As illustrated in Figure 2-4, the number of 1-pound weights placed on a scale totals 11 pounds, which gives a total weight or pressure of 11 psi if resting on 1 square inch of surface. Likewise, a column of

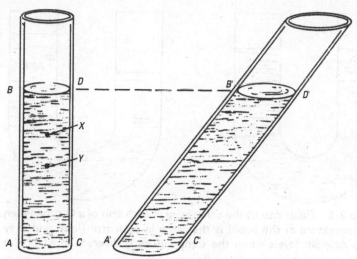

**Figure 2-3**  The principle that the pressure on any portion of a liquid having uniform density is proportional to its depth below the surface.

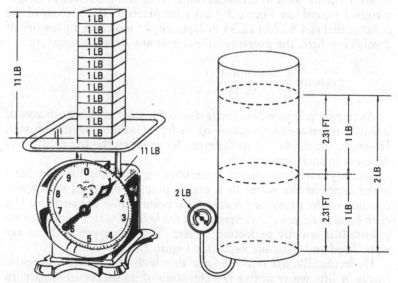

**Figure 2-4**  The principle that the pressure of water at any point below the surface is proportional to the depth of the point below the surface. Pressure in psi is Illustrated (left), and the relation between water pressure at any point below the surface and the depth in feet is shown (right).

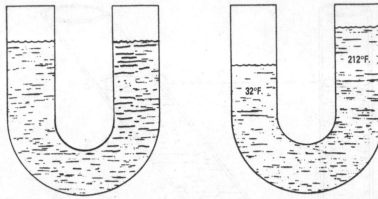

**Figure 2-5** Fluids rise to the same level in each arm of a U-tube when the temperature of the liquid is the same in each arm (left), and they rise to different levels when the temperature is different in each arm (right).

water 1 square inch in cross-sectional area and 2.31 feet in height weighs 1 pound (see Figure 2-4). A gage placed at the bottom of the column that is 4.62 feet (2.31 × 2), in height indicates a pressure of 2 psi. Therefore, the pressure of water at any depth is equal to:

$$\text{pressure} = \frac{\text{depth in feet}}{2.31}$$

As shown in Figure 2-5, *fluids rise to the same level in each arm of a U-tube when the temperature of the liquid is the same in each arm.* However, the fluids rise to different levels when the temperature is different in each arm (right).

Primary considerations in hydraulics are head and lift. The *head* is the depth of the water in a vessel, pipe, or conduit, which is a measure of the pressure on any given point below the surface. The term *head* indicates the difference in the level of water between two points. It is usually expressed in feet. The two types of head are *static head* and *dynamic head* (see Figure 2-6).

Hydraulically, *lift* is the height to which atmospheric pressure forces or lifts water above the elevation of its source of supply. In respect to pump operation, the height, measured from the elevation of the source of supply to the center point of the inlet opening of the pump is termed *lift* (see Figure 2-7 and Figure 2-8). The two types of lift are *static lift* and *dynamic lift*.

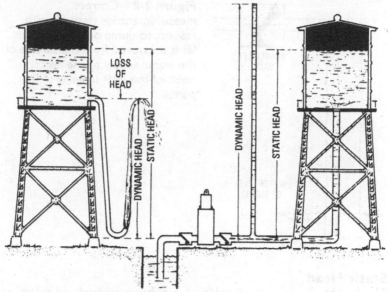

**Figure 2-6**  Static head and dynamic head.

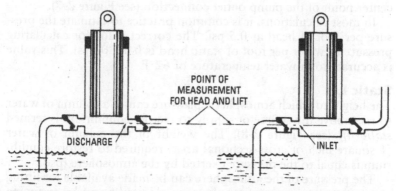

**Figure 2-7**  Correct points of measurement for head and lift.

## Hydrostatics

The branch of hydraulics dealing with the pressure and equilibrium (at rest) of water and other liquids is known as *hydrostatics*. Water is the liquid that is most often considered in studying the basic principles of hydraulics, but other liquids are also included.

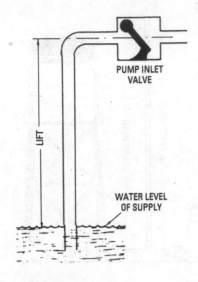

**Figure 2-8** Correct measurement for static lift with respect to pump operation. Static lift is measured from the surface of the water supply to the center point of the inlet opening of the pump.

## Static Head

The height above a given point of a column or body of water at rest (the weight of the water causing pressure) is termed *static head*. With respect to pump operation, the head is measured from the center point of the pump outlet connection (see Figure 2-7).

In most calculations, it is common practice to estimate the pressure per foot of head at 0.5 psi. The correct value for calculating pressure of water per foot of static head is 0.43302 psi. This value is accurate for a water temperature of 62°F.

## Static Lift

The height to which atmospheric pressure causes a column of water to rise above the source of supply to restore equilibrium is termed *static lift* (see Figure 2-8). The weight of the column of water (1 square inch of cross-sectional area) required to restore equilibrium is equal to the pressure exerted by the atmosphere (psi).

The pressure of the atmosphere can be made available for lifting water from the source of supply to an elevated pump by removing air from the inlet pump (see Figure 2-9). In the left-hand diagram, the inlet pipe (before connection with the pump) presses downward on the surface of the water with equal pressure (psi) on both the inside and outside of the pipe, because the pipe is open at the top. If the end of the pipe is connected to a pump that removes air from the pipe to create a partial vacuum, the water rises to a height *A* (see Figure 2-9) that is determined by the available atmospheric pressure.

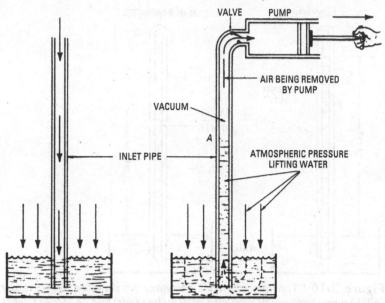

**Figure 2-9** The method of making the atmospheric pressure available for lifting water from the source to an elevated pump.

When the barometer indicates 30 inches of mercury at sea level, the atmospheric pressure at sea level is 14.74 psi, which is the atmospheric pressure that can maintain or balance a 34.042-foot column of water, if the column is completely devoid of air and the water temperature is 62°F. Thus, the atmospheric pressure lifts the water to a height that establishes equilibrium between the weight of the water and the pressure of the atmosphere.

When the water temperature is warmer than 62°F, the height to which the water can be lifted decreases because of the increased vapor pressure. A boiler-feed pump taking water at 153°F, for example, cannot produce a vacuum higher than 21.78 inches, because the water begins to boil at that point and the pump chamber fills with steam. Therefore, the theoretical lift is:

$$34 \times \frac{21.78}{30} = 24.68 \text{ ft (approximately)}$$

The result is approximate because no correction is made for the 34-foot column of water at 62°F. At 153°F, of course, the column length would be increased slightly.

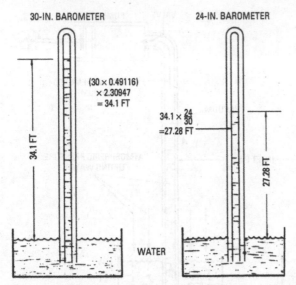

**Figure 2-10** The maximum lift of water at standard temperature (62°F) when barometric pressure is 30 inches (left) and 24 inches (right).

The maximum height to which water at standard temperature (62°F) can be lifted is determined by the barometric pressure. As shown in Figure 2-10, the water rises to a height of 34.1 feet when the barometric pressure is 30 inches, and it rises to 27.28 feet when the barometric pressure is 24 inches (using 0.49116 psi and 2.30947 feet as head of water for each psi).

## Displacement

The weight of the water pushed aside (displaced) by the flotation of a vessel is termed *displacement*. The term *draft* is used to indicate the depth to which an object or vessel sinks into the water. The depth at which the weight of the water displaced is equal to the weight of the object or vessel.

Figure 2-11 shows displacement and draft. If a box made of paper, a cigar box, and a solid block of wood (all of same dimensions) are placed in a pan of water, each sinks to a depth proportional to its weight. Each object sinks until the weight of the water that it displaces is equal to the weight of the object. Thus, the cigar box (2 pounds) sinks to twice the depth of the paper box (1 pound) and the wooden block (3 pounds) sinks to three times the depth that the paper box (1 pound) sinks.

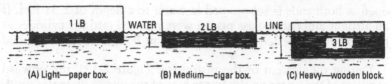

(A) Light—paper box.    (B) Medium—cigar box.    (C) Heavy—wooden block.

**Figure 2-11**   Displacement or draft of objects placed in a liquid.

The resultant pressure of a fluid on an immersed body acts upward vertically through the center of gravity of the displaced fluid, and it is equal to the weight of the fluid displaced. This is known as *Archimedes' Principle*. The center point of pressure for any plane surface acted upon by a fluid is the point of action of the resultant pressure acting upon the surface.

The draft of a boat is less in salt water than it is in fresh water because salt water is heavier than fresh water.

## Buoyancy

The power or tendency of a liquid to keep a vessel afloat is termed *buoyancy*. It is the upward pressure exerted by a fluid on a floating body. Figure 2-12 illustrates buoyancy. In the diagram, a cylinder

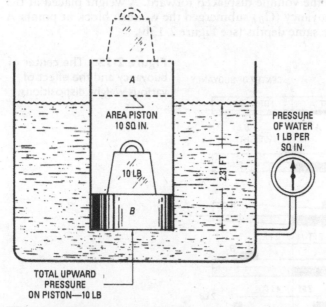

**Figure 2-12**   Buoyancy is illustrated by submerging a cylinder containing a frictionless piston in water.

open at both ends is submerged in water to a depth of 2.31 feet. If an airtight and frictionless piston were inserted into the cylinder at point A and released, it would sink to point B and remain suspended at that point.

During the descent of the piston (see Figure 2-12), the upward pressure of the water on the lower face of the piston increases gradually. The piston does not descend past point B, because the total upward pressure on the piston at B is equal to its weight, and the entire system (piston and displaced water) is in a state of equilibrium, as shown here:

Weight of piston = 10 lb

Pressure of water = 1 psi

Area of piston = 10 sq in

Total pressure on piston (1 × 10) = 10 lb

The center of gravity of the liquid displaced by the immersed body is termed the *center of buoyancy*. As illustrated in Figure 2-13a, a rectangular block of wood placed in water floats evenly (that is, on an even keel), because the volume of water displaced aft is proportional to the volume displaced forward. A weight placed at the center of buoyancy ($C_B$) submerged the wooden block at points A and B to the same depths (see Figure 2-13b).

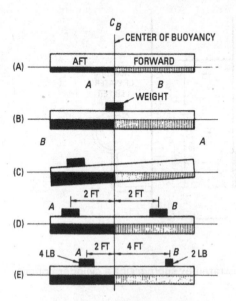

**Figure 2-13** The center of buoyancy and the effect of various weight dispositions.

If the weight is placed in an aft position (see Figure 2-13c), the block is immersed to a greater depth than the forward position. Thus, the center of buoyancy shifts to a point that depends on the position of the weight and its size.

If equal weights (*A* and *B* in Figure 2-13d) are placed at equal distances, the center of buoyancy is the center point and the block remains level. In actual practice, it is not practical to place equal weights at equal distances from the center of buoyancy.

Thus, a 4-pound weight (*A* in Figure 2-13e) is placed 2 feet aft, and a 2-pound weight *B* is placed 4 feet forward to prevent shifting the center of buoyancy. This is also illustrated in Figure 2-14, in which the same wooden block is pivoted through its center of buoyancy $C_B$. The pivot forms the fulcrum, or *origin of moments.*

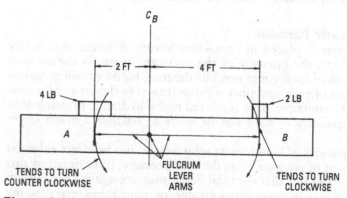

**Figure 2-14** Equilibrium with reference to the center of buoyancy (CB) and proper distribution of weights.

A *moment* is the measure of a force (or weight) by its effect in producing rotation about a fixed point or fulcrum, and it is measured in pounds per feet (lb/ft) when the force is measured in pounds and the distance is measured in feet. The 4-pound weight placed on a 2-foot lever arm (see Figure 2-14) tends to rotate the wooden block counterclockwise. This is with a force of 8 lb/ft (4 lb × 2 ft). Opposed to this force, the 2-pound weight on a 4-foot lever arm tends to turn the block clockwise with a force of 8 lb/ft (2 lb × 4 ft). Since the moments are then equal and opposite, there is no resultant tendency to rotate the wooden block. It is in a state of equilibrium.

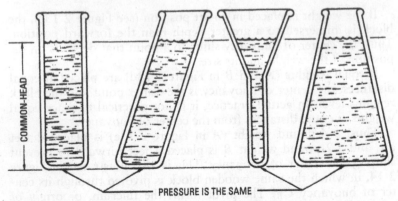

**Figure 2-15** The hydraulic principle that the pressure, in lb. per sq. in., is identical for containers that have different shapes.

## Hydrostatic Paradox

When water is placed in containers having different shapes (see Figure 2-15), the intensity of the pressure, in psi, is the same at the bottom of each container, but the total liquid pressure against the bottom of each container is proportional to the area of the bottom of the container. This is related to the hydraulic principle that *a small quantity of fluid can be made to balance a much larger weight.*

The quantity of liquid, or its total weight, has no effect either on the intensity of pressure or on the total pressure, if the head remains the same. The fact that the total liquid pressure against the bottom of a vessel may be many times greater (or many times less) than the total weight of the liquid is termed a *hydrostatic paradox.*

## Hydrostatic Balance

Archimedes' Principle states that *a body immersed in a fluid loses an amount of weight that is equivalent to the weight of the fluid displaced.* When a body is immersed in a liquid, it is acted upon by two forces:

- *Gravity*, which tends to lower the body.
- *Buoyancy*, which tends to raise the body.

Figure 2-16 illustrates this principle. An assembly consisting of a hollow brass cylinder and a solid cylinder of the same size is suspended from one pan of the balance, and a counterweight is placed on the other pan to balance the assembly. If the hollow cylinder

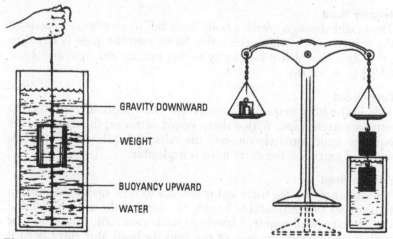

**Figure 2-16** Illustrating Archimedes' Principle that a body immersed in a fluid loses an amount of weight that is equivalent to the weight of the fluid displaced.

is filled with water, equilibrium is disturbed. However, if the balance is lowered to submerge the solid cylinder, equilibrium is restored. Thus, when the solid cylinder is submerged, a portion of its weight equal to the weight of the water in the hollow cylinder is lost.

## Hydrodynamics

The branch of physics dealing with the motion and action of water and other liquids is called *hydrodynamics*. Various forces act upon a liquid, causing it to be in a state of motion.

### Dynamic Head

The *dynamic head* of water is an equivalent or virtual head of water in motion, which represents the resultant pressure necessary to force the water from a given point to a given height and to overcome all frictional resistance. The dynamic head operating to cause *flow* of a liquid is divided into three parts:

- Velocity head
- Entry head
- Friction head

### Velocity Head

The height through which a body must fall in a vacuum to acquire the velocity with which the water flows into the pipe is equal to $(v^2 \div 2g)$, in which $v$ is velocity in feet per second (fps) and $2g = 64.32$ ($g = 32.16$).

### Entry Head

This is the head required to overcome the frictional resistance to entrance to the pipe. With a sharp-edged entrance, the entry head is equal to approximately one-half the velocity head; with a smooth, rounded entrance, the entry head is negligible.

### Friction Head

This is caused by the frictional resistance to flow inside the pipe. In most pipes of considerable length, the sum of the entry head and the velocity head required barely exceeds one foot. In a long pipe with a small head, the sum of the velocity head and entry head is usually so small that it can be omitted. The loss of head caused by the friction of water in pipes and elbows of various sizes and the various rates of flow can be obtained from tables that are used in pump calculations.

## Dynamic Lift

The *dynamic lift* of water is an equivalent or virtual lift of water in motion, which represents the resultant pressure necessary to lift the water from a given point to a given height and to overcome all frictional resistance (see Figure 2-17). The practical limit of actual lift in pump operation ranges from 20 to 25 feet. The practical limit of actual lift is reduced by longer inlet lines, by a larger number of elbows, and by pipes that are too small. Higher altitudes also reduce the practical limit of lift.

Figure 2-18 illustrates the operation of a common vacuum-type lift pump. When the mercury in the barometer $O$ is at 30 inches (corresponding to atmospheric pressure at 14.74 pounds) the head of water is 2.30947 feet for each psi. Therefore, if water, rather than mercury, were used in the barometer, the height of the column of water (at atmospheric pressure of 14.74 pounds, or 30 inches of mercury) would be 34.042 feet (14.74 × 2.30947). The piston of the pump in the illustration is located at 31.76 feet above the water level, and an attached mercurial gage $H$ would read 28 inches, leaving a margin of only 2 inches of available pressure to overcome friction and to lift the foot valve. The distance 31.76 feet is the approximate height that the water can be lifted at atmospheric pressure of

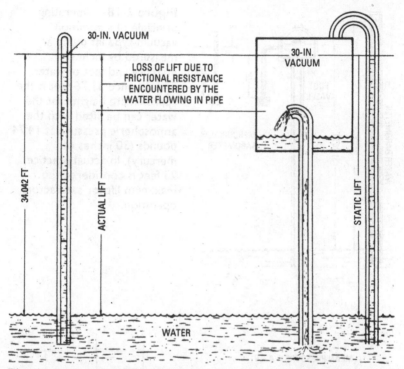

**Figure 2-17** Theoretical lift (left) for a pump, which corresponds to static lift for a given barometer reading, but is not obtained in actual practice; the dynamic lift (right) is actual lift, plus all frictional resistance.

14.74 pounds (30 inches of mercury). The maximum practical lift for satisfactory pump operation is approximately 25 feet.

The term *negative lift* is applied when the level of the water supply is higher than the pump inlet; or it is the vertical distance from the water supply level to the pump inlet at a lower level. This is sometimes called *suction head* (see Figure 2-19).

As shown in Figure 2-19, the standard measurement for negative lift is from the surface of the water supply to the lower center point of the pump inlet, but the pressure caused by the negative lift varies, depending on the position on the piston. It should be noted that the column to the piston balances a portion of the total column. Therefore, the actual amount of negative lift at a given instant is the difference between the two columns.

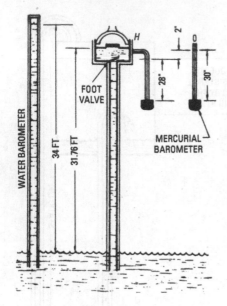

**Figure 2-18** Operating principle of a common vacuum-type lift pump as measured by inches of mercury and feet of water. The distance 31.76 feet is the approximate height that the water can be lifted with the atmospheric pressure at 14.74 pounds (30 inches of mercury). In actual practice, 25 feet is considered the maximum lift for satisfactory operation.

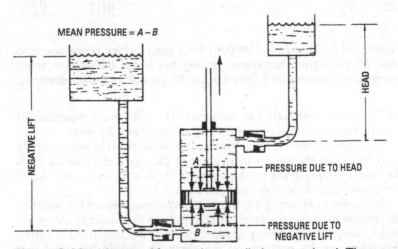

**Figure 2-19** Negative lift, sometimes called suction head. The pump shown is double-acting, but the second set of valves is not shown for the sake of simplicity.

### Table 2-1  Theoretical Lift for Various Temperatures (Leakage not Considered)

| Temp. (°F) | Absolute Pressure of Vapor (lb/in²) | Vacuum (in. of mercury) | Lift (ft.) | Temp. (°F) | Absolute Pressure of Vapor (lb/in²) | Vacuum (in. of mercury) | Lift (ft.) |
|---|---|---|---|---|---|---|---|
| 102.1 | 1 | 27.88 | 31.6 | 182.9 | 8 | 13.63 | 15.4 |
| 128.3 | 2 | 25.85 | 29.3 | 188.3 | 9 | 11.60 | 13.1 |
| 141.6 | 3 | 23.83 | 27.0 | 193.2 | 10 | 9.56 | 10.8 |
| 153.1 | 4 | 21.78 | 24.7 | 197.8 | 11 | 7.52 | 8.5 |
| 162.3 | 5 | 19.74 | 22.3 | 202.0 | 12 | 5.49 | 6.2 |
| 170.1 | 6 | 17.70 | 20.0 | 205.9 | 13 | 3.45 | 3.9 |
| 176.9 | 7 | 15.67 | 17.7 | 209.6 | 14 | 1.41 | 1.6 |

### Effect of Temperature on Theoretical Lift

When the water is warm, the height to which it can be lifted decreases because of the increased vapor pressure. A boiler-feed pump, receiving water at 201.96°F, for example, could not produce a vacuum greater than 5.49 inches, because the water at the temperature begins to boil, filling the pump chamber with steam. Therefore, the corresponding theoretical lift is:

$$34 \times \frac{5.49}{30} = 6.22 \text{ ft}$$

Table 2-1 shows the theoretical maximum lift for various temperatures (leakage not considered). A rise in temperature causes expansion. The column of water that can be supported by atmospheric pressure is lengthened (see Figure 2-20).

At 62°F, the atmosphere can support a 34-foot column of water at a barometer reading of 30 inches of mercury. As the temperature rises, the water expands, which lengthens the column of water.

As shown in Figure 2-20, a 34-foot column of water at a temperature of 60°F is placed in a tube that is closed at the bottom and is open at the top. If the water is heated to 180°F, the weight of the water per cubic foot decreases from 62.36 pounds, at 60°F, to 60.57 pounds at 180°F (from Table 1-4 in Chapter 1). Thus, from expansion of the water, the length of the 34-foot column (see Figure 2-20) becomes:

$$34 \times \frac{62.36}{60.57} = 35 \text{ ft}$$

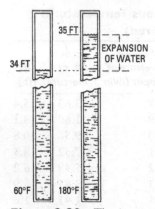

**Figure 2-20** The increase in maximum theoretical lift with an increase in water temperature.

Table 2-2 shows the properties of water with respect to temperature changes. Pressure, specific volume, density, and specific heat all vary with temperature changes.

When a liquid is placed in a closed chamber (which is otherwise empty and at a uniform temperature), evaporation occurs more or less rapidly at first. After a time, however, the space outside the liquid becomes partially filled with stray molecules that have escaped through the surface film. These molecules move about inside the chamber, and they are deflected from its walls and from each other. Some molecules may return to the surface of the liquid, and they may be attracted to the interior portion of the liquid. Ultimately, as many molecules may be returned to the liquid as are leaving it, and an equilibrium may be attained, at which stage evaporation may be said to be have ceased. There is no further loss to the liquid or gain to the vapor outside of it. However, a continued exchange of molecules occurs as new molecules are projected from the surface and other molecules are falling into the liquid in equal numbers. Thus, the chamber is filled with saturated vapor. The vapor is then *saturated*. In any state before this final stage is arrived at, the vapor is said to be *nonsaturated*. A saturated vapor is a vapor that is in equilibrium with its own liquid.

**Effect of Temperature on Dynamic Lift**
Pumps handling water at high temperature must work on reduced actual lift, because the boiling point is related to the pressure. At 212°F, a pump cannot lift any water, because the cylinder fills with steam on the admission stroke.

Theoretically, a pump (with no leakage) can draw or lift water to a height of 34.042 feet. This is when the barometer reads 30 inches of mercury, but the pump cannot attain a perfect vacuum. That is because of valve leakage, air in the water, and water vapor. The actual height of the water is usually less than 20 feet, and it is considerably less for warm or hot water.

When the water is warm, the height that it can be lifted decreases, because of increased vapor pressure. A boiler feed pump receiving water at 153°F, for example, can produce a vacuum not larger than

## Table 2-2 Properties of Water with Respect to Change in Temperature

| Temp. (°F) | Pressure, (lb/in²) | Specific Volume, (cu. ft. per lb.) | Density, (lb. per cu. ft) | Specific Heat |
|---|---|---|---|---|
| 20 | 0.06 | 0.01603 | 62.37 | 1.0168 |
| 30 | 0.08 | 0.01602 | 62.42 | 1.0098 |
| 40 | 0.12 | 0.01602 | 62.43 | 1.0045 |
| 50 | 0.18 | 0.01602 | 62.42 | 1.0012 |
| 60 | 0.26 | 0.01603 | 62.37 | 0.9990 |
| 70 | 0.36 | 0.01605 | 62.30 | 0.9977 |
| 80 | 0.51 | 0.01607 | 62.22 | 0.9970 |
| 90 | 0.70 | 0.01610 | 62.11 | 0.9970 |
| 100 | 0.95 | 0.01613 | 62.00 | 0.9967 |
| 110 | 1.27 | 0.01616 | 61.86 | 0.9970 |
| 120 | 1.69 | 0.01620 | 61.71 | 0.9974 |
| 130 | 2.22 | 0.01625 | 61.55 | 0.9979 |
| 140 | 2.89 | 0.01629 | 61.38 | 0.9986 |
| 150 | 3.71 | 0.01634 | 61.20 | 0.9994 |
| 160 | 4.74 | 0.01639 | 61.00 | 1.0002 |
| 170 | 5.99 | 0.01645 | 60.80 | 1.0010 |
| 180 | 7.51 | 0.01651 | 60.58 | 1.0019 |
| 190 | 9.34 | 0.01657 | 60.36 | 1.0029 |
| 200 | 11.52 | 0.01663 | 60.12 | 1.0039 |
| 210 | 14.13 | 0.01670 | 59.88 | 1.0050 |
| 220 | 17.19 | 0.01677 | 59.63 | 1.0070 |
| 230 | 20.77 | 0.01684 | 59.37 | 1.0090 |
| 240 | 24.97 | 0.01692 | 59.11 | 1.012 |
| 250 | 29.82 | 0.01700 | 58.83 | 1.015 |
| 260 | 35.42 | 0.01708 | 58.55 | 1.018 |
| 270 | 41.85 | 0.01716 | 58.26 | 1.021 |
| 280 | 49.18 | 0.01725 | 57.96 | 1.023 |
| 290 | 57.55 | 0.01735 | 57.65 | 1.026 |
| 300 | 67.00 | 0.01744 | 57.33 | 1.029 |
| 310 | 77.67 | 0.01754 | 57.00 | 1.032 |
| 320 | 89.63 | 0.01765 | 56.66 | 1.035 |
| 330 | 103.00 | 0.01776 | 56.30 | 1.038 |
| 340 | 118.00 | 0.01788 | 55.94 | 1.041 |
| 350 | 135.00 | 0.01800 | 55.57 | 1.045 |
| 360 | 153.00 | 0.01812 | 55.18 | 1.048 |
| 370 | 173.00 | 0.01825 | 54.78 | 1.052 |
| 380 | 196.00 | 0.01839 | 54.36 | 1.056 |
| 390 | 220.00 | 0.01854 | 53.94 | 1.060 |
| 400 | 247.00 | 0.01870 | 53.50 | 1.064 |
| 410 | 276.00 | 0.01890 | 53.00 | 1.068 |
| 420 | 308.00 | 0.01900 | 52.60 | 1.072 |
| 430 | 343.00 | 0.01920 | 52.20 | 1.077 |
| 440 | 381.00 | 0.01940 | 51.70 | 1.082 |

21.78 inches, because the water begins to boil at that point and the pump chamber is filled with steam. The corresponding theoretical lift then is:

$$34 \times \frac{21.78}{30} = 24.68 \text{ ft (approximately)}$$

The result is approximate because no correction has been made for the 34-foot column of water at 62°F, which is lengthened slightly. In addition, the practical lift is considerably less.

## Total Column
The term *total column* has been coined to avoid the term *total head*. The total column is head plus lift. The term *total head* means head plus lift. Lift should never be called head. The two types of total column are *static* and *dynamic*.

### Static Total Column
The static lift plus the static head, or the height or distance from the level of supply to the level in the tank, in feet, is termed the *static total column*. The column is causing pressure resulting from its weight (see Figure 2-21). Thus, in the illustration:

static lift *AB* + static head *CD* = static column *EF*

### Dynamic Total Column
The dynamic lift plus the dynamic head (or the equivalent total column of water in motion) is termed the *dynamic total column*. It represents the pressure resulting from the static total column, plus the resistance to flow caused by friction (see Figure 2-22).

### Problem
If static lift is 20 feet, lift friction is 10 percent; static head is 200 feet; and head friction is 20 percent, what is the dynamic total column? What is the corresponding pressure?

### Solution

$$\text{dynamic lift} = 20 \times 1.10 = 22 \text{ ft}$$
$$\text{dynamic head} = 200 \times 1.20 = \underline{240 \text{ ft}}$$
$$\text{total dynamic column} = 262 \text{ ft}$$
$$\text{corresponding pressure} = 262 \times 0.43302 = 113.5 \text{ psi}$$

## Friction of Water in Pipes
In the plumbing trade, there has been wide misunderstanding concerning the laws governing rates of discharge of water from faucets

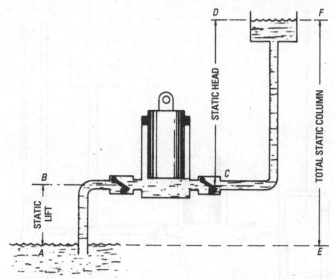

**Figure 2-21**  Static total column (EF), which is static lift (AB) plus static head (CD), extending vertically from the level of the water supply to the point of discharge or surface level of the water in the tank.

and the relations of pressure and discharge. This misunderstanding has produced waste. Table 2-3 can be valuable in calculating the loss of pressure caused by friction in pipes of various sizes that deliver various volumes of water per minute.

In addition to the friction encountered in pipes, friction occurs in the faucets. The friction varies with the type and make of faucet. The friction losses in Table 2-4 refer to a single type and make of faucet. Allowances should be made for other types of faucets.

A plumber should be able to estimate the losses in pressure caused by the several factors that are involved, and thus be able to determine the pressure required at the fixtures to deliver water at a given rate of flow. The following problems are used to illustrate methods of calculation.

### Problem
What pressure is required for a flow of 10 gallons per minute through a ³/₄-inch pipeline that is 350 feet long?

### Solution
From Table 2-3, the drop in pressure is 16.5 pounds per 100 feet of ³/₄-inch pipe for a flow of 10 gallons per minute. For 350 feet of

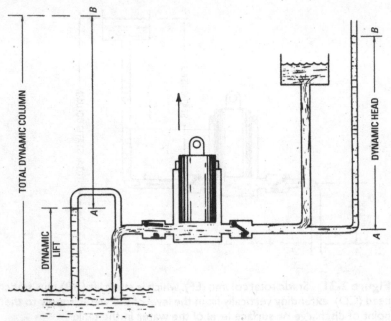

**Figure 2-22** The dynamic total column, which is dynamic lift plus dynamic head. Note that the length of AB (left) is equal to the dynamic head AB (right).

pipe, the required pressure is:

$$\frac{16.5 \times 350}{100} = 16.5 \times 3.5 = 57.8 \text{ lb}$$

**Problem**

The height of the water in a windmill tank is 60 feet. How many gallons per minute can flow through 500 feet of $1\frac{1}{2}$-inch pipe with a vertical rise of 23 feet?

**Solution**

2.3 ft head = 1 lb pressure:

Hence

pressure due to tank elevation = $60 \div 2.3 = 26.1$ lb

pressure loss due to 23-ft rise = $23 \div 2.3 = \underline{10.0}$ lb

total pressure available to cause flow = $16.1$ lb

pressure available per 100 ft of pipe = $16.1 \div 5 = 3.2$ lb

**Table 2-3   Loss in Pressure by Friction per 100-Ft Length of Wrought Iron Pipe (in lb. per sq. in.)**

| Gallons per Minute | Nominal Diameter in Inches | | | | | |
|---|---|---|---|---|---|---|
| | $\frac{1}{2}$ | $\frac{3}{4}$ | 1 | $1\frac{1}{4}$ | $1\frac{1}{2}$ | 2 |
| 1 | 0.9 | 0.2 | — | — | — | — |
| 2 | 3.2 | 0.8 | — | — | — | — |
| 3 | 6.9 | 1.8 | 0.6 | — | — | — |
| 4 | 11.7 | 3.2 | 0.9 | 0.3 | 0.1 | — |
| 5 | 17.8 | 4.5 | 1.4 | 0.4 | 0.2 | — |
| 6 | 24.8 | 6.4 | 2.0 | 0.5 | 0.3 | 0.1 |
| 8 | 42.6 | 10.9 | 3.4 | 0.9 | 0.4 | 0.1 |
| 10 | 64.0 | 16.5 | 5.1 | 1.4 | 0.6 | 0.2 |
| 12 | 90.0 | 23.1 | 7.1 | 1.9 | 0.9 | 0.3 |
| 14 | 120.0 | 30.4 | 9.6 | 2.5 | 1.2 | 0.4 |
| 16 | — | 40.0 | 12.2 | 3.2 | 1.5 | 0.5 |
| 18 | — | 48.7 | 15.2 | 4.0 | 1.8 | 0.7 |
| 20 | — | 59.2 | 18.3 | 4.8 | 2.3 | 0.8 |
| 30 | — | 120.0 | 38.7 | 10.2 | 4.8 | 1.7 |
| 40 | — | — | 66.0 | 17.4 | 8.2 | 2.9 |
| 50 | — | — | 98.0 | 26.1 | 12.4 | 4.3 |

**Table 2-4   Loss in Pressure of Water through a $\frac{1}{2}$-in. Commercial Faucet**

| Rate of Flow (gal. per min.) | Loss in Pressure (lb. per sq. in.) |
|---|---|
| 4 | 2 |
| 5 | $3\frac{1}{2}$ |
| 8 | 9 |
| 10 | 15 |
| 15 | 33 |
| 20 | 60 |

From Table 2-3, the two values nearest 3.2 pounds for $1\frac{1}{2}$-inch pipe are: 2.3 pounds pressure for 20 gallons per minute, and 4.8 pounds pressure for 30 gallons per minute. Thus, by interpolation, the 3.2 pounds pressure corresponds to a flow of:

$$20 + \left[ \frac{(3.2 - 2.3)}{(4.8 - 2.3)} \times 10 \right] = 23.6 \text{ gal/min}$$

**Problem**

A $3/4$-inch pipe extends 75 feet with a vertical rise of 23 feet. What pressure is required at the water mains in the street to deliver 10 gallons per minute through a $1/2$-inch faucet?

**Solution**

From Table 2-3:

> Friction loss through 75 ft of $3/4$-inch pipe = $(16.5 \times 75/100) = 12.4$ lb
>
> Loss due to 23-ft rise $(23 \div 2.3) = 10.0$ lb

From Table 2-4:

> Loss in flow through faucet at 10 gallons per minute = 15.0 lb

Therefore,

> $12.4 + 10 + 15.0 = 37.4$ lb pressure needed

In most installations, a pressure loss caused by flow through the meter must be considered (Table 2-5). At a given rate of flow (30 gallons per minute), it may be noted that the loss in pressure in passing through the meter decreases rapidly as the size of the meter increases.

**Table 2-5 Pressure-Loss in Flow through Meter**

| Meter Size (in) | Gal. per Min. | Pressure Loss (lb) |
|---|---|---|
| $5/8$ | 10 | 2–7 |
| $5/8$ | 30 | 16–65 |
| $3/4$ | 30 | 14–30 |
| 1 | 30 | 4–7 |

**Problem**

Water in a $3/4$-inch pipe passes through a meter and extends 125 feet from a main to a faucet. The faucet is 23 feet above the street main, and must deliver 10 gallons per minute. What pressure is required at the main?

**Solution**

From Table 2-3:

> Friction loss in 125 ft of pipe = $(16.5 \times 1.25) = 20.6$ lb
>
> Pressure loss due to 23-ft rise $(23 \div 2.3) = 10.0$ lb
>
> From Table 2-4: friction loss through faucet = 15.0 lb
>
> From Table 2-5: pressure loss through meter = 3.0 lb

Therefore,

Total pressure required at street main = 48.6 lb

In the previous calculations, loss of pressure caused by the fittings has not been considered. Unless numerous fittings have been used, this loss may be either omitted or estimated. Roughly, the loss in pressure caused by an elbow in small pipes may be estimated at $\frac{1}{4}$ to $\frac{1}{2}$ pound, which indicates that it is desirable to use as few fittings as possible. Table 2-6 provides the loss in head caused by friction in terms of head (in feet) for various sizes of smooth 90° elbows and for 100 feet of smooth pipe. Table 2-7 gives the approximate resistance to water flow of other common fittings, in equivalent feet of straight pipe.

## Flow of Water

The quantity of water discharged through a pipe is determined by the following:

- Head
- Length of pipe
- Character of interior surface
- Number and sharpness of bends

The head is measured vertically between the surface level at the pipe inlet and the level at the center point of the discharge end. Flow is independent of the position of the pipe, regardless of whether it is horizontal or inclined.

## Measurement of Water Flow

The most accurate method that can be used to measure water flow is to measure the volume, or to weigh the liquid delivered. The chief disadvantage of these methods lies in the fact that they can be used only in measuring small quantities of water. Large quantities of water can be measured with the following devices:

- Weir
- Pitot tubes
- Venturi meter

### Weir

The weir is a device that is commonly used to measure water flow. It consists of a notch in the vertical side of a tank or reservoir through which water may flow to be measured. The weir is especially adaptable for measuring the flow of small streams (see Figure 2-23).

## Table 2-6 Loss of Head Due to Friction

### Loss of Head in Feet in Various Sizes of Smooth 90° Elbows

| Gallons Per Min. Delivered | Pipe Sizes, Inches–Inside Diameter | | | | | |
|---|---|---|---|---|---|---|
| | 1 | 1¼ | 1½ | 2 | 2½ | 3 |
| 20 | 2.52 | 0.89 | 0.42 | 0.146 | 0.067 | 0.038 |
| 25 | 3.84 | 1.33 | 0.62 | 0.218 | 0.101 | 0.057 |
| 30 | 5.44 | 1.88 | 0.88 | 0.307 | 0.142 | 0.083 |
| 35 | 7.14 | 2.50 | 1.18 | 0.408 | 0.189 | 0.107 |
| 40 | 9.12 | 3.20 | 1.50 | 0.528 | 0.242 | 0.137 |
| 45 | — | 4.00 | 1.86 | 0.656 | 0.308 | 0.173 |
| 50 | — | 4.80 | 2.27 | 0.792 | 0.365 | 0.207 |
| 70 | — | 9.04 | 4.24 | 1.430 | 0.683 | 0.286 |
| 75 | — | — | 4.80 | 1.670 | 0.781 | 0.458 |
| 90 | — | — | 6.72 | 2.320 | 0.991 | 0.600 |
| 100 | — | — | 8.16 | 2.860 | 1.320 | 0.744 |
| 125 | — | — | — | 4.320 | 2.060 | 1.140 |
| 150 | — | — | — | 6.080 | 2.810 | 1.580 |
| 175 | — | — | — | 8.160 | 3.720 | 2.100 |
| 200 | — | — | — | 10.320 | 4.740 | 2.670 |
| 250 | — | — | — | — | 7.260 | 4.080 |

### Loss of Head per 100 Feet of Smooth Pipe

| Gallons Per Min. Delivered | Pipe Sizes, Inches—Inside Diameter | | | | | |
|---|---|---|---|---|---|---|
| | 1 | 1¼ | 1½ | 2 | 2½ | 3 |
| 20 | 42.0 | 11.1 | 5.2 | 1.82 | 0.61 | 0.25 |
| 25 | 64.0 | 16.6 | 7.8 | 2.73 | 0.92 | 0.38 |
| 30 | 89.0 | 23.5 | 11.0 | 3.84 | 1.29 | 0.54 |
| 35 | 119.0 | 31.2 | 14.7 | 5.10 | 1.72 | 0.71 |
| 40 | 152.0 | 40.0 | 18.8 | 6.60 | 2.20 | 0.91 |
| 45 | — | 50.0 | 23.2 | 8.20 | 2.80 | 1.15 |
| 50 | — | 60.0 | 28.4 | 9.90 | 3.32 | 1.38 |
| 70 | — | 113.0 | 53.0 | 18.40 | 6.21 | 2.57 |
| 75 | — | — | 60.0 | 20.90 | 7.10 | 3.05 |
| 90 | — | — | 84.0 | 29.40 | 9.81 | 4.01 |
| 100 | — | — | 102.0 | 35.80 | 12.00 | 4.96 |
| 125 | — | — | — | 54.00 | 18.20 | 7.60 |
| 150 | — | — | — | 76.00 | 25.50 | 10.50 |
| 175 | — | — | — | 102.00 | 33.80 | 14.00 |
| 200 | — | — | — | 129.00 | 43.10 | 17.80 |
| 250 | — | — | — | — | 66.00 | 27.20 |

**Table 2-7   Approximate Resistance of Common Fittings to Water Flow, in Equivalent Feet of Straight Pipe**

| Nominal Diameter, In. | Standard L | Medium L | Long Sweep L | 45° L | Gate Valve, Open | Globe Valve, Open |
|---|---|---|---|---|---|---|
| $1/2$ | 1.0 | 0.9 | 0.7 | 0.6 | 0.23 | 5.0 |
| $3/4$ | 1.51 | 1.3 | 1.0 | 0.80 | 0.40 | 10.0 |
| 1 | 1.8 | 1.5 | 1.3 | 1.0 | 0.55 | 13.0 |
| $1\,1/4$ | 2.7 | 2.4 | 2.1 | 1.5 | 0.80 | 18.0 |
| $1\,1/2$ | 3.0 | 2.6 | 2.3 | 1.6 | 0.90 | 20.0 |
| 2 | 4.1 | 3.5 | 3.0 | 2.3 | 1.3 | 29.0 |
| $2\,1/2$ | 5.5 | 4.8 | 3.8 | 2.8 | 1.4 | 35.0 |
| 3 | 7.0 | 6.0 | 4.6 | 3.5 | 1.9 | 45.0 |
| 4 | 10.0 | 8.5 | 6.6 | 5.0 | 2.6 | 60.0 |
| 6 | 16.0 | 14.0 | 12.0 | 8.0 | 4.5 | 110.0 |
| 8 | 21.0 | 18.0 | 15.0 | 12.0 | 6.0 | 150.0 |
| 10 | 26.0 | 22.0 | 18.0 | 15.0 | 8.0 | 200.0 |
| 14 | 40.0 | 35.0 | 31.0 | 23.0 | 12.0 | 280.0 |
| 18 | 55.0 | 45.0 | 40.0 | 30.0 | 16.0 | 380.0 |
| 20 | 60.0 | 55.0 | 42.0 | 34.0 | 17.0 | 420.0 |

To construct a weir, a notched board is placed across a small stream at some point, allowing a small pond to form. The notch in the board should be beveled on both side edges and on the bottom. The bottom of the notch is called the *crest* of the weir. The crest should be level and sides should be vertical.

A stake should be driven near the bank in the pond above the weir at a distance less than the width of the notch. The top of the stake should be level with the crest. The depth of the water above the top of the stake can be measured with a graduated rule (see Figure 2-23), or it can be measured with a hook gage for precision (see Figure 2-24).

The *hook gage* consists of a graduated slide arranged to slide in the frame. It can be finely adjusted by means of the screw $T$ (see Figure 2-24) that passes through a lug $L$ with a milled nut. A vernier $V$ is also provided.

The hook gage is first set to zero, at which point the hook is level with the crest, and the slide is raised until the hook pierces the surface of the water. The hook gage is located behind the weir to avoid the curvature effect as the water approaches the weir.

**Figure 2-23** A typical weir. Here it is used to measure the water flow in a small stream.

Not all formulas for weirs obtain the same results. The Francis formula has been used widely. It was derived from a series of experiments on 10-foot weirs as follows:

$$Q = 3.33 (L - 0.14) H \frac{3}{2}, \text{ for one end contracted}$$

$$Q = 3.33 (L - 0.24) H \frac{3}{2}, \text{ for both ends contracted}$$

$$Q = 3.33 \left( L - H \frac{3}{2} \right), \text{ without end contractions}$$

where the following is true:

    $Q$ is cubic feet per second
    $L$ is length of weir, in feet
    $H$ is head over crest, in feet

A formula used for the V-notched weir (for small flows) is:

$$Q = 2.544 \, H \, \frac{5}{2}$$

where the following is true:

H is head, in feet, above the apex of the triangle.

A weir table (Table 2-8) can be used to determine the amount of water flowing over the weir. The first vertical column represents the depth of flow over the notch, and the fractions in the column heads represent parts of an inch. The body of the table indicates the cubic feet of water per minute per inch (or fractional inch) of weir. The result for a 1-inch weir must be multiplied by the total horizontal length of the weir.

### Pitot Tube

A pitot tube is a bent tube that is used to determine the velocity of running water by placing the curved end underneath the surface of the water

**Figure 2-24**  Hook gage, used to measure the depth of water above a weir.

and observing the height to which the water rises in the tube. It is a type of current meter, and the basic feature is the thin-edged orifice at the curved end of the tube (see Figure 2-25). The basic principle of the pitot tube is that when it is placed in running water with the orifice turned upstream, the impact of the fluid causes an excess pressure in the tube that is equal to the velocity head.

The pitot tube formula is derived as follows. The head of the water in the tube due to impact is $v^2/g$, the head caused by velocity is $v$, and the water should rise a height $h$ above the surface. Experiments have shown that the actual height that the water rises is more nearly equal to the velocity head $v^2/2g$ than to $v^2/g$. Thus, the head $h$ is usually considered to be:

$$h = \frac{cv^2}{2g}$$

### Table 2-8   Weir Table for Ft³ of Water per Inch of Depth above Notch

| Inches | | $^1/_8$ | $^1/_4$ | $^3/_8$ | $^1/_2$ | $^5/_8$ | $^3/_4$ | $^7/_8$ |
|---|---|---|---|---|---|---|---|---|
| 0 | 0.00 | 0.01 | 0.05 | 0.09 | 0.14 | 0.19 | 0.26 | 0.32 |
| 1 | 0.40 | 0.47 | 0.55 | 0.64 | 0.73 | 0.82 | 0.92 | 1.02 |
| 2 | 1.13 | 1.23 | 1.35 | 1.46 | 1.58 | 1.70 | 1.82 | 1.95 |
| 3 | 2.07 | 2.21 | 2.34 | 2.48 | 2.61 | 2.76 | 2.90 | 3.05 |
| 4 | 3.20 | 3.35 | 3.50 | 3.66 | 3.81 | 3.97 | 4.14 | 4.30 |
| 5 | 4.47 | 4.64 | 4.81 | 4.98 | 5.15 | 5.33 | 5.51 | 5.69 |
| 6 | 5.87 | 6.06 | 6.25 | 6.44 | 6.62 | 6.82 | 7.01 | 7.21 |
| 7 | 7.40 | 7.60 | 7.80 | 8.01 | 8.21 | 8.42 | 8.63 | 8.83 |
| 8 | 9.05 | 9.26 | 9.47 | 9.69 | 9.91 | 10.13 | 10.35 | 10.57 |
| 9 | 10.80 | 11.02 | 11.25 | 11.48 | 11.71 | 11.94 | 12.17 | 12.41 |
| 10 | 12.64 | 12.88 | 13.12 | 13.36 | 13.60 | 13.85 | 14.09 | 14.34 |
| 11 | 14.59 | 14.84 | 15.09 | 15.34 | 15.59 | 15.85 | 16.11 | 16.36 |
| 12 | 16.62 | 16.88 | 17.15 | 17.41 | 17.67 | 17.94 | 18.21 | 18.47 |
| 13 | 18.74 | 19.01 | 19.29 | 19.56 | 19.84 | 20.11 | 20.39 | 20.67 |
| 14 | 20.95 | 21.23 | 21.51 | 21.80 | 22.08 | 22.37 | 22.65 | 22.94 |
| 15 | 23.23 | 23.52 | 23.82 | 24.11 | 24.40 | 24.70 | 25.00 | 25.30 |
| 16 | 25.60 | 25.90 | 26.20 | 26.50 | 26.80 | 27.11 | 27.42 | 27.72 |
| 17 | 28.03 | 28.34 | 28.65 | 28.97 | 29.28 | 29.59 | 29.91 | 30.21 |
| 18 | 30.54 | 30.86 | 31.18 | 31.50 | 31.82 | 32.15 | 32.47 | 32.80 |
| 19 | 33.12 | 33.45 | 33.78 | 34.11 | 34.44 | 34.77 | 35.10 | 35.44 |
| 20 | 35.77 | 36.11 | 36.45 | 36.78 | 37.12 | 37.46 | 37.80 | 38.15 |

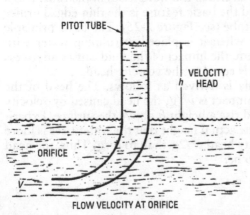

**Figure 2-25**   Basic principle of a pitot tube, used for measuring the flow of water.

The coefficient $c$ is fairly constant for any given tube. The quantity of fluid discharged can be calculated by the following formula:

$$Q = ca\sqrt{2gh}$$

in which the following is true:

Q is quantity, in cubic feet per second
$c$ is coefficient of discharge for orifice
$a$ is area of the orifice
$h$ is velocity head, in feet

### Venturi Meter

A venturi meter is an instrument (similar to an hourglass) used for accurately measuring the discharge of fluid or gas through a pipe. It consists of a conical nozzle-like reducer followed by a more gradual enlargement to the original size, which is the size of the pipe in which the meter is laid (see Figure 2-26).

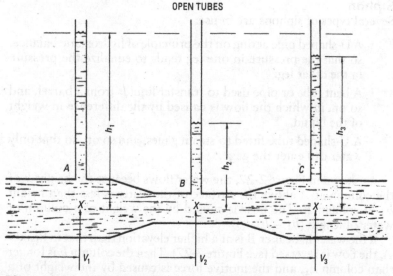

**Figure 2-26**   Basic principle of a venturi meter showing relative pressure heads at points A, B, and C.

In the illustration, the pressure heads ($h_1$, $h_2$, and $h_3$) are shown by the vertical tubes $A,B,C$ as they appear when measured by open water columns. The pressure head $h_2$ is less than $h_1$ by approximately the same difference as the difference in velocity heads as:

$$\frac{(V_2)^2}{2g} - \frac{(V_1)^2}{2g}$$

The pressure $h_3$ at tube C is approximately the same as that at tube A, or $h_1$, being slightly less because of loss caused by friction of the flowing water. An equation for discharge past the meter is:

$$Q = (cAa\sqrt{2gh}) \div \sqrt{A^2 - a^2}$$

in which the following is true:

Q is discharge, in cubic feet per second

A is area at A, in square feet

a is area at B, in square feet

c is coefficient (varies between 0.97 and 1.0)

## Siphon

Several types of siphons are in use:

- A U-shaped pipe acting on the principle of hydrostatic balance, so that the pressure in one leg tends to equalize the pressure in the other leg.

- A bent tube or pipe used to transfer liquids from a barrel, and so on, in which the flow is caused by the difference in weight of the liquid.

- A U-shaped tube fitted to steam gages, and so on, so that only water can enter the gage.

As shown in Figure 2-27, the water flows because the height $h$ of the column at A above the surface of the water is less than height $h_1$ of column B. The flow is caused by the weight of a column of water of length $(h_1 - h)$.

If the water in beaker B is at a higher elevation than that in beaker A, the flow is reversed (see Figure 2-27). Then the column $h$ is longer than column $h_1$, and the motive force is caused by the weight of a column having the length $(h - h_1)$. The atmosphere presses equally on the surface of the water in the beaker, tending to force the water upward in the columns. The unequal weight of water in the two columns upsets equilibrium and causes the water to flow from the beaker having the shorter column to the beaker having the longer column.

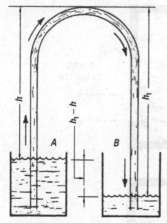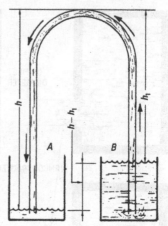

**Figure 2-27** Basic principle of a siphon. Flow is from the shorter column to the longer column (left), and is reversed when the column length is reversed (right).

## Flow through Orifices

An orifice is cut through either the flat side or the bottom of a vessel, leaving sharp edges (see Figure 2-28). The stream lines of force set up in the water approach the orifice in all directions. The direction of flow of the water particles converges. This is with the exception of those near the center-point that produce a contraction of the jet when the orifices are not shaped properly.

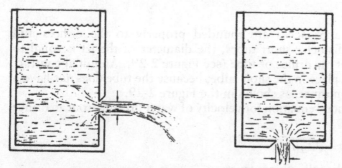

**Figure 2-28** Directions of flow of water particles through orifices in the side (left) and bottom (right) of a vessel to form a contraction of the jet.

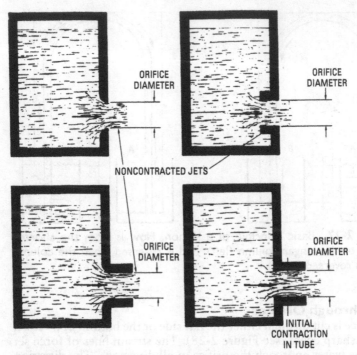

**Figure 2-29** Noncontraction jets in flow of water through properly rounded approaches to orifices (left) and a constant-diameter short tube (upper right). The diameter of the jet equals the area of the orifice or tube. Initial contraction occurs inside the tube (lower right), because the tube does not have a rounded approach.

If the approaches are rounded properly to the orifices and constant-diameter short tubes, the diameter of the jet is equal to the area of the orifice or tube (see Figure 2-29). An initial contraction takes place in the short tube, because the tube does not have a rounded entrance, as shown in the Figure 2-29.

The general equation for velocity of water flowing from an orifice or tube is:

$$v = \sqrt{2gH}$$

in which the following is true:

$v$ is theoretical velocity, in feet per second, corresponding to head $H$

$H$ is head of water, in feet, on centerline of jet

$g$ is acceleration due to gravity (32.2 feet per second per second)

If friction and contraction of the jet are considered, the general equation for discharge of water is:

$$Q = CA\sqrt{2gH}$$

in which the following is true:

$Q$ is discharge, in cubic feet per second

$C$ is coefficient of discharge, which is the product of coefficient of friction $C_1$ and coefficient of contraction $C_2$

$A$ is area, in square feet, of the orifice or tube

The path of a jet issuing from a horizontal orifice or a tube describes a *parabola* (see Figure 2-30). A parabola is a plane curve; every point on the curve is equidistant from a fixed point called

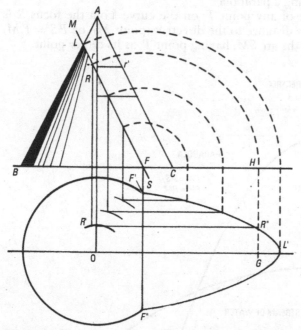

**Figure 2-30**  A parabola. The path described by a jet of liquid issuing from a horizontal orifice or jet is illustrated by a parabola.

the *focus* and from a fixed line called the *directrix*. A parabola is generated by a plane cutting a cone parallel to one of its elements.

As shown in Figure 2-30, the plane *MS* cuts the element *AB* at point *L* and at point *F* on the base. The point *F* is projected downward, cutting the curve at the two points *F'* and *F"*. With point *F* as the center point and radius *LF*, swing point *L* around and project downward to the axis *OG*, obtaining point *L'*, on the curve.

Any other point, *R*, for example, can be obtained as follows: swing point *R* around with point *F* as center point and project downward with line *HG*. Describe an arc (radius = rr' of cone at elevation of point *R*), and where the arc cuts the projection of point *R* at R', project the point R' to line *HG* to obtain point *R"*, which is a point on the curve. The other points can be obtained in a similar manner. The curve can be traced through points F', R", L', and so on, and similar points on the opposite side of the axis, ending at point F" to form the curve, or parabola.

As shown in Figure 2-31, a jet of water issuing from the side of a vessel is drawn downward on leaving the orifice by the force of gravity, describing a parabola.

The distance of any point *P* on the curve from the focus *S* is equivalent to its distance to the directrix (point *M*), or $PS = PM$, as indicated by the arc *SM*, having point *P* as its center point.

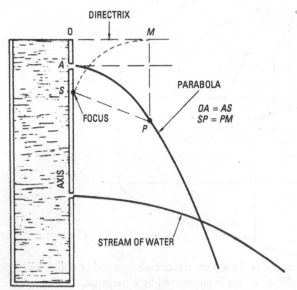

**Figure 2-31** The flow of water through an orifice with a horizontal axis.

If the side $AD$ of the vessel is inclined to vertical at an angle $\theta$ (see Figure 2-32), the jet issues normally with respect to side $AD$, rises to a highest point $C$, and then curves downward. If the distances $x$ and $y$ are the horizontal abscissa and vertical ordinate, measured from the orifice $A$, the equation for the curve, which is also the equation for a common parabola, is:

$$y = x \tan\theta - \frac{x^2 \sec^2 \theta}{4h}$$

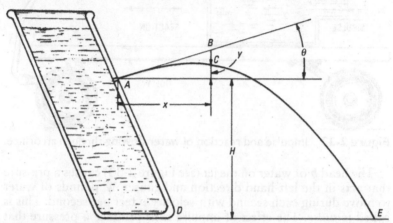

**Figure 2-32** The flow of water through an orifice whose axis is inclined to the horizontal.

When a stream of water impinges on a solid surface, it presses on the surface with a force equal and opposite to the force, which changes the velocity and direction of motion of the water (see Figure 2-33). When the orifice is opened at point $A$, the head $h$ of water causes a pressure $P$ that acts in the direction indicated by the arrow. This causes the cart supporting the vessel to move in the same direction.

The equation for the reaction of the jet is:

$$P = 2ahw$$

in which the following is true:

 $P$ is reaction

 $a$ is area of orifice

 $h$ is head of water on orifice

 $w$ is weight per unit volume of water

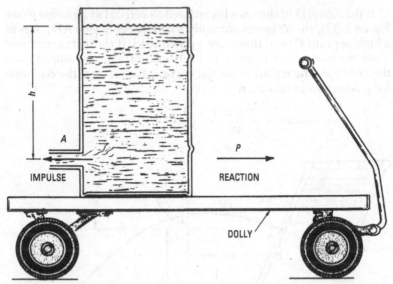

**Figure 2-33** Impulse and reaction of water flowing through an orifice.

The head *h* of water on the jet (see Figure 2-33) causes a pressure that acts in the left-hand direction and causes *W* pounds of water to move during each second with velocity *v* feet per second. This is called *impulse*. The effect of impulse is to produce a pressure that reacts in the right-hand direction, causing the dolly supporting the vessel to move in a direction opposite the direction of jet flow, which is called *reaction*.

**Specific Gravity**
The ratio of the weight of a given volume of a substance to that of an equal volume of another substance, which is used as a standard of comparison (water for liquids and solids; air or hydrogen for gases), is known as *specific gravity* (sp. gr.). Water is the standard that is considered here.

When dealing with solids, the specific gravity is the ratio of the weight in the air of a given substance to the weight of an equal volume of water. That is, the specific gravity (sp. gr.) is a number that indicates how many times a given volume of a substance is heavier than an equal volume of water.

Since the weight of water varies with temperature, comparisons should be made with water at 62°F. One cubic inch of pure water weighs 0.0361 pounds at 62°F. Therefore, if the specific gravity of a material is known, its weight per cubic inch can be calculated by multiplying its specific gravity by 0.0361. To calculate the weight of one cubic foot of a given material, multiply its specific gravity by 62.35 (the weight of 1 cubic foot of water at 62°F). This can be illustrated by a typical problem.

## Problem
If the specific gravity of wrought iron is 7.85, what is its weight per cubic inch?

## Solution
Since water at 62°F weighs 0.0361 pounds, the weight of 1 cubic inch of wrought iron is:

$$7.85 \times 0.036 = 0.2826 \, lb$$

The specific gravity of a liquid indicates the weight of a given volume of the liquid in comparison with the weight of an equal volume of water at 62°F (see Table 2-9). Since the *density* of a substance is mass per unit of volume and weight is equivalent to mass, then

### Table 2-9   Specific Gravities, Degrees Baume and Degrees API (at 60°F.)

| Degrees Baume | Specific Gravity | Degrees A.P.I. | Specific Gravity |
|---|---|---|---|
| 10 | 1.0000 | 10 | 1.0000 |
| 15 | 0.9655 | 15 | 0.9659 |
| 20 | 0.9933 | 20 | 0.9340 |
| 25 | 0.9032 | 25 | 0.9042 |
| 30 | 0.8750 | 30 | 0.8762 |
| 35 | 0.8485 | 35 | 0.8498 |
| 40 | 0.8235 | 40 | 0.8251 |
| 45 | 0.8000 | 45 | 0.8017 |
| 50 | 0.7777 | 50 | 0.7796 |
| 55 | 0.7568 | 55 | 0.7587 |
| 60 | 0.7368 | 60 | 0.7389 |
| 65 | 0.7179 | 65 | 0.7201 |
| 70 | 0.7000 | 70 | 0.7022 |
| 75 | 0.6829 | 75 | 0.6852 |
| 80 | 0.6666 | 80 | 0.6690 |
| 85 | 0.6511 | 85 | 0.6536 |
| 90 | 0.6363 | 90 | 0.6388 |

density can be defined as weight per unit of volume. This is indicated in the illustration of the density of steam at different pressures (see Figure 2-34), in which the weight of steam increases with the pressure.

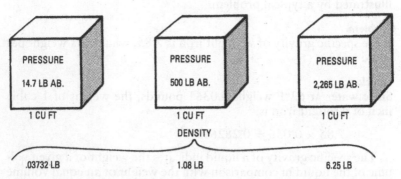

**Figure 2-34** Density of steam at various pressures. Density increases as pressure increases.

Figure 2-35 shows the effect of temperature on density of water. The water expands with a rise in temperature. As the temperature rises, the original volume at lower temperature increases and occupies more space, becoming lighter per unit volume, or in density.

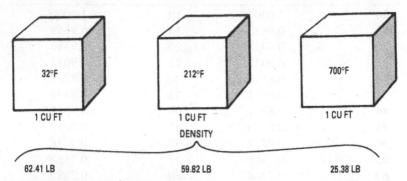

**Figure 2-35** Density of water at various temperatures. Density is less as temperature increases.

A *hydrometer* is an instrument that is used for determining the specific gravity of liquids. It is a closed glass tube containing air, with a weighted bulb at one end causing it to float upright in the liquid. The depth to which the hydrometer sinks can be read on the graduated scale to obtain the specific gravity. The lighter (or less dense) the liquid, the lower the tube sinks.

Baume scale hydrometers for testing the specific gravity of oil are for liquids lighter than water, and marked accordingly. Baume scales are also available for liquids heavier than water. The comparison is made with distilled water at 60°F (the oil is also tested at 60°F).

Either a Baume scale hydrometer or an American Petroleum Institute (API) hydrometer can be used to test the specific gravity of fuel oil or lubricating oil. Temperature is extremely important in determining specific gravity.

Specific gravity reading can be converted to a Baume reading by the formula:

$$\text{degrees Baume} = \frac{140}{\text{sp. gr.}} - 130$$

Baume reading can be converted to specific gravity by the formula:

$$\text{specific gravity} = \frac{140}{130 + \text{degrees Baume}}$$

**Example**

If the Baume reading for oil is 26°, the specific gravity of the oil can be calculated by the formula:

$$\text{sp. gr.} = \frac{140}{130 + 26} = 0.897$$

A rule for roughly converting the specific gravity at any temperature to the standard 60°F is: *For every 10°F above 60°F, subtract 1° from the Baume reading, and for every 10°F add 1°.* The rule can be applied as follows:

$$\text{degrees Baume (60°F)}$$
$$= \text{degrees Baume} - \frac{(\text{indicated °F} - 60\text{°F})}{10}$$

**Example**

If a hydrometer indicates a specific gravity of 27.5° Baume at an oil temperature of 75°F, what is the Baume reading at 60°F?

$$\text{degrees Baume (60°F)} = 27.5\text{°F} - \frac{(75\text{°F} - 60\text{°F})}{10}$$

$$= 27.5° - 1.5° = 26° \text{ Baume}$$

**Example**

If the hydrometer indicates 24° Baume at an oil temperature of 40°F, what is the Baume reading at 60°F?

$$\text{degrees Baume } (60°F) = 24° - \frac{(40°F - 60°F)}{10}$$

$$= 24° - (-2°)$$

$$= 24° + 2° = 26° \text{ Baume}$$

The *American Petroleum Institute (API)* specific gravity measurement is similar to Baume, except in scale divisions. Specific gravity can be converted to degrees API by the formula:

$$\text{degrees A.P.I.} = \frac{141.5}{\text{sp. gr.}} - 131.5$$

Degrees API can be converted to specific gravity by the formula:

$$\text{sp. gr.} = \frac{141.5}{131.5 + \text{degrees A.P.I.}}$$

## Summary

Hydraulics is the branch of physics that deals with the mechanical properties of water and other liquids and with the application of these properties in engineering. The basic principles of hydraulics are:

- The pressure exerted by a liquid on a surface is proportional to the area of the surface.
- A smaller quantity of water can be made to balance a much larger weight.
- The pressure on any portion of a fluid having uniform density is proportional to its depth below the surface.
- Fluids rise to the same level in each arm of a U-tube when the temperature of the liquid is the same throughout.

The branch of hydraulics dealing with the pressure and equilibrium (at rest) of water and other liquids is known as hydrostatics. Water is most often considered in the study of hydraulics, but other liquids are included.

Static head is the height of a column or body of water at rest above a given point—the weight of the water causing pressure. The head is measured from the center point of the pump outlet connection. In most calculations, the pressure per foot of static head can be estimated at 0.5 psi. For accurate calculations, the pressure of water per foot of static head is 0.43302 psi at 62°F.

The height to which atmospheric pressure causes a column of water to rise above the source of supply to restore equilibrium is termed static lift. The weight of the column of water (1 square inch of cross-sectional area) required to restore equilibrium is equal to the pressure exerted by the atmosphere (psi). The maximum height to which water at standard temperature (62°F) can be lifted is determined by the barometric pressure. Water rises to a height of 34.1 feet when the barometric pressure is 30 inches.

The weight of the water pushed aside (displaced) by the flotation of a vessel is termed displacement. The term draft indicates the depth to which the vessel or object sinks in the water. This is the depth whereby the weight of the water displaced is equal to the weight of the vessel or object.

Buoyancy is the power or tendency of a liquid to keep a vessel afloat. It is the upward pressure exerted by a fluid on a floating body. The center of gravity of the liquid displaced by the body immersed in it is called the center of buoyancy.

When water is placed in containers having different shapes, the intensity of the pressure in psi is the same at the bottom of each container. The fact that the total liquid pressure against the bottom of a vessel may be many times greater or less than the total weight of the liquid is termed a hydrostatic paradox.

Archimedes established that *a body immersed in a fluid loses an amount of weight that is equivalent to the weight of the fluid displaced.* When a body is immersed in a liquid, it is acted upon by two forces:

• Gravity, which tends to lower the body.
• Buoyancy, which tends to raise the body.

The branch of physics dealing with the *motion and action* of water and other liquids is termed hydrodynamics. Various forces act on a liquid, causing it to be a state of motion.

The dynamic head of water is an equivalent or virtual head of water in motion. It represents the pressure necessary to force the water from a given point to a given height and to overcome all frictional resistance. The dynamic head causing flow of a liquid is divided into three parts: (1) velocity head, (2) entry head, and (3) friction head.

The dynamic lift of water is an equivalent or virtual lift of water in motion that represents the resultant pressure necessary to lift the water from a given point to a given height and to overcome all frictional resistance. The practical limit of actual lift in pump operation ranges from 20 to 25 feet. When the water is warm, the height to which it can be lifted decreases, because of increased vapor pressure.

The term total column means head plus lift. The static total column is the static lift plus the static head. The height of distance from the level of supply to the level in the tank (in feet) is termed the static total column. The dynamic total column is the dynamic lift plus the dynamic head, or the equivalent total column of water in motion. It represents the pressure resulting from the static total column plus the resistance to flow caused by friction.

Friction of water in pipes results in loss of pressure and reduced volume of water per minute delivered. Also, friction in the faucets, in the fitting, and in the flow through the meter must be considered.

Water flow can be measured by means of the weir, pitot tube, and venturi meter. Actual measurements of the volume and of the weight of the liquid delivered may be more accurate, but they can be used only when the quantity of water delivered is small.

The general equation for velocity of water flowing from an orifice or tube is:

$$V = \sqrt{2gh}$$

The general equation for discharge of water through an orifice is:

$$Q = ca\sqrt{2gh}$$

The path of a jet of water issuing from a horizontal orifice or a tube describes a parabola. The jet is drawn downward on leaving the orifice by the force of gravity.

Specific gravity is the ratio of the weight of a given substance to the weight of an equal volume of another substance, which is used as a standard of comparison (water for liquids and solids; air or hydrogen for gases). One cubic inch of pure water weighs 0.0361 pounds at 62°F. Therefore, if the specific gravity of a material is known, its weight per cubic inch can be calculated by multiplying its specific gravity by 0.0361. To calculate the weight of 1 cubic foot of a given material, multiply its specific gravity by 62.35 (the weight of 1 cubic foot of water at 62°F).

A hydrometer is an instrument that is used to determine the specific gravity of liquids. Either a Baume scale hydrometer or an American Petroleum Institute (API) hydrometer can be used to test the specific gravity of fuel oil or lubricating oil.

A specific gravity reading can be converted to a Baume reading by the following formula:

$$\text{degrees Baume} = \frac{140}{\text{sp. gr.}} - 130$$

Specific gravity reading can be converted to degrees API by the following formula:

$$\text{degrees A.P.I.} = \frac{141.5}{\text{sp. gr.}} - 131.5$$

Degrees API can be converted to specific gravity by the following formula:

$$\text{sp. gr.} = \frac{141.5}{131.5 + \text{degrees A.P.I.}}$$

## Review Questions

1. What is hydraulics?
2. List four important principles of hydraulics.
3. What is meant by a *head* of water?
4. What is meant by *lift*, hydraulically?
5. What is the difference between *static head* and *dynamic head*?
6. What is the difference between *static lift* and *dynamic lift*?
7. What is the maximum height to which atmospheric pressure can lift water?
8. What factor determines the maximum height to which water is lifted at standard temperature (62°F)?
9. To what depth does an object sink in water?
10. What is *buoyancy*?
11. What is the practical limit of pump operation? Why?
12. What is meant by *total column* of water?
13. What is the effect of friction in pipes, fittings, and so on?
14. Name three instruments for measuring the flow of water.
15. Explain how a siphon operates.
16. What is meant by *specific gravity* of a substance?
17. Explain the basic principle of a hydrometer.

18. When did the fundamental law evolve for the science of hydraulics?
19. What is *hydrostatics*?
20. Define *displacement*.
21. Define *draft*.
22. What is Archimedes' Principle?
23. Define *hydrostatic paradox*.
24. When is hydrostatic balance achieved?
25. What is that branch of physics dealing with the motion and action of water and other liquids called?
26. What is meant by the term *negative lift*? What is another name for this term?
27. What causes a difference between *actual lift* and *static lift*?
28. Why must pumps handling water at high temperature work on reduced actual lift?
29. If you have a ½-inch commercial faucet with a rate of flow of 10 gallons per minute what would be the loss in pressure caused by the faucet?
30. What four factors determine the quantity of water discharged through a pipe?
31. What is the approximate resistance to water flow of a 45°F L fitting that has a nominal diameter of 3 inches?
32. What is a *hook gage*? How does it work?
33. Sketch a venturi meter and identify the various points.
34. How does a siphon work?
35. What is the specific gravity at 35° Baume?
36. What is the specific gravity at 60°API?
37. What does *API* stand for?
38. What is the general equation for velocity of water flowing from an orifice or tube?
39. What is the term *parabola* used to describe?

# Part II

# Pumps

# Chapter 3

## Centrifugal Pumps

All types of centrifugal pumps depend on *centrifugal force* for their operation. Centrifugal force acts on a body moving in a circular path, tending to force it farther away from the axis or center point of the circle described by the path of the rotating body.

### Basic Principles

The rotating member inside the casing of a centrifugal pump provides rapid rotary motion to the mass of water contained in the casing. That means the water is forced out of the housing through the discharge outlet by means of *centrifugal force*. The vacuum created thereby enables atmospheric pressure to force more water into the casing through the inlet opening. This process continues as long as motion is provided to the rotor, and as long as a supply of water is available. In the centrifugal pump, *vanes* or *impellers* rotating inside a close-fitting housing draw the liquid into the pump through a central inlet opening, and by means of centrifugal force the liquid is thrown outward through a discharge outlet at the periphery of the housing.

Figure 3-1 and Figure 3-2 show the basic principle of centrifugal pump operation. If a cylindrical can with vanes A and C (for rotating the liquid when the can is rotated) is mounted on a shaft with a pulley for rotating the can at high speed, centrifugal force acts on the water (rotating at high speed) to press the water outward to the walls of the can. This causes the water to press outward sharply. Since it cannot move beyond the walls of the can, pressure forces the water upward, causing it to overflow as the water near the center of the can is drawn downward. Atmospheric pressure forces the water downward, since a vacuum is created near the center as the water moves outward toward the sides of the can. It can be noted in Figure 3-1 that the water has been lifted a distance DD'.

Since the water that spills over the top has a high velocity that is equal to the rim speed, the kinetic energy that has been generated is wasted, unless an arrangement is made to catch the water and an additional supply of water is provided (see Figure 3-2). In Figure 3-2, a receiver catches the water as it spills over, and a supply tank is connected with the hollow shaft to supply water to the can. Instead of rotating the can, only the vanes can be rotated to obtain the same result.

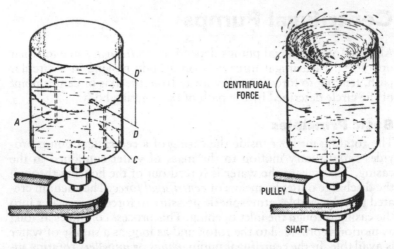

**Figure 3-1** Basic principle of a centrifugal pump. The radial vanes A and C cause the liquid to revolve when the cylinder is rotated (left). Centrifugal force pushes the liquid outward toward the walls of the cylinder and then upward, causing it to overflow when the cylinder is rotated at high speed (right).

## Pumps Having Straight Vanes
In the first practical centrifugal pump, the rotor was built with straight (radial) vanes (see Figure 3-3). The essential parts of a centrifugal pump are:

- *Impeller*, or rotating member.
- *Case*, or housing surrounding the rotating member.

In the centrifugal pump, water enters through the inlet opening in the center of the impeller, where it is set in rotation by the revolving blades of the impeller. The rotation of the water, in turn, generates centrifugal force, resulting in a pressure at the outer diameter of the impeller; when flow takes place, the water passes outward from the impeller at high velocity and pressure into the gradually expanding passageway of the housing and through the discharge connection to the point where it is used.

## Pump Having Curved Vanes
Curved vanes were first used by Appold in England in 1849. Figure 3-4 shows the cover and inner workings of a centrifugal pump having curved vanes and casing (commonly known as a *volute*

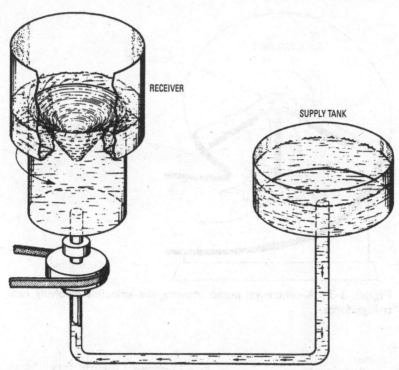

RECEIVER

SUPPLY TANK

**Figure 3-2**   A basic centrifugal pump with supply tank and receiver.

*pump*). An inlet pipe connection (*A*) to the cover directs the water to the eye (*B*) of the rotating impeller. The curved vanes (*C*) of the impeller direct the water from the eye to the discharge edge (*D*), moving the water in a spiral path. As the impeller revolves, the water moves toward the discharge edge, and then enters the volute-shaped passageway (*E*) where it is collected from around the impeller and directed to the discharge connection (*F*).

## The Volute

A *volute* is a curve that winds around and constantly recedes from a center point. The volute is a spiral that lies in a single plane (in contrast with a conical spiral). It is the shape of the periphery of the case or housing that surrounds the impeller of a volute-type centrifugal pump.

The volute-type casings or housings form a progressively expanding passageway into which the impeller discharges the water.

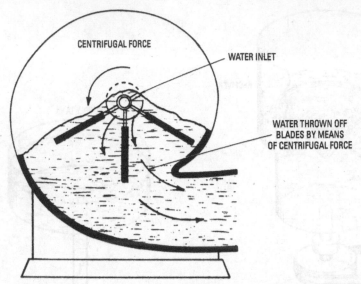

**Figure 3-3** A centrifugal pump showing the principle involving centrifugal force.

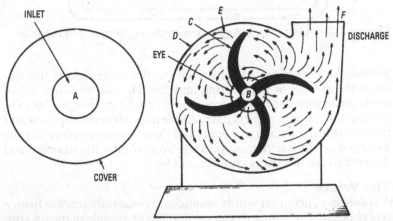

**Figure 3-4** Basic parts of a centrifugal pump showing cover (left) and section of a centrifugal pump, commonly termed a volute pump because of the shape of its housing.

The volute-shaped passageway collects the water from the impeller and directs it to the discharge outlet.

The volute-shaped housing is proportioned to produce equal flow velocity around the circumference, and to reduce gradually the velocity of the liquid as it flows from the impeller to the discharge outlet. The objective of this arrangement is to change velocity head to pressure head.

### Curvature of the Impeller Vanes

If the vanes could be curved to a shape that is mathematically correct, a different curve would be required for each change in working conditions. However, this is not practical because a large number of patterns in stock would be required. Figure 3-5 shows a simple method for describing the curve of the vanes for impellers of the larger diameters and for lifts of 60 feet or more as follows:

- Divide the circle into a number of arms (six, for example).
- Bisect each radius.
- Using the point $B$ as a center point and a radius $BC$, describe the curves that represent the working faces of the vanes:

$$BC = AB + \frac{1}{6}AB$$

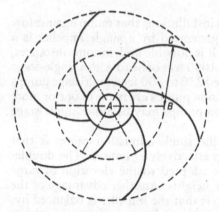

**Figure 3-5**  Layout for the curved vanes of an impeller.

### Basic Classification

The centrifugal pump is a device for moving liquids and gases. The two major parts of the device are the impeller (a wheel with vanes) and the circular pump casing around it. The volute centrifugal pump is the most common type. In it, the fluid enters the pump at high speed near the center of the rotating impeller. Here it is thrown against the casing by the vanes. The centrifugal pressure forces the fluid through an opening in the casing. This outlet widens progressively in a spiral that reduces the speed of the fluid and increases the pressure. Centrifugal pumps produce a continuous flow of fluid at high pressure. The pressure can be increased by linking several impellers together in one system. In such a multistage pump, the outlet for each impeller casing serves as the inlet to the next impeller. Centrifugal pumps are used for a wide variety of purposes, such as pumping liquids for water supply, irrigation, and sewage disposal systems. Such devices are also utilized as gas compressors.

The basic designs of centrifugal pumps correspond to the various principles of operations. Centrifugal pumps are designed chiefly with respect to the following:

- Intake, as single-admission or double-admission.
- Stage operation, as single-stage or multistage.
- Output, as large-volume (low-head), medium-volume (medium-head), and small-volume (high-head).
- Impeller, as type of vanes, number of blades, housing, and so on.

## Single-Stage Pump
This type of pump is adapted to installations that pump against low to moderate heads. The head generated by a single impeller is a function of its tangential speed. It is possible and, in some instances, practical to generate as much as 1000 feet of head with a single-stage impeller, but for heads that exceed 250 to 300 feet, multistage pumps are generally used. Single-admission pumps are made in one or more stages, and the double-admission pumps may be either single-stage or multistage (see Figure 3-6).

The chief disadvantage of the single-admission pump is that the head at which it can pump effectively is limited. The double-admission single-stage pump is adapted to the elevation of large quantities of water to moderate heights. Another advantage of the double-admission type of pump is that the impeller is balanced hydraulically in an axial direction, because the thrust from one admission stream is counteracted by the thrust from the other admission stream.

Single-stage centrifugal pumps are widely used in portable pumping applications as well as in stationary pumping applications. These portable pumps can be operated by air motors, electric motors, gasoline engines, and diesel engines. Many are used in the

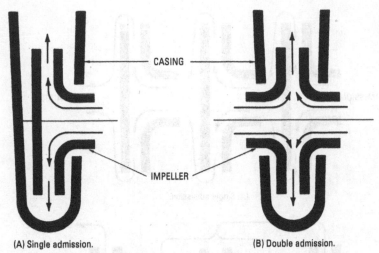

(A) Single admission.                    (B) Double admission.

**Figure 3-6**   Single-admission (left) and double-admission (right) types of impeller for single-stage centrifugal pumps, illustrating flow path of the liquid.

contracting field. Stationary single-stage centrifugal pumps are very popular in home use in both shallow and deep wells for supplying the home water supply.

## Multistage Pump

The multistage centrifugal pump is essentially a high-head or high-pressure pump. It consists of two or more stages. The number of stages depends on the size of head against which the pump is to work. Each stage is essentially a separate pump. However, these stages are located in the same housing and the impellers are attached to the same shaft. Up to eight stages may be found in a single housing.

The initial or first stage receives the water directly from the source through the admission pipe, builds the pressure up to the correct single-stage pressure, and passes it onward to the succeeding stage. In each succeeding stage, the pressure is increased or built up until the water is delivered from the final stage at the pressure and volume that the pump is designed to deliver. Figure 3-7 shows water flow in single-admission and double-admission multistage pumps.

Stationary-mounted centrifugal pumps are found on many machine tools for delivering coolant to the cutting tools. Multiple-stage centrifugal pumps are used in home and industry where large volumes of water at high pressure are required.

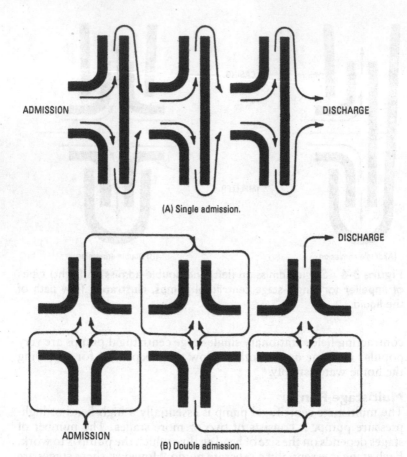

(A) Single admission.

(B) Double admission.

**Figure 3-7** Flow path of liquid from admission to discharge for single-admission (upper) and double-admission (lower) types of impellers in multistage pumps.

## Impellers

The efficiency of a centrifugal pump is determined by the type of impeller. The vanes and other details are designed to meet a given set of operating conditions. The number of vanes may vary from one to eight, or more, depending on the type of service, size, and so on.

Figure 3-8 shows a single-vane *semi-open* impeller. This type of vane is adapted to special types of industrial pumping problems that require a rugged pump for handling liquids containing fibrous

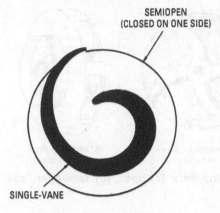

**Figure 3-8** A single-vane, semi-open impeller.

materials and some solids, sediment, or other foreign materials in suspension.

The *open* type of vane is suited for liquids that contain no foreign matter or material that may lodge between the impeller and the stationary side plates. Liquids containing some solids (such as those found in sewage or drainage, where there is a limited quantity of sand or grit) may be handled by the open type of vane.

In addition to the open and semi-open types of impellers, the *enclosed* or shrouded type of impeller may be used (see Figure 3-9), depending on the service, desired efficiency, and cost. The enclosed type of impeller is designed for various types of applications. The shape and the number of vanes are governed by the conditions of service. It is more efficient, but its initial cost is also higher. Shrouded impellers do not require wearing plates. The enclosed impeller reduces wear to a minimum, ensures full-capacity operation with initially high efficiency for a prolonged period of time, and does not clog, because it does not depend on close operating clearances.

The *axial-flow* type of impeller (see Figure 3-10) is used to obtain a flow of liquid in the direction of the axis of rotation. These propeller-type impellers are designed to handle a large quantity of water at no lift and at low head in services such as irrigation, excavation, drainage, sewage, etc. The pumping element must be submerged at all times. This type of pump is not suitable for pumping that involves a lift.

The *mixed-flow* type of impeller (see Figure 3-11) is used to handle a large quantity of water at low head. High-capacity low-head pumps are designed on the mixed-flow principle to increase the

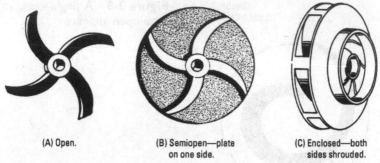

(A) Open.

(B) Semiopen—plate on one side.

(C) Enclosed—both sides shrouded.

**Figure 3-9** Several types of impellers: (a) open, (b) semi-open, and (c) enclosed.

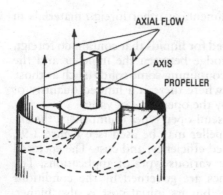

AXIAL FLOW

AXIS

PROPELLER

**Figure 3-10** Propeller-type impeller used to obtain axial flow.

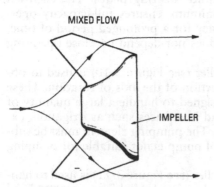

MIXED FLOW

IMPELLER

**Figure 3-11** Mixed-flow type of impeller.

rotational speeds, reduce the size and bulk of the pump, and to increase efficiency.

## Balancing

The centrifugal pump is inherently an unbalanced machine. These pumps are subject to end thrust, which means that some method must be devised to counteract this load. In a single-admission type of impeller, an unbalanced hydraulic thrust is directed axially toward the admission side, because the vacuum in the admission side causes atmospheric pressure to produce a thrust on the impeller. Various methods of balancing have been tried:

- *Natural balancing*, as by opposing impellers.
- *Mechanical balancing*, as by a balancing disk, and so on.

### Natural Balancing

This method of balancing is generally used in double-admission, single-stage pumps and in single- and double-admission multistage pumps. Figure 3-12 shows back-to-back single-admission impellers and a double-admission impeller. In the diagrams, the two thrusts $P$ and $P'$ act in axial directions and oppose each other.

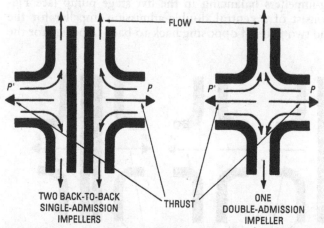

**Figure 3-12** Natural balancing achieved in single-stage pumps by means of two back-to-back single-admission types of impellers (left) and one double-admission type of impeller (right).

The opposing-impellers method is applied to multistage pumps by various impeller arrangements. The three-stage pump arrangement (see Figure 3-13) consists of a central double-admission impeller and two opposing single-admission end impellers. The arrows indicate the opposing thrusts.

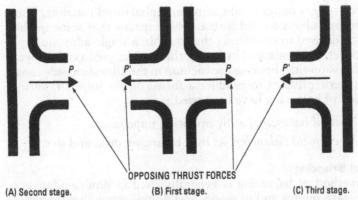

OPPOSING THRUST FORCES

(A) Second stage.          (B) First stage.          (C) Third stage.

**Figure 3-13** Natural balancing achieved in a three-stage pump by means of two opposed single-admission types of impellers (left and right) and one double-admission type of impeller (center).

Opposing-impellers balancing in the five-stage pump (see Figure 3-14) consist of a central double-admission impeller for the first stage and two pairs of opposing back-to-back impellers for the

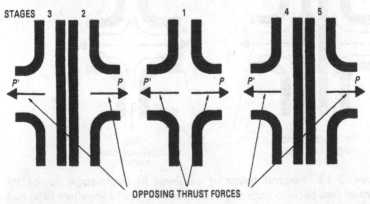

STAGES    3      2              1              4      5

OPPOSING THRUST FORCES

**Figure 3-14** Natural balancing achieved in a five-stage pump by means of an assembly by two pairs of opposed single-admission back-to-back type impellers (left and right) and one double-admission type of impeller (center).

second and third stages and for the fourth and fifth stages. The method is also illustrated in the six-stage pump shown in Figure 3-15.

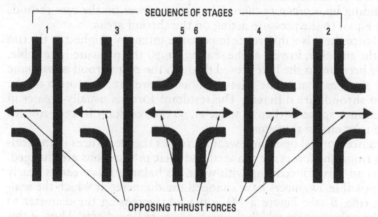

SEQUENCE OF STAGES

1   3   5 6   4   2

OPPOSING THRUST FORCES

**Figure 3-15** Natural balancing achieved in a six-stage centrifugal pump by means of an assembly of four single-admission type impellers (left and right) and one double-admission type of impeller (center).

### Mechanical Balancing

The liquid enters through the eye of the impeller in an axial direction and leaves in a radial direction, thereby creating an end thrust that also results from the fact that the liquid in the clearance spaces is under pressure. These forces are not present in open impellers, since there are no shrouds upon which the forces act.

In a single-admission enclosed-impeller pump (see Figure 3-16), the liquid from the housing, being under pressure, leaks backward

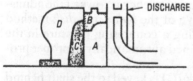

DISCHARGE

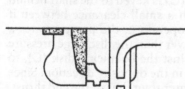

**Figure 3-16** Mechanical balancing in a centrifugal pump by means of a balancing disk.

through the clearance spaces $A$ and $D$, past the sealing rings $C$ and $B$, to the inlet. The impeller is usually cored in the rear shroud to permit the accumulating leakage to pass onward to the inlet without building up a pressure there. Therefore, forces on the two shrouds are equal to the pressure acting on the shroud areas.

Since there is a difference in pressure intensity (highest at the rim of the impeller, lowest at the sealing rings) the pressure is variable. The pressure at the holes cored through the rear shroud is not quite the same as that in the inlet chamber. Therefore, the forces on the two shrouds are different. The resultant force is usually greater in clearance space $D$ than in space $A$, so the resultant force is toward the inlet end of the pump.

Since normal operating wear increases the clearances in the sealing rings, the forces are changed and their proportions are changed, the end-thrust increasing with wear. To balance this wear as nearly as possible, engineers have changed the diameter at which the sealing ring $B$ (see Figure 3-16) is placed, increasing the diameter to reduce the area on which the high-pressure liquid acts. Thus, if this ring were moved outward to the rim of the impeller, the force on the rear shroud could be reduced to change the resultant end thrust to the opposite direction or away from the admission.

To provide the pump with a minimum of end thrust during its useful life, the ring is placed outward at a distance far enough to reverse the thrust while the pump is new and while the clearances in the sealing rings are small. Clearances increase with use. Thrust is reduced gradually. It finally reverses itself near the admission end. The proportions are so fixed that, by the time the thrust becomes large enough to cause an undue load on the thrust bearing on the shaft, leakage through the increased clearance spaces is sufficient to affect seriously the efficiency of the pump. Then the clearance rings can be replaced at a small cost to restore the pump to its original condition.

As previously explained, the thrust tendency is toward the admission side of the pump. The objective of the *balancing disk* method is to balance this thrust by providing a countering pressure in the opposite direction, which is maintained automatically in proper proportions against the balancing disk.

In Figure 3-16, the balancing disk $(C)$ is keyed to the shaft behind the last-stage impeller and runs with a small clearance between it and the balance seat $(B)$. While in operation, the pump creates a pressure in space $A$ that is slightly lower than the discharge pressure of the pump. This pressure acts against the balancing disk $(C)$, to counterbalance the end thrust that is in the opposite direction. Since the pressure against the disk $C$ is greater than that of the end thrust, the complete rotating element of the pump, together with disk C, is caused to move slightly, so that disk C is moved away from seat B.

This action results in a small leakage into the balance chamber (D), thereby reducing the pressure in space A, which causes the rotating element to return to a position where leakage past disk C and seat B enables the pressure in space A to balance the thrust—thrust and balance pressure are in equilibrium. The leakage between disk C and seat B is slight and not large enough to affect the efficiency or capacity of the pump for a considerable period.

## Construction of Pumps

In its primitive form, the centrifugal pump was inefficient and was intended only for pumping large quantities of water at low heads. This type of pump is now highly developed, and many types of these pumps are available for a wide variety of service requirements.

Although the earlier pumps were adapted only to low heads, this has been overcome by connecting two or more units on a single shaft and operating them in series—passing the water through each unit in succession, with the total head pumped against divided between the units (multistage pump).

The earlier multistage assemblies were bulky, because they consisted of separate units coupled. However, several stages are now housed in a single casing or housing. The centrifugal pump usually gives best results when it is designed for specific operating conditions.

### Casing or Housing

The casing is usually a two-piece casting split on a horizontal or diagonal plane, with inlet and discharge openings cast integrally with the lower portion. Centrifugal pumps are either single-inlet or double-inlet types. The double-inlet type of pump is usually preferred, because the end thrusts are equalized when variations in pressure occur on either the discharge or the inlet side.

The diagonally split casing or housing (see Figure 3-17) permits easy removal of internal parts. The discharge and suction piping need not be disturbed.

Figure 3-18 shows the offset-volute design of casing. This type of casing also features top-centerline discharge. It also has self-venting and back pullout.

### Impeller

Impeller design varies widely to meet a wide variety of service conditions. The selection of the proper type of impeller is of prime importance in obtaining satisfactory and economical pump operation.

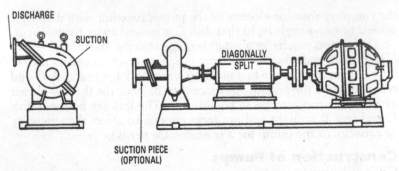

**Figure 3-17** View of a paper-stock centrifugal pump. The diagonally-split casing permits easy removal of the interior parts without disturbing the discharge of the suction piping. *(Courtesy Buffalo Forge Company)*

**Figure 3-18** An offset-volute design of casing (left) and cover (right). The casing is designed for top centerline discharge, self-venting, and back pullout. *(Courtesy Buffalo Forge Company)*

A high degree of efficiency can be obtained with the open-type impeller under certain conditions by carefully proportioning the curvature of the blades and by reducing the side clearances to a minimum with accurate machining of impeller edges and side plates. The open-type impeller is often used to handle large quantities of water at low heads, such as those encountered in irrigation, drainage, storage of water, and circulation of water through condensers.

The enclosed-type impeller (see Figure 3-19) is generally considered to be a more efficient impeller. The vanes are cast integrally on both sides and are designed to prevent packing of fibrous materials between stationary covers and the rotating impeller.

**Figure 3-19**   Open-type (left) and enclosed-type (right) impellers.
(Courtesy Buffalo Forge Company)

Figure 3-20 shows an enclosed-type double-inlet impeller and wearing ring. The impeller is cast in a single piece of bronze, although some liquids require that the impeller be made of chrome, Monel, nickel, or a suitable alloy.

## Stuffing-Box Assembly
The assembly shown in Figure 3-21 is equipped with a five-ring packing and seal cage. The seal cage is split and glass-filled with material that is resistant to corrosion and heat. Mechanical seals can be used, and they are available in materials suitable for corrosive and non-corrosive applications.

## Bearings and Housings
Most pumps are equipped with ball bearings (see Figure 3-22). Typical construction consists of single-rows of deep-grooved ball bearing

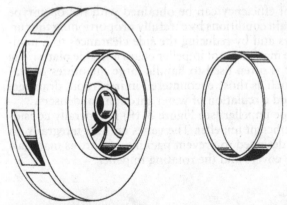

**Figure 3-20** Enclosed-type double-inlet impeller and wearing ring. Some liquids require that the impeller be made of special metals, such as chrome, Monel, nickel, or a suitable alloy. *(Courtesy Buffalo Forge Company)*

**Figure 3-21** Stuffing-box assembly for a centrifugal pump.
*(Courtesy Buffalo Forge Company)*

**Figure 3-22** Bearings for a centrifugal pump. The outboard thrust bearing is a double-row deep-grooved ball bearing, and the inboard guide bearing is a single-row deep-grooved ball bearing. *(Courtesy Buffalo Forge Company)*

of ample size to withstand axial and radial loads. The bearing housing may be of the rotating type, so that the entire rotating element can be removed from the pump without disturbing the alignment or exposing the bearings to water or dirt. The bearing housing may be positioned by means of dowel pins in the lower portion of the casing and securely clamped by covers split on the same plane as the pump casing. Then the entire bearing can be removed from the shaft without damage by using the sleeve nut as a puller. However, single ball bearings are an exception. Most pumps today are provided with double ball bearings.

## Shaft Assembly

The shaft (see Figure 3-23) is machined accurately to provide a precision fit for all parts, including the impeller and bearings. The shaft in most centrifugal pumps must be protected against corrosive or abrasive action by the liquid pumped, by such methods as using cast bronze sleeves that fit against the impeller hub and are sealed by a thin gasket.

**Figure 3-23**  A solid one-piece stainless-steel shaft for a centrifugal pump. The shaft is machined accurately to provide a precision fit for all parts. *(Courtesy Buffalo Forge Company)*

## Drive

Centrifugal pumps are driven by direct drive or by a belt and pulley. A large subbase is usually provided for direct-drive connection to an engine or motor, and the two units are connected by a suitable coupling (see Figure 3-24). A disassembly of a dredging pump with a base or pedestal for belt-and-pulley drive is shown in Figure 3-25.

Vertical pumps are generally made with a single inlet. Since the weight of the impeller and shaft requires a thrust bearing, this weight can be proportioned to take care of the unbalanced pressure caused by the single-inlet characteristic. Figure 3-26 shows a vertical motor-mount dry-pit centrifugal pump. Gearing for vertical pumps is seldom advisable, except where bevel gears may be needed to transmit power from a horizontal shaft to a vertical shaft. Figure 3-27 shows a diagram with dimensions of a horizontal angle-flow centrifugal pump. Prior to introduction of the angle-gear drive, most deep-well turbine engines were driven by electric motors. However, this

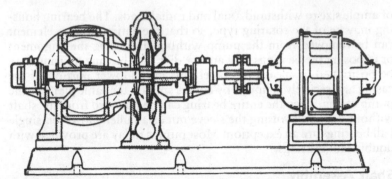

**Figure 3-24** Centrifugal pump and motor placed on a large subbase and connected by a suitable coupling for direct-drive.
*(Courtesy Buffalo Forge Company)*

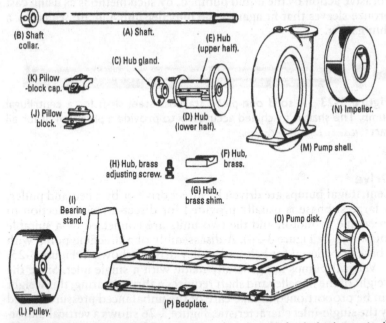

**Figure 3-25** Disassembly view showing parts of a belt-drive single-stage centrifugal pump. The parts shown are: (A) shaft; (B) shaft collar; (C) hub gland; (D) hub [lower half]; (E) hub [upper half]; (F) hub, brass; (G) hub, brass shim; (H) hub, brass adjusting screw; (I) bearing stand; (J) pillow block; (K) pillow-block cap; (L) pulley; (M) pump shell; (N) impeller; (O) pump disk; and (P) bedplate. *(Courtesy Buffalo Forge Company)*

**Figure 3-26** A vertical motor-mount centrifugal pump.
(Courtesy Deming Division, Crane Co.)

development has resulted in the use of more gasoline and diesel power units.

## Installation

When the correct type of centrifugal pump has been selected for the service requirements, it must be installed properly to give satisfactory service and to be reasonably trouble-free. Several important factors must be considered for proper pump installation, depending on the size of the pump.

## Location

The pump should be located where it is accessible and where there is adequate light for inspection of the packing and bearings. A centrifugal pump requires a relatively small degree of attention, but if it

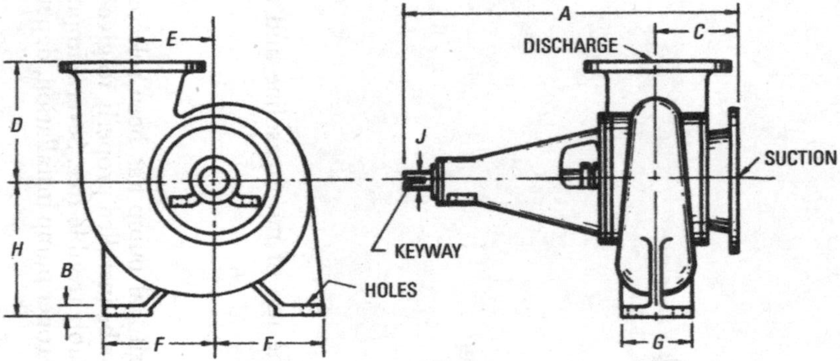


| PUMP SIZE | A | B | C | D | E | F | G | H | J | KEYWAY | L |
|---|---|---|---|---|---|---|---|---|---|---|---|
| 10" | 4'1$\frac{1}{2}$" | 1$\frac{1}{8}$" | 16$\frac{1}{2}$" | 14" | 10$\frac{1}{2}$" | 14" | 9" | 17$\frac{1}{2}$" | 2$\frac{1}{4}$" | $\frac{5}{8}$" × $\frac{5}{16}$" × 4" | 41$\frac{1}{4}$" |
| 12" | 4'2" | 1$\frac{1}{8}$" | 17" | 15" | 10$\frac{3}{4}$" | 15" | 13$\frac{1}{2}$" | 18" | 2$\frac{1}{4}$" | $\frac{5}{8}$" × $\frac{5}{16}$" × 4" | 41$\frac{1}{8}$" |
| 16" | 5'6$\frac{1}{4}$" | 1$\frac{1}{4}$" | 24" | 20" | 14$\frac{1}{2}$" | 20" | 18" | 24" | 3" | $\frac{3}{4}$" × $\frac{1}{4}$" × 4$\frac{1}{2}$" | 41$\frac{3}{8}$" |

**Figure 3-27** Horizontal angle-flow type of centrifugal pump with tabulated dimensions.

is inaccessible, it probably receives no attention until a breakdown requiring major repairs occurs.

The height of lift must also be considered. The lift is affected by temperature, height above sea level, and pipe friction, foot valve, and strainer losses. The elevation of the pump with respect to the liquid to be pumped should be at a height that is within the practical limit of the dynamic lift.

Piping also affects the location of the pump. The pump should be located so that the piping layout is as simple as possible.

## Foundation

The foundation should be rugged enough to afford a permanent rigid support to the entire base area of the bedplate and to absorb normal strains and shocks that may be encountered in service. Concrete foundations are usually the most satisfactory.

Foundation bolts of the specified size should be located in the concrete according to drawings submitted prior to shipment of the unit. If the unit is mounted on steelwork or other type of structure, it should (if possible) be placed directly above the main members, beams, and walls, and it should be supported in such a way that the base plate cannot be distorted or the alignment disturbed by a yielding or springing action of the structure or the base plate. The bottom portion of the bedplate should be located approximately $3/4$ inch above the top of the foundation to leave space for grouting.

## Leveling

Pumps are usually shipped already mounted, and it is usually unnecessary to remove either the pump or the driving unit from the base plate for leveling. The unit should be placed above the foundation and supported by short strips of steel plate and wedges near the foundation bolts. A $3/4$-inch to 2-inch space between the base plate and the foundation should be allowed for grouting. The *coupling bolts should be removed* before proceeding with leveling the unit and aligning the coupling halves.

When scraped clean, the projecting edges of the pads supporting the pump and motor feet can be used for leveling, employing a spirit level. If possible, the level should be placed on an exposed part of the pump shaft, sleeve, or planed surface of the casing.

The wedges underneath the base plate can be adjusted until the pump shaft is level, and the flanges of the suction and discharge nozzles are either vertical or horizontal, as required. At the same time, the pump should be placed at the specified height and location. Accurate alignment of the unbolted coupling halves between the pump and the driver shafts must be maintained while proceeding with the leveling of the pump and base.

To check the alignment of the pump and driver shafts, place a straightedge across the top and side of the coupling, checking the faces of the coupling halves for parallelism by means of a tapered thickness gage or feeler gage at the same time (see Figure 3-28).

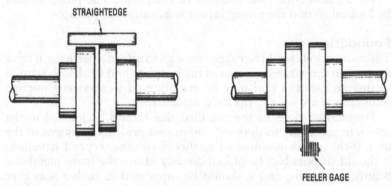

**Figure 3-28** Use of a straightedge and feeler gage to align the pump and driver shafts.

If the coupling halves are true circles, have the same diameter, and have flat faces, true alignment exists when the distances between the faces are equal at all points and when a straightedge lies squarely across the rim at all points. The test for parallelism is to place a straightedge across the top and side of the coupling, checking the faces of the coupling halves for parallelism by means of a tapered thickness gage or feeler gage at the same time.

### Turbine Drive

If the pump is driven by a steam turbine, final alignment should be made with the driver at operating temperature. If this is impossible, an allowance in the height of the turbine and shaft while cold should be made. In addition, if the pump is to handle hot liquids, allowance should be made for elevation of the shaft when the pump expands. In any event, the alignment should be checked while the unit is at operating temperature and adjusted as required, before placing the pump in actual service. The application of heat to the steam and exhaust piping results in expansion. The turbine nozzles should not be subjected to piping strains in the installation.

## Motor Drive

An allowance for heat is unnecessary for electric motors. The motor alone should be operated (if possible) before aligning the pump, so that the magnetic center of the rotor can be determined. If this is impossible, the rotor of the motor should be turned over and reversed to determine collar clearances, and then placed in the middle position for aligning. If the faces are not parallel, the thickness gage (or feelers) varies at different points. If one coupling is higher than the other, the distance can be determined by means of the straightedge and feeler gages.

## Space between Coupling Faces

The clearance between the faces of the couplings of the pin-and-buffer type, and the ends of the shafts in other types of couplings, should be set so that they cannot touch, rub, or exert a force on either the pump or the driver. The amount of clearance may vary with the size and type of coupling used. Sufficient clearance for the unhampered endwise movement of the shaft of the driving element to the limit of its bearing clearance should be allowed. On motor-driven units, the magnetic center of the motor determines the running position of the half-coupling of the motor. This can be checked by operating the motor while it is disconnected.

## Grouting

The grouting process involves pouring a mixture of cement, sand, and water into the voids of the stone, brick, or concrete work, either to provide a solid bearing or to fasten anchor bolts, dowels, and so on. The usual grouting mixture consists of one part cement, two parts sand, and enough water to cause the mixture to flow freely underneath the bedplate.

A wooden form is built around the outside of the bedplate to contain the grout and to provide sufficient head for ensuring a flow of the mixture beneath the entire head plate. The grout should be allowed to set for 48 hours. Then the hold-down bolts should be tightened and the coupling halves rechecked.

## Inlet Piping

In a new installation, it is advisable to flush the inlet pipe with clear water before connecting it to the pump. Except for misalignment, most problems with individual centrifugal pump installations can be traced to faults in the inlet lines. It is extremely important to install the inlet piping correctly.

The diameter of the inlet piping should not be smaller than the inlet opening, and it should be as short and direct as possible. If a long inlet cannot be avoided, the size of the piping should be increased. Air pockets and high spots in an inlet line invariably cause trouble. Preferably, there should be a continual rise (without high spots) from the source of supply to the pump.

When the supply liquid is at its lowest level, the end of the pipe should be submerged to a depth equal to four times its diameter (for large pipes). Smaller pipes should be submerged to a depth of 2 to 3 feet. After installation is completed, the inlet piping should be blanked off and tested hydrostatically for air leaks before the pump is operated.

A strainer should be attached to the end of the inlet pipe to prevent lodging of foreign material in the impellers. The clear and free opening of the strainer should be equal to three or four times the area of the inlet pipe. If the strainer is likely to become clogged frequently, the inlet pipe should be placed where it is accessible. A foot valve for convenience in priming may be necessary where the pump is subjected to intermittent service. The size and type of foot valve should be selected carefully to avoid friction loss through the valve.

## Discharge Piping
The discharge piping (like the inlet piping) should also be as short and as free of elbows as possible, to reduce friction. Check valves and gate valves should be placed near the pump. The check valve protects the casing of the pump from breakage caused by water hammer, and it prevents the pump operating in reverse if the driver should fail. The gate valve can be used to shut off the pump from the discharge piping when inspection or repair is necessary.

## Pumps Handling High-Temperature Liquids
Special types of multistage pumps are constructed for handling high-temperature liquids, with a key and keyway on the feet and base of the lower half of the casing. One end of the pump is bolted securely, but the other end (on some units) is bolted with spring washers positioned underneath the nuts on the casing feet, permitting one end of the casing to move laterally as the casing expands. Some special types of hot-liquid multistage pumps are doweled at the inboard end, and other types are doweled at the thrust-bearing end, sometimes with the dowels placed crosswise. Dowels at the opposite end (if used) are fitted parallel to the pump shaft, to allow the casing to expand at high temperatures.

When handling hot liquids, the nozzle flanges should be disconnected after the unit has been placed in service, to determine whether the expansion is in the proper direction.

### Jacket Piping
Multistage pumps use jacketed or separately-cooled thrust bearings. If hot liquids are being handled, care must be taken to be certain that the jacket or oil-cooler water piping is connected.

### Drain Piping
To determine whether the water is flowing and to regulate the amount of flow, it is good practice to pipe the discharge from the jacket or cooler into a funnel connected to a drain. All drain and drip connections should be piped to a point where leakage can be disposed of.

## Operation
Before starting the operation of a centrifugal pump, the driver should be tested for its direction of rotation (see Figure 3-29), with the coupling halves disconnected. The arrow on the pump casing indicates the direction of rotation.

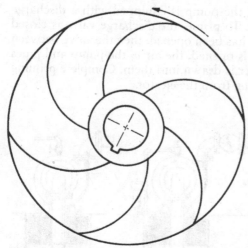

**Figure 3-29**  The direction of rotation of the impeller on a centrifugal pump.

The ball bearings should be supplied with the grade of lubricant recommended by the pump manufacturer. Oil-lubricated bearings should be filled level with the overflow.

The cooling water piping to the thrust-bearing housing also requires attention. Cooling water should not be used on bearings that are warm to the hand only. Use only sufficient water to keep the lubricant at a safe working temperature.

Periodically, the water supply should be flushed freely to remove particles of scale and similar particles that may stop the flow on a throttled valve. Final inspection of all parts should be made carefully before starting the pump. It should be possible to rotate the rotor by hand.

## Priming
A centrifugal pump should not be operated until it is filled with water. If the pump is run without liquid, there is danger of damage to liquid-lubricated internal parts. Some specially-constructed types of centrifugal pumps have been designed to start dry. Liquid from an external source is used to seal the stuffing boxes and to lubricate the impeller wearing rings and shaft sleeves in the same manner as with a stuffing-box packing.

### Ejector Method
As shown in Figure 3-30, the pump is equipped with a discharge valve and a steam ejector. To prime, the discharge valve is closed after the steam-inlet valve has been opened; then the valve between the ejector and the pump is opened, the air in the pump and pipes is exhausted, and the water is drawn into them. Complete priming is indicated by water issuing from the ejector.

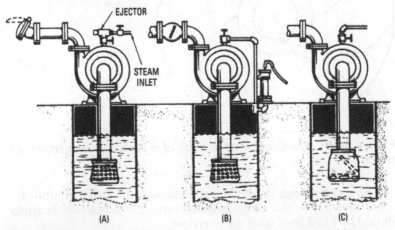

**Figure 3-30** Three methods of priming centrifugal pumps: (a) ejector; (b) lift-type pump; and (c) foot valve with top discharge.

The ejector is shut off by first closing the valve between the ejector and the pump, and then closing the steam-inlet valve. If it is inconvenient to place the ejector near the pump, the air pipe can be extended, using an air pipe that is slightly larger than when the ejector is placed near the pump.

### Hand-Type Lift Pump (or Powered Air Pump) Method

A check valve can be used instead of the discharge valve, and either a hand-type old-fashioned lift pump (see Figure 3-30) or a powered air pump can be substituted for the steam ejector. A valve should be placed in the air pipe. The valve should be closed before starting. In the older installations, the kitchen-type lift pumps were often used for priming. Sometimes the kitchen-type pumps themselves require a small quantity of water to water-seal them, making them an effective air pump.

### Foot-Valve and Top-Discharge Method

When the foot valve is used (see Figure 3-30), the centrifugal pump and inlet pipe are filled with water through the discharge or top portion of the pump from either a small tank or a hand-type pump. If the inlet pipe is long, at least 5 feet of discharge head on the pump is required to prevent water from being thrown out before the water in the inlet line has begun to move, causing failure in starting. When check or discharge valves are used, a vacuum gage placed on the air-priming pump at the head of the well or pit indicates that the pump is primed. A steam-type air ejector may be operated with water if 30 to 40 pounds of water pressure is available. However, a special type of ejector is required.

For automatic priming, a pressure regulator can be connected into the discharge line. The regulator automatically starts an air pump if the main pump loses its prime. It stops the priming pump when the priming action is completed.

In the float-controlled automatic priming system (see Figure 3-31), a priming valve is connected between the top of the casing of the pump and a float-controlled air valve. The air valve is connected to a priming switch, with the contacts in series with the control circuit of the main motor starter. A check valve in the discharge line operates a switch that shunts the priming switch.

In actual operation, the priming valve and air valve are in the indicated positions so long as the pump is not primed. When the float switch closes, the priming pump is started. It continues in operation until the main pump is primed and until the water is high enough in the float chamber to close the priming switch. The closing of the priming switch starts the main pump, and the priming pump is

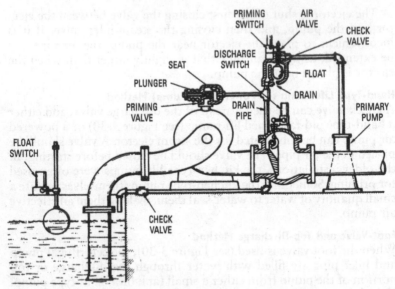

**Figure 3-31**  Float-controlled automatic priming system.

stopped by a contact that is opened when the pump motor control circuit is energized. When the main pump is running, the discharge check valve is held open, and the contacts on its switch are closed to complete a holding circuit for the pump motor contactor around the priming switch. This switch permits the priming switch to open, without shutting down the pump.

A number of small holes in the priming valve plunger permit air to pass freely during the priming operation. These small holes are so proportioned that the pressure developed by the pump forces the plunger to its seat, thereby cutting off communication to the priming pump when priming has been completed and the pump has started. When the priming line is sealed, the water drains from the float chamber, and the priming switch opens. However, the pump motor contactor is held closed by a circuit through the discharge-valve switch. When the float switch opens, the main pump shuts down. It should be noted that the float switch starts the priming pump when it closes, and it stops the centrifugal pump when it opens.

Figure 3-32 shows another type of automatic priming system in which a vacuum tank is used as a reserve to keep the main pump primed. This system consists of a motor-driven air pump and a vacuum tank connected between the section of the air pump and the priming connections on the centrifugal pump. The air tank serves

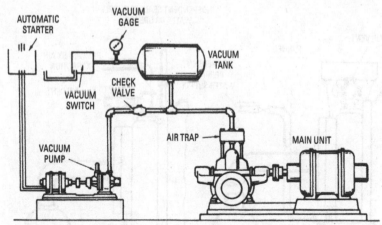

**Figure 3-32** Automatic priming system in which a vacuum tank is used as a reserve to keep the main pump primed.

as a reserve on the system, so that the vacuum producer needs starting only intermittently. A vacuum switch starts and stops the vacuum producer at the predetermined limits of vacuum. An air trap in the line between the pump to be primed and the vacuum tank prevents water rising into the vacuum system after the pump is primed. Air is drawn from the pump suction chamber whether the pump is idle, under vacuum, or in operation. The priming pipes and air-trap valves are under vacuum to ensure that the centrifugal pump remains primed at all times. Modifications of this type of system are used for pumps that handle sewage, paper stock, sludge, or other fluids that carry solids in suspension. Figure 3-33 shows other diagrams of piping for priming.

## Starting the Pump

Prior to starting a pump having oil-lubricated bearings, the rotor should be turned several times, either by hand or by momentarily operating the starting switch (with the pump filled with water), if the procedure does not overload the motor. This starts a flow of oil to the bearing surfaces. The pump can be operated for a few minutes with the discharge valve closed without overheating or damage to the pump.

Various items should be checked before starting the pump. On some installations, an extended trial run is necessary. The vent valves should be kept open to relieve pocketed air in the pump and system during these trials. This circulation of water prevents overheating the pump.

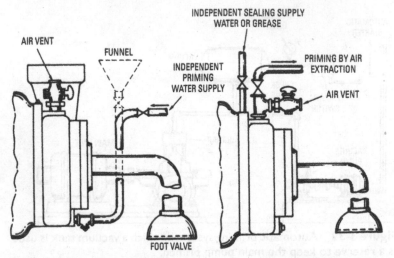

**Figure 3-33** Piping for priming systems using independent priming water supply (left) and priming by extraction of air (right).

Then, to cut the pump into the lines, close the vent valves and open the discharge valve slowly. The pump is started with the discharge valve closed, because the pump operates at only 35 to 50 percent of full load when the discharge valve is closed. If the liquid on the upper side of the discharge check valve is under sufficient head, the pump can be started with the discharge valve in open position.

Gland packing should not be too tight, because heat may cause it to expand, thereby burning out the packing and scoring the shaft. A slight leakage is desirable at first; then the packing can be tightened after it has been warmed and worn. A slight leakage also indicates that the water seal is effective without undue binding, keeping the gland and shaft cool.

A specific type of packing is required for high temperatures and for some liquids (see Figure 3-34). Graphite-type packing is recommended for either hot or cold water. Rapid wear of the sleeves may result if flax packing or metallic packing is used on centrifugal pumps with bronze shaft sleeves. Each ring of packing should be inserted separately and pushed into the stuffing box as far as the gland permits. The split openings of the successive packing rings should be positioned at 90-degree intervals.

As mentioned, the centrifugal pump should not be operated for long periods at low capacity because of the possibility of overheating. However, a pump may be operated safely at low capacity if a

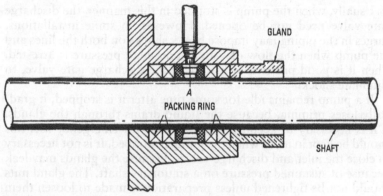

**Figure 3-34** Water-seal cage of a centrifugal pump.

permanent bypass from the discharge to the inlet (equal to one-fifth the size of the discharge pipe) is installed.

To start a hot-condensate pump, open the inlet and discharge valves on the independent stuffing-box seals before operating. Usually, the air-extraction apparatus is in service before the hot-condensate pump is started, and the main turbine is heated at the same time. This provides an accumulation of water in the hot well. If allowed to collect above the level of the hot-condensate gage glass, the steam jets or other extracting apparatus do not entrain water, which renders them inoperative. As soon as a supply of condensate begins flowing to the hot well, the pump can be started and the air valves on top of the pump opened as the pump reaches full speed. Then the air valves should be closed.

Except for the bearings and the glands, a centrifugal pump does not require attention once it is operating properly. These pumps should operate for long periods without attention other than to observe that there is a drip of liquid from the glands, that the proper oil is supplied to the bearings, and that the oil is changed at regular intervals.

## Stopping the Pump

Normally, when there is a check valve near the pump in the discharge line, the pump is shut down by stopping the motor, securing it until needed again by closing the valves in the following order:

1. Discharge
2. Inlet
3. Cooling water supply
4. At all points connecting to the system

Usually, when the pump is stopped in this manner, the discharge gate valve need not be opened. However, in some installations, surges in the piping may impose heavy shocks on both the lines and the pump when the flow of water under high pressure is arrested. Then it is good practice to first close the discharge gate valve, to eliminate shock.

If a pump remains idle for some time after it is stopped, it gradually loses priming, because the liquid drains through the glands. If it is necessary to keep the pump primed for emergency use, this should be kept in mind when the pump is stopped. It is not necessary to close the inlet and discharge valves, because the glands may leak because of sustained pressure on a stationary shaft. The gland nuts should not be tightened unless preparation is made to loosen them again when starting the pump.

## Abnormal Operating Conditions

Centrifugal pumps should run smoothly and without vibration when they are operating properly. The bearings operate at a constant temperature, which may be affected by the location of the units. This temperature may range as low as 100°F, but the operating temperature is usually maximum temperature at minimum flow, varying with pump capacity.

If, for some reason, a pump either does not contain liquid or becomes vapor-bound, vibration will occur because of contact between the stationary and the rotating parts, and the pump may become overheated. Vapor may be blown from the glands and, in extreme instances, the thrust bearings may suddenly increase in temperature; damage may result from the rotor being forced in a single direction.

If the pump is overheated because of a vaporized condition and the rotor has not seized, open all vents and prime or flood liquid into the pump. A low-temperature liquid should not be admitted suddenly to a heated pump, because fracture or distortion of its parts may result. An overheated pump should not be used unless an emergency exists (to save a boiler from damage, for example).

Vibration also may result from excessive wear on the pump rotor or in the shaft bearings, which causes the pump and motor shafts to become misaligned. These should be corrected at the first opportunity. If a rotor has seized, it is necessary to dismantle it completely and to rectify the parts by filing, machining, and so on.

## Troubleshooting

Many difficulties may be experienced with centrifugal pumps. Location of these troubles and their causes are discussed here.

## Reduced Capacity or Pressure and Failure to Deliver Water

When the capacity or pressure of the pump is reduced and the pump fails to deliver water, any of the following may be the cause:

- Pump is not primed
- Low speed
- Total dynamic head is higher than the pump rating
- Lift is too high (normal lift is 15 feet)
- Foreign material is lodged in the impeller
- Opposite direction of rotation
- Excessive air in water
- Air leakage in inlet pipe or stuffing boxes
- Insufficient inlet pressure for vapor pressure of the liquid
- Mechanical defects (such as worn rings, damaged impeller, or defective casing gasket)
- Foot valve is either too small or restricted by trash
- Foot valve or inlet pipe is too shallow

## Loses Water after Starting

If the pump starts and then loses water, the cause may be any the following:

- Air leak in inlet pipe
- Lift too high (over 15 feet)
- Plugged water-seal pipe
- Excessive air or gases in water

## Pump Overloads Driver

If the driver is overloaded by the pump, the cause may be any of the following:

- Speed too high
- Total dynamic head lower than pump rating (pumping too much water)
- Specific gravity and viscosity of pumped liquid different from pump rating
- Mechanical defects

## Pump Vibrates

The causes of pump vibration may be the following:

- Misalignment
- Foundation not rigid enough
- Foreign material causes impeller imbalance
- Mechanical defects, such as bent shaft, rotating element rubbing against a stationary element, or worn bearings

Centrifugal pumps can be operated only in one direction. The arrow on the casing indicates the direction of rotation. During pump operation, the stuffing boxes and bearings should be inspected occasionally. The pump should be disassembled, cleaned, and oiled if it is to be idle for a long period. If the pump is to be exposed to freezing temperatures, it should be drained immediately after stopping.

## Pointers on Pump Operation

Various abnormal conditions may occur during pump operations. Following are some suggestions:

- If the pump discharges a small quantity of water during the first few revolutions and then churns and fails to discharge more water, air is probably still in the pump and piping, or the lift may be too great. Check for a leaky pipe, a long inlet pipe, or lack of sufficient head.
- If the pump produces a full stream of water at first and then fails, the cause is failure of the water supply, or the water level receding below the lift limit. This can be determined by placing a vacuum gage on the inlet elbow of the pump. This problem can be remedied by lowering the pump to reduce the lift.
- If the pump delivers a full stream of water at the surface or pump level, but fails at a higher discharge point, then the pump speed is too low.
- If the pump delivers a full stream of water at first and the discharge decreases slowly until no water is delivered, an air leak at the packing gland is the cause.
- If a full stream of water is delivered for a few hours and then fails, then the inlet pipe or the impeller is obstructed if the flow from the water supply is unchanged.
- If heavy vibration occurs while the pump is in operation, then the shaft has been sprung, the pump is out of alignment, or an obstruction has lodged in one side of the impeller.

- If the bearings become hot, then the belt is unnecessarily tight, the bearings lack oil, or there is an end thrust.

- If hot liquids are to be pumped, the lift should be as small as possible, because the boiling point of the liquid is lowered under vacuum and the consequent loss of priming is caused by the presence of vapor.

- If the water is discharged into a sump or tank near the end of the inlet pipe, there is great danger of entraining air into the inlet pipe.

- If a pump is speeded up to increase capacity beyond its maximum rating, a waste of power results.

- If a pump remains idle for some time, its rotor should be rotated by hand once each week. For long periods, it should be taken apart, cleaned, and oiled.

## Maintenance and Repair

Liberal clearances are provided between the rotating and the stationary pump parts to allow for small machining variations, and for the expansion of the casing and rotor when they become heated. Stationary diaphragms, wearing rings, return channels, and so on (which are located in the casing or other stationary parts) are slightly smaller in diameter than the bore of the casing, which is machined with a $1/32$-inch gasket between the flanges. When the flange nuts are tightened, the casing must not bind on the stationary parts. Operating clearances (such as those at the wearing rings) depend on the actual location of the part, the type of the material, and the span of the bearing.

### Lateral End Clearances

Lateral movement between the rotor and stator parts is necessary for mechanical and hydraulic considerations, and for expansion variations between the casing and the rotor. End movement is limited to $1/64$ inch in small pumps and certain other types of pumps, and it may be as much as $1/2$ inch on larger units and on those handling hot liquids. On pumps handling cold liquids, the clearance is divided equally when the thrust bearing is secured in position and the impellers are centralized. The recommendations of the pump manufacturer should be consulted for the designed lateral clearances before proceeding with an important field assembly.

### Parts Renewals

Since the pump casings are made from castings, it is sometimes necessary to favor variations in longitudinal dimensions on the casings by making assembly-floor adjustments to the rotor, so that designed lateral clearances can be preserved and so that the impellers can be placed in their correct positions with respect to the diaphragms, diffusers, and return channels.

When a rotor with its diffusers is returned to the factory for repair, the repaired rotor can be placed in the pump with no adjustment if it is still possible to calibrate dimensions on the worn parts. When it is impossible to obtain complete details on the used parts, replacement parts are made to standard dimensions.

When the assembled rotor (with stage pieces, and so on) is placed in the lower portion of the casing, the total lateral clearance should be checked. With the thrust bearing assembled and the shaft in position, the total clearance should be divided properly and the impellers centered in their volutes. Final adjustments are made by manipulation of the shaft nuts.

The casing flange gaskets should be replaced with the same type and thickness of material as the original gaskets. The inner edge of the gasket must be trimmed accurately along the bore of the stuffing box. The edges must be trimmed squarely and neatly at all points. This is especially so where the gasket abuts on the outer diameter and sides of the stationary parts between the stages. It should be overlapping sufficiently for the upper portion of the casing to press the edges of the gasket against the stator parts while the casing is being tightened. This ensures proper sealing between the stages. In the trimming-operation, the gasket should first be cemented to the lower portion of the casing with shellac. Then a razor blade can be used to cut the gasket edges squarely, overlapping at the same time.

## Pointers on Assembly
Unnecessary force should not be used in tightening the impeller and shaft sleeve nuts. Bending of the shaft may result, and the concentricity of the rotor parts operating in close clearances with the stator parts may be destroyed, causing rubbing and vibration.

### Locking Screws
A dial test indicator should be used to determine whether the shaft has been bent when securing the safety-type locking screws.

Design considerations require that all parts be mounted on the shaft in their original order. *Opposed impellers* are both right-hand and left-hand types in the same casing. Diaphragms, wearing rings, and stage bushings are individually fitted, and their sealing flanges between the stages are checked for each location. Stage bushings with stop pieces are not interchangeable with each other, because the stop pieces may locate at various positions. All stationary parts assembled on the rotor (such as stuffing-box bushings, diaphragms, and so on) possess stops that consist of either individual pins or flange halves in the lower casing only, to prevent turning. These parts must be in position when lowering the rotor into the casing, so that all stops are in their respective recesses in the lower casing. Otherwise, the upper casing may foul improperly-positioned parts during the mounting procedure.

### Deep-Well Pump Adjustments

Before operating these pumps, the impeller or impellers must be adjusted to their correct operating positions by either raising or lowering the shaft by means of the adjuster nut provided for the purpose. The shaft is raised to its uppermost position by turning the adjusting nut downward. Measure the distance the shaft has been raised above its lowest position, and then back off the adjusting nut until the shaft has been lowered one-third the total distance that it was raised. Lock the nut in position with the key, setscrew, or locknut provided. With either a key or a setscrew, it is usually necessary to turn the adjusting nut until it is possible to insert the key or setscrew in place through both the adjusting nut and motor clutch.

### Installing Increaser on Discharge Line

Hydraulic losses can be reduced by installing an increaser on the discharge line. Figure 3-35 shows the hookup with the check and discharge valves. The discharge line should be selected with reference to friction losses, and it should never be smaller than the pump discharge outlet—preferably, one to two sizes larger. The pump should not be used to support heavy inlet or discharge piping. Pipes or

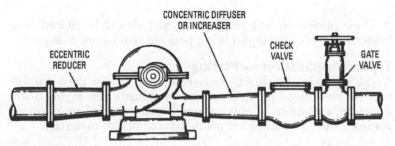

**Figure 3-35** Installation of a diffuser or increaser on the discharge line.

fittings should not be forced into position with the flange bolts, because the pump alignment may be disturbed. Independent supports should be provided for all piping. When piping is subjected to temperature changes, expansion and contraction should not exert a strain on the pump casing. In hotels, apartment buildings, hospitals, and so on, where noise is objectionable, the discharge pipe should not be attached directly to the steel structural work, hollow walls, and so on, because vibration may be transmitted to the building. Preferably, the discharge line should be connected to the pump discharge outlet through a flexible connection.

## Corrosion-Resisting Centrifugal Pumps

Centrifugal pumps are designed for liquid transfer, recirculation, and other applications where no suction lift is required. These pumps produce high flow rates under moderate head conditions. However, centrifugal pumps are not recommended for pumping viscous fluids. Table 3-1 is a viscosity chart for liquids up to 1500 Saybolt-seconds universal (ssu). *Saybolt-seconds universal (ssu)* is the time required for a gravity flow of 60 cubic centimeters. Performance decrease and horsepower adjustment are shown in the table.

### Table 3-1   Viscosity Chart

| Viscosity ssu | 30 | 100 | 250 | 500 | 750 | 1000 | 1500 |
|---|---|---|---|---|---|---|---|
| Flow Reduction Percent (in G.P.M.) | — | 3 | 8 | 15 | 20 | 25 | 30 |
| Head Reduction Percent (in feet) | — | 2 | 5 | 12 | 15 | 20 | 25 |
| Horsepower Increase Percent | — | 10 | 20 | 30 | 50 | 65 | 85 |

These pumps are not self-priming and should be placed at or below the level of the liquid being pumped (see Figure 3-36).

### Typical Application — Plating

A typical plating-shop installation serves to show how the corrosion-resisting centrifugal pump is utilized in industry (see Figure 3-37). There are a number of applications for this type of pump in the plating shop. The pumps are manufactured in two materials:

- Noryl
- Polypropylene

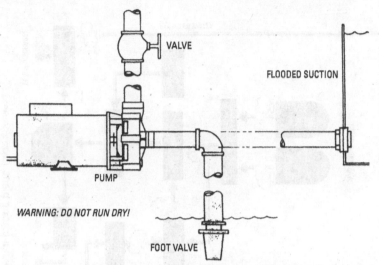

VALVE

FLOODED SUCTION

PUMP

WARNING: DO NOT RUN DRY!

FOOT VALVE

**Figure 3-36** Connecting the pump for proper operation. *(Courtesy Sherwood)*

Figure 3-37 shows a pump mounted on the end of an electric motor. The plastic pump is combined with a non-metallic seal that offers compatibility with a wide variety of corrosive chemicals used in the plating industry.

**Figure 3-37** Corrosion-resisting centrifugal pump. *(Courtesy Sherwood)*

## Plating

A typical plating line consists of a series of baths and rinses, each requiring circulation, transfer, and filtering pumps. Figure 3-38 shows a single plating line, whereas most installations involve multiple lines.

## Shaft Seal

Noncorrosive thermoplastic mechanical stationary shaft seals isolate all metal parts from liquid contact (see Figure 3-39). Stainless steel components are sealed in polypropylene. The seal guarantees reliability through simplicity and balance of design.

## Water Treatment

The waste-treatment system neutralizes spent plating solutions and waste acids throughout the total system.

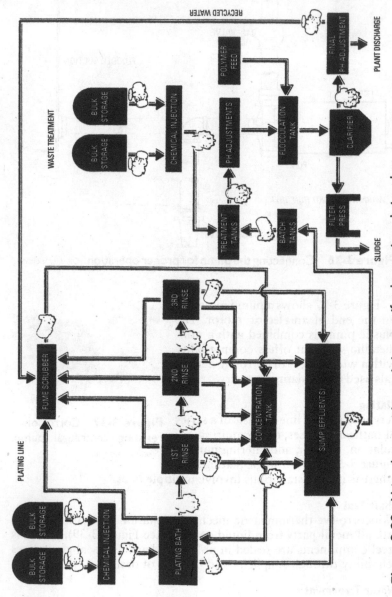

**Figure 3-38** Using the corrosion-resisting centrifugal pump in a plating plant. *(Courtesy Sherwood)*

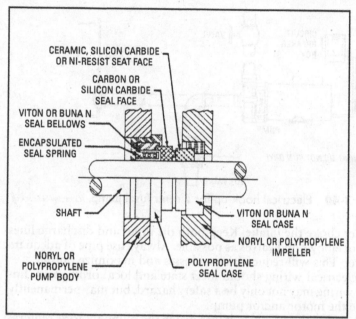

CERAMIC, SILICON CARBIDE
OR NI-RESIST SEAT FACE

CARBON OR
SILICON CARBIDE
SEAL FACE

VITON OR BUNA N
SEAL BELLOWS

ENCAPSULATED
SEAL SPRING

SHAFT

VITON OR BUNA N
SEAL CASE

NORYL OR POLYPROPYLENE
IMPELLER

NORYL OR
POLYPROPYLENE
PUMP BODY

POLYPROPYLENE
SEAL CASE

**Figure 3-39** Corrosion-resisting centrifugal pump seal. *(Courtesy Sherwood)*

## Fume Scrubbers

The actual plating process emits malodorous and noxious fumes. Removal of the most difficult fumes and odors is done by using air scrubbers. Pumps are used as circulation units in the removal of contaminant particles from fumes emitted in the process environments.

## Corrosion-Resisting Pump Installation

Locate the pump as near the source to be pumped as possible. A flooded suction situation is preferred. Since this type of pump is not self-priming, *it must never be allowed to run dry.* If the fluid level is below the pump, therefore, a foot valve must be installed and the pump primed prior to start-up (see Figure 3-40).

Mount the motor base to a secure, immobile foundation. Use only plastic fittings on both the intake and discharge ports, and seal the pipe connections with Teflon tape. These fittings should be self-supported and in a neutral alignment with each port. That is, fittings must not be forced into alignment that may cause premature line failure or damage to the pump volute.

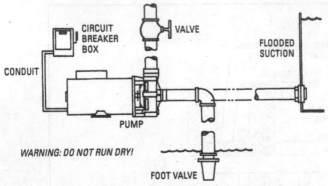

**Figure 3-40** Electrical hook-up for a centrifugal pump. *(Courtesy Sherwood)*

Never choke the intake. Keep both the input and discharge lines as free of elbows and valves as possible. Always use pipe of adequate diameter. This will reduce friction losses and maximize output.

All electrical wiring should meet state and local ordinances. Improper wiring may not only be a safety hazard, but may permanently damage the motor and/or pump.

Ensure that the supply voltages match the pump motor's requirements. Check the motor wiring and connect (according to instructions on the motor) to match the supply voltage. Be sure of proper rotation. *Improper rotation will severely damage the pump and void the warranty.*

The power cord should be protected by conduit or by cable, and should be of the proper gage or wire size. It should be no longer than necessary. Power should be drawn directly from a box with circuit breaker protection or with a fused disconnect switch.

Always switch off the power before repairing or servicing the pump or the motor. Check for proper rotation of three-phase (3Φ) motors.

The water flush will provide decontamination of chemicals or elastomers and seal and seat faces, while providing lubrication required for start-up. Water-flushed seals are recommended for abrasive solutions, high temperature service, or when pumps may be run dry or against dead head conditions. Where conditions cause the pumped liquid to form crystals, or if the pump remains idle for a period without adequate flushing, a water flush seal system is advised.

Two methods of water flushing can be used:

- **Direct Plumbing to City Water**—This provides the best possible approach to flushing the seal and seat faces. Caution must be taken to conform to local city ordinances that may require *backflow preventers*. These are a series of check valves required to prevent contamination of city water should there be a shutoff of the water supply. Also be aware of the addition of water into the chemicals pumped where some imbalance may be created altering the chemicals' formulation and aggressiveness.

• **Recirculation of Solution Pumped**—This system takes a bleed off the pump discharge and recirculates the solution in the seal chamber. Although not nearly as effective as the direct water flush, it will provide cooling to the seal and seat faces under operation. This system is not effective where crystallization occurs or for pumps in idle conditions. Figures 3-41 and 3-42 show the pump intake and discharge and plug locations.

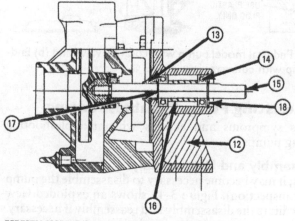

**PEDESTAL MODELS (Same Parts for All Pump Models)**

| | |
|---|---|
| 13 — PEDESTAL MOUNTING BRACKET | 17 — SPACER |
| 14 — SLINGER | 18 — INTERNAL RTG. RING |
| 15 — BALL BEARING (2) | 19 — EXTERNAL RTG. RING |
| 16 — SHAFT | |

**Figure 3-41**  Pedestal model pump, cutaway view. *(Courtesy Sherwood)*

## Maintenance

Lubricate the motor as per instructions on the motor. The rotary seal requires no lubrication after assembly. The pump must be drained before servicing, or if stored in below-freezing temperatures. Periodic replacement of seals may be required due to normal carbon wear.

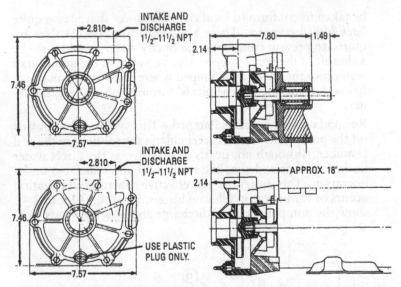

**Figure 3-42** (a) Pedestal model pump with cutaway view, and (b) End-mount model pump with cutaway view. *(Courtesy Sherwood)*

## Corrosion-Resisting Pump Troubleshooting
Table 3-2 shows symptoms and actions when troubleshooting corrosion-resisting pumps.

### Pump-End Assembly and Disassembly
In some instances, it may become necessary to disassemble the pump (for example, for inspection). Figure 3-43 shows an exploded view of the pump to facilitate the disassembly and reassembly if necessary.

#### Pump Disassembly
To disassemble the pump, follow these steps:

1. Shut off power to motor before disconnecting any electrical wiring from the back of the motor.
2. Disassemble the body-motor assembly from the volute. The volute may be left in-line if you wish.
3. Remove the cap covering the shaft at the back of the motor and with a large screwdriver, prevent the shaft from rotating while you unscrew the impeller.
4. Remove the ceramic piece from the impeller.

## Table 3-2  Troubleshooting Corrosion-Resisting Pumps

| Symptom | Action |
| --- | --- |
| Motor will not rotate | Check for proper electrical connections to the motor |
| | Check main power box for blown fuse, etc. |
| | Check thermal overload on the motor |
| Motor hums or will not rotate at correct speed | Check for proper electrical connections to the motor and the proper cord size and length |
| | Check for foreign material inside the pump |
| | Remove the bracket and check for impeller rotation without excessive resistance |
| | Remove the pump and check the shaft rotation for excessive bearing noise |
| | Have an authorized service person check the start switch and/or start capacitor on the motor |
| Pump operates with little or no flow | Check to insure that the pump is primed |
| | Check for a leaking seal |
| | Check for improper line voltage to the motor or incorrect rotation |
| | Check for clogged inlet port and/or impeller |
| | Check for defective check valve or foot valve |
| | Check inlet lines for leakage, of either fluid or air |
| Pump loses prime | Check for defective check valve or foot valve |
| | Check for leaking seal |
| | Check for inlet line air leakage |
| | Check for low fluid supply |
| Motor or pump overheats | Check for proper line voltage and phase, also proper motor wiring |
| | Check for binding motor shaft or pump parts |
| | Check for inadequate ventilation |
| | Check temperature of fluid being pumped—should not exceed 194°F (90°C) for extended periods |

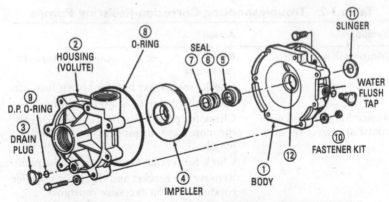

**Figure 3-43** Exploded view of corrosion-resisting centrifugal pump.
*(Courtesy Sherwood)*

| REF. NO. | DESCRIPTION | REF. NO. | DESCRIPTION |
|---|---|---|---|
| 1. | BODY | 7. | SEAT WASHER |
| 2. | HOUSING | | SEAL ASSEMBLY (5, 6, & 7) |
| 3. | DRAIN PLUG | 8. | O-RING GASKET |
| 4. | DRAIN PLUG | 9. | DRAIN PLUG O-RING |
| 5. | SEAL | 10. | FASTENER KIT |
| 6. | SEAL SEAT | 11. | SLINGER |
| | | 12. | SEAL O-RING |

5. Detach the body from the motor.

6. Remove the carbon-graphite seal from the body by pressing out from the back. *Do not dig out from the front.*

**Pump-End Assembly**
To assemble the pump-end assembly, follow these steps:

1. Clean and inspect all pump parts: O-ring, seal seats, motor shaft, and so on.

2. Apply lubricant to the body bore hole and O-ring for seal installation.

3. Press the carbon graphite seal into the body while taking care not to damage the carbon graphite face.

4. Place the slinger (rubber washer) over the motor shaft and mount the body to the motor.

5. Carefully grease the boot or O-ring around the ceramic piece and press it into the impeller. If the ceramic piece has an O-ring, the marked side goes in.

6. Sparingly lubricate carbon-graphite and ceramic sealing surfaces with lightweight machine oil. Do not use silicon lubricants or grease.

7. Thread the impeller onto the shaft and tighten. If required, remove the motor end-cap and use a screwdriver on the back of the motor shaft to prevent shaft rotation while tightening. Replace the motor end-cap.

8. Electrically, connect the motor so that the impeller will rotate counterclockwise (CCW) facing the pump with the motor toward the rear. Incorrect rotation will damage the pump and void the warranty.

## Note

For three-phase power, electrically check the rotation of the impeller with the volute disassembled from the bracket. If the pump-end is assembled and the rotation is incorrect, serious damage to the pump-end assembly will occur and invalidate the warranty. If the rotation is incorrect, simply exchange any two leads to reverse the direction of rotation.

9. Seat the O-ring in the volute slot and assemble the volute to the body.

10. Install the drain plug in the volute drain hole.

## Good Safety Practices

Following are some good safety tips:

- When wiring a motor, follow all local electrical and safety codes, as well as the *National Electrical Code* (NEC) and the *Occupational Safety and Health Act* (OSHA).

- Always disconnect the power source before performing any work on or near the motor or its connected load. If the power disconnect point is out of sight, lock the disconnect in the open position and tag it to prevent unexpected application of power. Failure to do so could result in electrical shock.

- Be careful when touching the exterior of an operating motor. It may be hot enough to be painful or cause injury. With modern motors, which are built to operate at higher temperatures, this condition is normal if operated at rated load and voltage.

- Do not insert any object into the motor.

- Pump rotates in one direction only—counterclockwise from the pump inlet end.

- Protect the power cable from coming in contact with sharp objects.
- Do not kink the power cable and never allow the cable to contact oil, grease, hot surfaces, or chemicals.
- Ensure that the power source conforms to the requirements of your equipment.
- Do not handle the pump with wet hands or when standing in water because electrical shock could occur. Disconnect the main power before handling the unit for any reason.
- The unit should run counterclockwise as viewed facing the shaft end. Clockwise rotation can result in damage to the pump motor, property damage, and/or personal injury.

## Impeller Design Considerations

Experience has been an important factor in determining the design of centrifugal impellers. Fundamentally, the centrifugal pump adds energy in the form of velocity to an already-flowing liquid; pressure is not added in the usual sense of the word. Kinetic energy is involved in all basic considerations of the centrifugal pump.

The design of the impeller of a centrifugal pump is based ultimately on the performance of other impellers. The general effect of altering certain dimensions is a result of modifying or changing the design of impellers that have been tested.

Calculations applicable to centrifugal pumps are based on the impeller diagram shown in Figure 3-44, in which the following symbols are used:

$V_2$ is tangential velocity at outer periphery

$V_1$ is tangential velocity at inner periphery

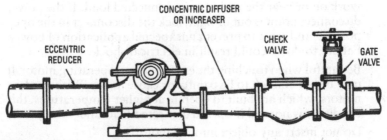

**Figure 3-44**  Impeller illustrating basis of calculations.

$Z_2$ is relative velocity of water at outlet

$Z_1$ is relative velocity of water at inlet

$C_2$ is absolute velocity of water at outlet

$J_2$ is radial velocity of water at outlet

$J_1$ is radial velocity of water at inlet

$W$ is tangential velocity of water at outlet

$d_2$ is outlet angle of impeller

$d_1$ is inlet angle of impeller

## Velocity of Impeller

In the Figure 3-44, water enters the impeller inlet with a radial velocity $J_1$, and it leaves the impeller at an absolute velocity $C_2$. The inner peripheral velocity $V_1$ and the outer peripheral velocity $V_2$ are in *feet per second*. If the theoretical head $H$ (in feet) represents the head against which the pump delivers water (with no losses), then:

$$H = \frac{(V_2)^2}{2g}; \quad \text{or } V_2 = \sqrt{2gH}$$

in which the following is true:

$g$ is the force of gravity = 32.2 feet per second

If the head $H$ against which the pump must work and the diameter of the impeller are known, the speed of the pump can be calculated by the preceding formulas.

### Problem

If a pump with an impeller diameter of $10^{3/4}$ inches is required to pump against a head of 100 feet, what is the required speed of the pump?

### Solution

By substituting in the previous formula:

$$V_2 = \sqrt{2gH}$$
$$= \sqrt{2 \times 32.2 \times 100}$$
$$= 80.25 \text{ ft/s}$$

Since the tangential velocity $V_2$ at the outer periphery of the impeller is 80.25 feet per second, this is equal to 4815 feet per minute

**164** Chapter 3

(80.25 × 60). Since the circumference ($\pi d$) of a $10^{3}/_4$-inch diameter impeller is 33.8 inches (3.14 × $10^{3}/_4$, or 2.8 feet, the impeller must rotate at a speed of 4815 feet per minute; this is equal to 1720 revolutions per minute (4815 ÷ 2.8).

## Total Hydraulic Load or Lift

As mentioned, the lift is the vertical distance measured from the level of the water to be pumped to the centerline of the pump inlet. If the water level is above the centerline of the pump, the pump is operating under *inlet head*, or negative lift. Under conditions of negative lift, the inlet head must be subtracted from the sum of the remaining factors.

The *discharge head*, in contrast to the inlet head, is the vertical distance measured between the centerline of the pump discharge opening and the level to which the water is elevated.

The loss of head because of friction must be considered. This loss can be obtained from Table 3-3. Other data for centrifugal pump calculations are given in Tables 3-4, 3-5, 3-6, 3-7, and 3-8.

## Velocity Head

The equivalent distance (in feet) through which a liquid must fall to acquire the same velocity is called *velocity head*. The velocity head can be determined from the following formula:

$$H_v = \frac{(V_2)^2}{2g} = \frac{(V_2)^2}{64.4}$$

in which

$$V = \frac{0.408 \times \text{gal/min}}{D^2}$$

Where $D$ equals the diameter of the pipe in inches.

The following example is used to illustrate calculation of the total hydraulic head that a pump may work against, using gage readings.

### Problem

Assuming that the distance $A$ (see Figure 3-45) or vertical distance from the centerline of the gage connection in the inlet pipe to the centerline of the pressure gage is 2 feet. The discharge pressure gage reading is 40 pounds. The vacuum gage reading is 15 inches (when discharging 1000 gallons of water per minute). The discharge pipe is

## Table 3-3  Loss of Head* Because of Friction in Pipes†

| Gallons per Min. | 1/2-Inch Pipe Vel.‡ | 1/2-Inch Pipe Fric.§ | 3/4-Inch Pipe Vel. | 3/4-Inch Pipe Fric. | 1 Inch Pipe Vel. | 1 Inch Pipe Fric. | 1 1/4-Inch Pipe Vel. | 1 1/4-Inch Pipe Fric. | 1 1/2-Inch Pipe Vel. | 1 1/2-Inch Pipe Fric. | 2-Inch Pipe Vel. | 2-Inch Pipe Fric. | 2 1/2-Inch Pipe Vel. | 2 1/2-Inch Pipe Fric. | 3-Inch Pipe Vel. | 3-Inch Pipe Fric. |
|---|---|---|---|---|---|---|---|---|---|---|---|---|---|---|---|---|
| 1 | 1.05 | 2.1 | — | — | — | — | — | — | — | — | — | — | — | — | — | — |
| 2 | 2.10 | 7.4 | 1.20 | 1.9 | — | — | — | — | — | — | — | — | — | — | — | — |
| 3 | 3.16 | 15.8 | 1.80 | 4.1 | 1.12 | 1.26 | — | — | — | — | — | — | — | — | — | — |
| 4 | 4.21 | 27.6 | 2.41 | 7.0 | 1.49 | 2.14 | 0.86 | 0.57 | 0.63 | 0.26 | — | — | — | — | — | — |
| 5 | 5.26 | 41.0 | 3.01 | 10.5 | 1.86 | 3.25 | 1.07 | 0.84 | 0.79 | 0.39 | — | — | — | — | — | — |
| 10 | 10.52 | 147.0 | 6.02 | 38.0 | 3.72 | 11.70 | 2.14 | 3.05 | 1.57 | 1.43 | 1.02 | 0.50 | 0.65 | 0.17 | 0.45 | 0.07 |
| 15 | — | — | 9.02 | 88.0 | 5.60 | 25.00 | 3.20 | 6.50 | 2.36 | 3.00 | 1.53 | 1.00 | 0.98 | 0.36 | 0.68 | 0.15 |
| 20 | — | — | 12.03 | 136.0 | 7.44 | 42.00 | 4.29 | 11.10 | 3.15 | 5.20 | 2.04 | 1.82 | 1.31 | 0.61 | 0.91 | 0.25 |
| 25 | — | — | — | — | 9.30 | 64.00 | 5.36 | 18.00 | 3.94 | 7.80 | 2.55 | 2.73 | 1.63 | 0.92 | 1.13 | 0.38 |
| 30 | — | — | — | — | 11.15 | 89.00 | 6.43 | 23.50 | 4.72 | 11.00 | 3.06 | 3.84 | 1.96 | 1.29 | 1.36 | 0.54 |
| 35 | — | — | — | — | 13.02 | 119.00 | 7.51 | 31.20 | 5.51 | 14.70 | 3.57 | 5.10 | 2.20 | 1.72 | 1.59 | 0.71 |
| 40 | — | — | — | — | 14.88 | 152.00 | 8.58 | 40.00 | 6.30 | 18.80 | 4.08 | 6.60 | 2.61 | 2.20 | 1.82 | 0.91 |
| 45 | — | — | — | — | — | — | 9.65 | 50.00 | 7.08 | 23.20 | 4.60 | 8.20 | 2.94 | 2.00 | 2.05 | 1.15 |
| 50 | — | — | — | — | — | — | 10.72 | 60.00 | 7.87 | 28.40 | 5.11 | 9.90 | 3.27 | 3.32 | 2.27 | 1.30 |

†Friction given per 100 ft. of 15-year-old iron pipe: for new and smooth iron pipe, use 0.71 of value shown in table

(continued)

## Table 3-3 (continued)

| Gallons per Min. | 1/2- Inch Pipe | | 3/4- Inch Pipe | | 1 Inch Pipe | | 1 1/4- Inch Pipe | | 1 1/2- Inch Pipe | | 2- Inch Pipe | | 2 1/2- Inch Pipe | | 3 Inch Pipe | |
|---|---|---|---|---|---|---|---|---|---|---|---|---|---|---|---|---|
| | Vel.‡ | Fric.§ | Vel. | Fric. | Vel. | Fric. | Vel. | Fric. | Vel. | Fric. | Vel. | Fric. | Vel. | Fric. | Vel. | Fric. |
| 70 | — | — | — | — | — | — | — | 113.00 | 11.02 | 53.00 | 7.15 | 18.40 | 4.58 | 6.20 | 3.18 | 2.57 |
| 90 | — | — | — | — | — | — | 15.01 | — | 14.17 | 84.00 | 9.19 | 28.40 | 5.88 | 9.00 | 4.09 | 4.08 |
| 100 | — | — | — | — | — | — | — | — | 15.74 | 102.00 | 10.21 | 35.00 | 6.54 | 12.00 | 4.54 | 4.96 |
| 120 | — | — | — | — | — | — | — | — | 18.89 | 143.00 | 12.25 | 50.00 | 7.84 | 16.80 | 5.45 | 7.00 |
| 140 | — | — | — | — | — | — | — | — | 22.04 | 199.00 | 14.30 | 67.00 | 9.15 | 23.30 | 6.35 | 9.20 |
| 160 | — | — | — | — | — | — | — | — | — | — | 16.34 | 86.00 | 10.46 | 29.00 | 7.26 | 11.00 |
| 180 | — | — | — | — | — | — | — | — | — | — | 18.38 | 107.00 | 11.76 | 36.70 | 8.17 | 14.80 |
| 200 | — | — | — | — | — | — | — | — | — | — | 20.42 | 129.00 | 13.07 | 43.10 | 9.08 | 17.80 |
| 220 | — | — | — | — | — | — | — | — | — | — | 22.47 | 154.00 | 14.38 | 52.00 | 9.99 | 21.30 |
| 240 | — | — | — | — | — | — | — | — | — | — | 24.51 | 182.00 | 15.69 | 61.00 | 10.89 | 26.10 |
| 260 | — | — | — | — | — | — | — | — | — | — | 26.55 | 211.00 | 16.99 | 70.00 | 11.80 | 29.10 |
| 280 | — | — | — | — | — | — | — | — | — | — | — | — | 18.30 | 81.00 | 12.71 | 33.40 |
| 300 | — | — | — | — | — | — | — | — | — | — | — | — | 19.61 | 92.00 | 13.62 | 38.00 |

* Loss of head given in feet
†Friction given per 100 ft. of 15-year-old iron pipe: for new and smooth iron pipe, use 0.71 of value shown in table
‡Velocity feet per second
§Friction head in feet

## Table 3-3 (continued)

| Gallons per Min. | 4-Inch Pipe | | 5-Inch Pipe | | 6-Inch Pipe | | 8-Inch Pipe | | 10-Inch Pipe | | 12-Inch Pipe | | 14-Inch Pipe | | 15-Inch Pipe | | 16-Inch Pipe | 20-Inch Pipe |
|---|---|---|---|---|---|---|---|---|---|---|---|---|---|---|---|---|---|---|
| | Vel.‡ | Fric.§ | Vel. | Fric. | Vel. | Fric. | Vel. | Fric. | Vel. | Fric. | Vel. | Fric. | Vel. | Fric. | Vel. | Fric. | | |
| 40 | 1.02 | 0.22 | — | — | — | — | — | — | — | — | — | — | — | — | — | — | — | — |
| 45 | 1.17 | 0.28 | — | — | — | — | — | — | — | — | — | — | — | — | — | — | — | — |
| 50 | 1.28 | 0.34 | — | — | — | — | — | — | — | — | — | — | — | — | — | — | — | — |
| 70 | 1.79 | 0.63 | 1.14 | 0.21 | — | — | — | — | — | — | — | — | — | — | — | — | — | — |
| 75 | 1.92 | 0.73 | 1.22 | 0.24 | — | — | — | — | — | — | — | — | — | — | — | — | — | — |
| 100 | 2.55 | 1.23 | 1.63 | 0.30 | 1.14 | 0.14 | — | — | — | — | — | — | — | — | — | — | — | — |
| 120 | 3.06 | 1.71 | 1.96 | 0.57 | 1.42 | 0.26 | — | — | — | — | — | — | — | — | — | — | — | — |
| 125 | 3.19 | 1.86 | 2.04 | 0.64 | 1.48 | 0.20 | — | — | — | — | — | — | — | — | — | — | — | — |
| 150 | 3.84 | 2.55 | 2.45 | 0.08 | 1.71 | 0.32 | — | — | — | — | — | — | — | — | — | — | — | — |
| 175 | 4.45 | 3.36 | 2.86 | 1.18 | 2.00 | 0.48 | — | — | — | — | — | — | — | — | — | — | — | — |
| 200 | 5.11 | 4.37 | 3.27 | 1.48 | 2.28 | 0.62 | — | — | — | — | — | — | — | — | — | — | — | — |
| 225 | 6.32 | 6.61 | 3.67 | 1.86 | 2.57 | 0.74 | — | — | — | — | — | — | — | — | — | — | — | — |
| 250 | 6.40 | 6.72 | 4.08 | 2.24 | 2.80 | 0.92 | 1.60 | 0.22 | — | — | — | — | — | — | — | — | — | — |
| 275 | 7.03 | 7.90 | 4.50 | 2.72 | 3.06 | 1.15 | 1.73 | 0.27 | — | — | — | — | — | — | — | — | — | — |
| 300 | 7.86 | 9.38 | 4.90 | 3.15 | 3.40 | 1.29 | 1.90 | 0.36 | — | — | — | — | — | — | — | — | — | — |
| 350 | 8.90 | 12.32 | 5.72 | 4.19 | 3.98 | 1.08 | 2.20 | 0.41 | — | — | — | — | — | — | — | — | — | — |
| 400 | 10.20 | 15.82 | 6.54 | 5.33 | 4.54 | 2.21 | 2.60 | 0.50 | 1.80 | 0.21 | — | — | — | — | — | — | — | — |
| 450 | 11.50 | 19.74 | 7.35 | 6.55 | 5.12 | 2.74 | 2.92 | 0.64 | — | — | — | — | — | — | — | — | — | — |

(continued)

# Table 3-3 (continued)

| Gallons per Min. | 4-Inch Pipe | | 5-Inch Pipe | | 6-Inch Pipe | | 8-Inch Pipe | | 10-Inch Pipe | | 12-Inch Pipe | | 14-Inch Pipe | | 15-Inch Pipe | | 16-Inch Pipe | | 20-Inch Pipe | |
|---|---|---|---|---|---|---|---|---|---|---|---|---|---|---|---|---|---|---|---|---|
| | Vel.‡ | Fric.§ | Vel. | Fric. | Vel. | Fric. | Vel. | Fric. | Vel. | Fric. | Vel. | Fric. | Vel. | Fric. | Vel. | Fric. | Vel. | Fric. | Vel. | Fric. |
| 475 | 12.30 | 22.96 | 7.88 | 7.22 | 5.55 | 3.21 | 3.10 | 0.79 | 1.94 | 0.25 | — | — | — | — | — | — | — | — | — | — |
| 500 | 12.77 | 24.08 | 8.17 | 8.12 | 5.60 | 3.26 | 3.20 | 0.81 | 2.04 | 0.28 | 1.42 | 0.11 | — | — | — | — | — | — | — | — |
| 550 | — | — | 8.99 | 9.66 | 6.16 | 3.93 | 3.52 | 0.98 | 2.25 | 0.33 | 1.57 | 0.14 | — | — | — | — | — | — | — | — |
| 600 | — | — | 9.80 | 11.34 | 6.72 | 4.79 | 3.84 | 1.16 | 2.46 | 0.39 | 1.71 | 0.15 | 1.37 | 0.09 | — | — | — | — | — | — |
| 650 | — | — | 10.62 | 13.16 | 7.28 | 5.50 | 4.16 | 1.34 | 2.66 | 0.46 | 1.85 | 0.19 | 1.47 | 0.10 | — | — | — | — | — | — |
| 700 | — | — | 11.44 | 15.12 | 7.64 | 6.38 | 4.46 | 1.54 | 2.86 | 0.52 | 2.00 | 0.22 | 1.58 | 0.11 | — | — | — | — | — | — |
| 750 | — | — | 12.26 | 17.22 | 8.50 | 7.00 | 4.80 | 1.74 | 3.06 | 0.59 | 2.13 | 0.24 | 1.68 | 0.13 | — | — | — | — | — | — |
| 800 | — | — | — | — | 9.08 | 7.90 | 5.12 | 1.97 | 3.28 | 0.67 | 2.27 | 0.27 | 1.79 | 0.14 | — | — | — | — | — | — |
| 850 | — | — | — | — | 9.58 | 8.75 | 5.48 | 2.28 | 3.48 | 0.75 | 2.41 | 0.31 | 1.89 | 0.16 | — | — | — | — | — | — |
| 900 | — | — | — | — | 10.30 | 10.11 | 5.75 | 2.46 | 3.68 | 0.83 | 2.56 | 0.34 | 2.00 | 0.17 | 1.73 | 0.12 | — | — | — | — |
| 950 | — | — | — | — | 10.72 | 10.71 | 6.06 | 2.87 | 3.88 | 0.91 | 2.70 | 0.35 | 2.10 | 0.19 | 1.82 | 0.14 | — | — | — | — |
| 1000 | — | — | — | — | 11.32 | 12.04 | 6.40 | 3.02 | 4.08 | 1.01 | 2.84 | 0.41 | 2.31 | 0.23 | 2.00 | 0.16 | — | — | — | — |
| 1100 | — | — | — | — | 12.50 | 14.31 | 7.03 | 3.51 | 4.50 | 1.29 | 3.13 | 0.49 | 2.52 | 0.26 | 2.18 | 0.19 | — | — | — | — |
| 1200 | — | — | — | — | 13.52 | 16.69 | 7.67 | 4.26 | 4.91 | 1.46 | 3.41 | 0.57 | — | — | — | — | — | — | — | — |
| 1500 | — | — | — | — | — | — | 9.60 | 6.27 | 6.10 | 2.09 | 4.20 | 0.85 | 3.15 | 0.39 | 2.73 | 0.28 | 2.39 | 0.24 | — | — |
| 2000 | — | — | — | — | — | — | 12.70 | 10.71 | 8.10 | 3.50 | 5.60 | 1.43 | 4.20 | 0.66 | 3.64 | 0.47 | 3.19 | 0.39 | — | — |
| 2500 | — | — | — | — | — | — | — | — | 10.10 | 6.33 | 7.00 | 2.18 | 5.25 | 1.01 | 4.55 | 0.72 | 3.99 | 0.56 | — | — |
| 3000 | — | — | — | — | — | — | — | — | 12.10 | 7.42 | 8.40 | 3.30 | 6.30 | 1.57 | 5.46 | 1.12 | 4.79 | 0.80 | 3.08 | 0.27 |

*Loss of head given in feet
†Friction given per 100 ft. of 15-year-old iron pipe: for new and smooth iron pipe, use 0.71 of value shown in table
‡Velocity feet per second
§Friction head in feet

### Table 3-4 Electric Current Consumption for Pumping

| Percent Efficiency of Pump, Motor, and Transformer | Consumption in Kilowatt-Hours per 24-hour Period (100 gallons/minute; 100 feet high) |
|---|---|
| 30 | 149.2 |
| 34 | 132.0 |
| 38 | 118.0 |
| 42 | 107.0 |
| 45 | 100.0 |
| 50 | 89.0 |
| 55 | 81.0 |
| 60 | 75.0 |
| 65 | 69.0 |
| 70 | 64.0 |
| 75 | 59.0 |

### Table 3-5 Weight and Volume of Water (Standard Gallons)

|  | Imperial, or English | U.S. |
|---|---|---|
| Cubic inches/gallon | 277.274 | 231.00 |
| Pounds/gallon | 10.00 | 8.3311 |
| Gallons/cubic foot | 6.232102 | 7.470519 |
| Pounds/cubic foot | 62.4245 | 62.4245 |

6 inches in diameter (at gage connection) and the inlet pipe is 8 inches in diameter (at gage connection). Calculate the total hydraulic load or lift.

### Solution

40 (psi) × 2.31 feet = 92.40 feet

15-inch vacuum (Table 3-7) = 17.01 feet

Distance (see Figure 3-37) = 2.00 feet

Velocity head in 6-inch pipe minus velocity

Head in 8-inch pipe, or (1.99 − 0.63) = 1.36 feet

*Total hydraulic load or lift* = 112.77 feet

(2.31 feet = height of water column exerting 1 psi)

### Table 3-6   Gallons of Water Per Minute Required to Feed Boilers (30 pounds, or 3.6 gallons, of water per horsepower, evaporated from 100° to 70 pounds per square inch of steam pressure)

| H.P Boiler | Feed Water (gallons) | H.P. Boiler | Feed Water (gallons) |
|---|---|---|---|
| 20 | 1.2 | 150 | 9.0 |
| 25 | 1.5 | 160 | 9.6 |
| 30 | 1.8 | 170 | 10.2 |
| 35 | 2.1 | 180 | 10.8 |
| 40 | 2.4 | 190 | 11.4 |
| 45 | 2.7 | 200 | 12.0 |
| 50 | 3.0 | 225 | 13.5 |
| 55 | 3.3 | 250 | 15.0 |
| 60 | 3.6 | 275 | 16.5 |
| 65 | 3.9 | 300 | 18.0 |
| 70 | 4.2 | 325 | 19.5 |
| 75 | 4.5 | 350 | 21.0 |
| 80 | 4.8 | 400 | 24.0 |
| 85 | 5.1 | 450 | 27.0 |
| 90 | 5.4 | 500 | 30.0 |
| 100 | 6.0 | 600 | 36.0 |
| 110 | 6.6 | 700 | 42.0 |
| 120 | 7.2 | 800 | 48.0 |
| 130 | 7.8 | 900 | 54.0 |
| 140 | 8.4 | 1000 | 60.0 |

If the inlet and discharge pumps are the same diameter at the gage connections, the velocity head is the same in both pipes, and a correction (such as was required in the previous solution) is not required. Also, the friction head in the inlet and discharge pipes is included in the gage readings.

When the discharge pipe is smaller in diameter than the inlet pipe, the difference between the velocity heads in both pipes should be added in calculating the total hydraulic load or lift; the difference should be subtracted if the inlet pipe is smaller than the discharge pipe.

Table 3-7 Capacity (in Gallons per Foot of Depth) of Round Tanks

| Inside Diameter Ft. | In. | Gallons One Foot in Depth | Inside Diameter Ft. | In. | Gallons One Foot in Depth | Inside Diameter Ft. | In. | Gallons One Foot in Depth |
|---|---|---|---|---|---|---|---|---|
| 1 | 0 | 5.87 | 7 | 6 | 330.38 | 13 | 9 | 1108.06 |
| 1 | 3 | 9.17 | 7 | 9 | 352.76 | | | |
| 1 | 6 | 13.21 | | | | 14 | 0 | 1151.21 |
| 1 | 9 | 17.98 | | | | 14 | 3 | 1192.69 |
| | | | 8 | 0 | 375.90 | 14 | 6 | 1234.91 |
| | | | 8 | 3 | 399.76 | 14 | 9 | 1277.86 |
| 2 | 0 | 23.49 | 8 | 6 | 424.36 | | | |
| 2 | 3 | 29.73 | 8 | 9 | 449.21 | 15 | 0 | 1321.54 |
| 2 | 6 | 36.70 | | | | 15 | 3 | 1365.96 |
| 2 | 9 | 44.41 | 9 | 0 | 475.80 | 15 | 6 | 1407.51 |
| | | | 9 | 3 | 502.65 | 15 | 9 | 1457.00 |
| 3 | 0 | 52.86 | 9 | 6 | 530.18 | | | |
| 3 | 3 | 62.03 | 9 | 9 | 558.45 | 16 | 0 | 1503.62 |
| 3 | 6 | 73.15 | | | | 16 | 3 | 1550.97 |
| 3 | 9 | 82.59 | 10 | 0 | 587.47 | 16 | 6 | 1599.06 |
| | | | 10 | 3 | 617.17 | 16 | 9 | 1647.89 |
| 4 | 0 | 93.97 | 10 | 6 | 653.69 | | | |
| 4 | 3 | 103.03 | 10 | 9 | 678.88 | 17 | 0 | 1697.45 |
| 4 | 6 | 118.93 | | | | 17 | 3 | 1747.74 |
| 4 | 9 | 132.52 | 11 | 0 | 710.69 | 17 | 6 | 1798.76 |
| | | | 11 | 3 | 743.36 | 17 | 9 | 1850.53 |
| 5 | 0 | 146.83 | 11 | 6 | 776.77 | | | |
| 5 | 3 | 161.88 | 11 | 9 | 810.91 | 18 | 0 | 1903.02 |
| 5 | 6 | 177.67 | | | | 18 | 3 | 1956.25 |
| 5 | 9 | 194.19 | 12 | 0 | 848.18 | 18 | 6 | 2010.21 |
| | | | 12 | 3 | 881.39 | 18 | 9 | 2064.91 |
| 6 | 0 | 211.44 | 12 | 6 | 917.73 | | | |
| 6 | 3 | 229.43 | 12 | 9 | 954.81 | 19 | 0 | 2121.58 |
| 6 | 6 | 248.15 | | | | 19 | 3 | 2176.68 |
| 6 | 9 | 267.61 | 13 | 0 | 992.62 | 19 | 6 | 2233.52 |
| | | | 13 | 3 | 1031.17 | | | |
| 7 | 0 | 287.80 | 13 | 6 | 1070.45 | 20 | 0 | 2349.46 |
| 7 | 3 | 308.72 | | | | | | |

### Table 3-8 Conversion of Vacuum Gage Reading (Inches of Mercury) to Lift in Feet (To convert to feet, multiply by 1.13)

| Inch Vac. | Feet | Inch Vac. | Feet | Inch Vac. | Feet | Inch Vac. | Feet |
|---|---|---|---|---|---|---|---|
| 1/4 | 0.28 | 8 1/4 | 9.35 | 16 1/4 | 18.42 | 24 1/4 | 27.50 |
| 1/2 | 0.56 | 1/2 | 9.64 | 1/2 | 18.71 | 1/2 | 27.78 |
| 3/4 | 0.85 | 3/4 | 9.92 | 3/4 | 18.99 | 3/4 | 28.07 |
| 1 | 1.13 | 9 | 10.21 | 17 | 19.28 | 25 | 28.35 |
| 1/4 | 1.41 | 1/4 | 10.49 | 1/4 | 19.56 | 1/4 | 28.63 |
| 1/2 | 1.70 | 1/2 | 10.77 | 1/2 | 19.84 | 1/2 | 28.91 |
| 3/4 | 1.98 | 3/4 | 11.06 | 3/4 | 20.13 | 3/4 | 29.20 |
| 2 | 2.27 | 10 | 11.34 | 18 | 20.41 | 26 | 29.48 |
| 1/4 | 2.55 | 1/4 | 11.62 | 1/4 | 20.70 | 1/4 | 29.76 |
| 1/2 | 2.84 | 1/2 | 11.90 | 1/2 | 20.98 | 1/2 | 30.05 |
| 3/4 | 3.12 | 3/4 | 12.19 | 3/4 | 21.27 | 3/4 | 30.33 |
| 3 | 3.41 | 11 | 12.47 | 19 | 21.55 | 27 | 30.62 |
| 1/4 | 3.69 | 1/4 | 12.75 | 1/4 | 21.83 | 1/4 | 30.90 |
| 1/2 | 3.98 | 1/2 | 13.04 | 1/2 | 22.11 | 1/2 | 31.19 |
| 3/4 | 4.26 | 3/4 | 13.32 | 3/4 | 22.40 | 3/4 | 31.47 |
| 4 | 4.54 | 12 | 13.61 | 20 | 22.68 | 28 | 31.75 |
| 1/4 | 4.82 | 1/4 | 13.89 | 1/4 | 22.96 | 1/4 | 32.03 |
| 1/2 | 5.11 | 1/2 | 14.18 | 1/2 | 23.24 | 1/2 | 32.32 |
| 3/4 | 5.39 | 3/4 | 14.46 | 3/4 | 23.53 | 3/4 | 32.60 |
| 5 | 5.67 | 13 | 14.74 | 21 | 23.81 | 29 | 32.89 |
| 1/4 | 5.95 | 1/4 | 15.02 | 1/4 | 24.09 | 1/4 | 33.17 |
| 1/2 | 6.23 | 1/2 | 15.31 | 1/2 | 24.38 | 1/2 | 33.46 |
| 3/4 | 6.52 | 3/4 | 15.59 | 3/4 | 24.66 | 3/4 | 33.74 |
| 6 | 6.80 | 14 | 15.88 | 22 | 24.95 | 30 | — |
| 1/4 | 7.08 | 1/4 | 16.16 | 1/4 | 25.23 | | |
| 1/2 | 7.37 | 1/2 | 16.45 | 1/2 | 25.51 | | |
| 3/4 | 7.65 | 3/4 | 16.73 | 3/4 | 25.80 | | |
| 7 | 7.94 | 15 | 17.01 | 23 | 26.08 | | |
| 1/4 | 8.22 | 1/4 | 17.29 | 1/4 | 26.36 | | |
| 1/2 | 8.50 | 1/2 | 17.57 | 1/2 | 26.65 | | |
| 3/4 | 8.79 | 3/4 | 17.86 | 3/4 | 26.93 | | |
| 8 | 9.07 | 16 | 18.14 | 24 | 27.22 | | |

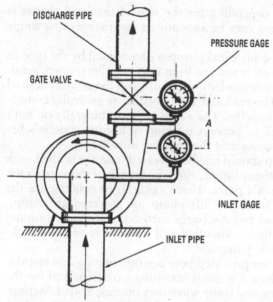

**Figure 3-45** Centrifugal pump, showing location of gages for taking readings to calculate total hydraulic load.

## Summary

The rotating member (impeller) inside the casing of a centrifugal pump provides rapid rotary motion to the mass of water inside the casing, forcing the water out of the casing through the discharge outlet by means of centrifugal force. The vacuum created thereby enables atmospheric pressure to force more water into the casing through the inlet opening.

The volute-type casing or housing forms a progressively expanding passageway into which the impeller discharges the water. The housing is proportioned to produce equal flow velocity around the circumference and to gradually reduce the velocity of the liquid as it flows from the impeller to the discharge outlet—the object being to change velocity head to pressure head.

Single-admission centrifugal pumps are made in one or more stages, and the double-admission pumps may be either single-stage or multistage pumps. The double-admission single-stage pump can elevate large quantities of water to moderate heights. The multistage pump is essentially a high-head or high-pressure pump consisting of two or more stages, depending on the size of the head that it is to pump against. There may be as many as eight stages in a single casing.

The efficiency of a centrifugal pump is determined by the type of impeller. Three types of vanes are found on impellers: open, semi-open, and enclosed (or shrouded), depending on the service, desired efficiency, and cost. The enclosed-type impeller is generally considered a more efficient impeller. The vanes are cast integrally on both sides and are designed to prevent packing of fibrous materials between the stationary cover and the rotating impeller.

A centrifugal pump should not be operated unless it is filled with liquid. If it is run without liquid, there is danger of damage to the liquid-lubricated internal parts. Three methods of priming are the ejector method, the hand-type lift pump (or powered air pump), and the foot-valve and top-discharge method. Automatic priming systems, such as the float-controlled and the reserve vacuum tank-types, are used on some pumps.

Liberal clearances are provided between the rotating and stationary pump parts to allow for small matching variations and for the expansion of the casing and rotor when they become heated. Stationary diaphragms, wearing rings, return channels, and so on (which are located in the casing) are slightly smaller in diameter than the casing bore, which is machined for a gasket between the flanges.

If the level of the water supply is above the centerline of the pump, the pump is operating under inlet head. The discharge head is the vertical distance measured between the centerline of the pump and the level to which the water is elevated.

## Review Questions

1. What is the basic operating principle of the centrifugal pump?
2. How is the volute-type casing proportional?
3. List three types of impeller vanes.
4. What is the chief advantage of the double-admission type of single-stage pump?
5. For what type of service is the multistage pump adapted?
6. For what types of service is the axial-flow type of impeller adapted?
7. In what direction does centrifugal force move objects?
8. What is the difference between a vane and impeller?
9. What is meant by the *eye of the impeller*?

10. Where does the volute pump get its name?

11. What is the difference between a single- and double-admission pump?

12. What determines the efficiency of a centrifugal pump?

13. For what is the open type of vane best suited?

14. How is natural balancing achieved in single-stage pumps?

15. How is mechanical balancing in a centrifugal pump achieved?

16. What is a *stuffing box* on a pump?

17. What type of drive is used on centrifugal pumps?

18. What type of thrust bearings do multistage pumps use?

19. Why is it important to make sure the direction of rotation on a centrifugal pump impeller is correct?

20. List three methods used to prime centrifugal pumps.

21. What is *gland packing*?

22. Where is graphite packing utilized?

23. What causes pump vibration?

24. List four reasons why a pump may have a reduced capacity, a reduced pressure, or a failure to deliver water.

25. What causes a pump to lose water after starting?

# Chapter 4

## Rotary Pumps

The rotary pump is used primarily as a source of fluid power in hydraulic systems. It is widely used in machine-tool, aircraft, automotive, press, transmission, and mobile-equipment applications.

A set of symbols has been established for pumps and other hydraulic components, approved by the American National Standard Institute, Inc., and known as ANS Graphic Symbols. These symbols are an aid to the circuit designer and to the maintenance personnel who might be checking out a hydraulic system. In the photos of each pump shown in this chapter will be the ANS Graphic Symbol for that pump.

### Principles of Operation

The rotary pump continuously scoops the fluid from the pump chamber, whereas the centrifugal pump imparts velocity to the stream of fluid. The rotary pump is a *positive-displacement* pump with a circular motion; the centrifugal pump is a *nonpositive-displacement* pump.

Rotary pumps are classified with respect to the impelling element as follows:

- *Gear*-type
- *Vane*-type
- *Piston*-type pumps

### Gear-Type Pumps

The gear-type pump is a power-driven unit having two or more intermeshing gears or lobed members enclosed in a suitably shaped housing (see Figure 4-1). Gear-type pumps are positive-delivery pumps, and their delivery rate can be changed only by changing the speed at which the pump shaft revolves. The efficiency of a gear-type pump is determined largely by the accuracy with which the component parts are machined and fitted.

Gear pumps are available in a wide range of ratings, from less than 1 gallon per minute (gpm) to more than 100 gpm and for pressures from less than 100 psi to more than 3000 psi.

In Figure 4-2 the two gears fit closely inside the housing. The hydraulic oil is carried around the periphery of the two gears, and it is then forced through the outlet port by the meshing of the two

**Figure 4-1**   Gears for gear-type pumps and motors.

(A)                    (B)                    (C)

**Figure 4-2**   Movement of a liquid through a gear-type hydraulic pump: (A) liquid entering the pump; (B) liquid being carried between the teeth of the gears; and (C) liquid being forced into the discharge line.

gears at their point of tangency. The gear-type pump is also classified as to the type of gears used:

- Spur-gear
- Helical-gear
- Herringbone-gear
- Special-gear such as the *Gerotor* principle

**Spur-Gear**

Two types of spur-gear rotary pumps are used:

- The *external* type
- The *internal* type

In operation of the external type of gear pump, vacuum spaces form as each pair of meshing teeth separates, and atmospheric pressure forces the liquid inward to fill the spaces. The liquid filling the space between two adjacent teeth follows along with them. This happens as they revolve and it is forced outward through the discharge opening. That is because the meshing of the teeth during rotation forms a seal that separates the admission and discharge portions of the secondary chamber (see Figure 4-2).

In operation of the internal-gear type of rotary pump, power is applied to the rotor and transferred to the idler gear with which it meshes. As the teeth come out of mesh, an increase in volume creates a partial vacuum. Liquid is forced into this space by atmospheric pressure and remains between the teeth of the rotor and idler until the teeth mesh to force the liquid from these spaces and out of the pump. The internal-gear type of rotary pump shown in Figure 4-3 and Figure 4-4 possesses only two moving parts: the precision-cut rotor and idler gears. As shown in the cross-sectional view of the pump, the teeth of the internal gear and idler gear separate at the suction port and mesh again at the discharge port.

Shaft rotation will determine which port is suction and which port is discharge. A look at Figure 4-3 will show how rotation determines which port is which. As the pumping elements (gears) come out of mesh, point A, liquid is drawn into the suction port; as the gears come into mesh, point B, the liquid is forced out the discharge port. Reversing the rotation reverses the flow through the pump. When determining shaft rotation, always look from the shaft end of the pump. Unless otherwise specified, rotation is assumed clockwise (CW), which makes the suction port on the right side of the pump.

Internal and external gear-shaped elements are combined in the Gerotor mechanism (see Figure 4-5). In the Gerotor mechanism, the tooth form of the inner Gerotor is generated from the tooth form of the outer Gerotor, so that each tooth of the inner Gerotor is in sliding contact with the outer Gerotor at all times, providing continuous fluid-tight engagement. As the teeth disengage, the space between them increases in size, creating a partial vacuum into which the liquid flows from the suction port to which the enlarging tooth space is exposed. When the chamber reaches its maximum volume, it is exposed to the discharge port. As the chamber diminishes in size because of meshing of the teeth, the liquid is forced from the pump. The various parts of the Gerotor pump are shown in Figure 4-6.

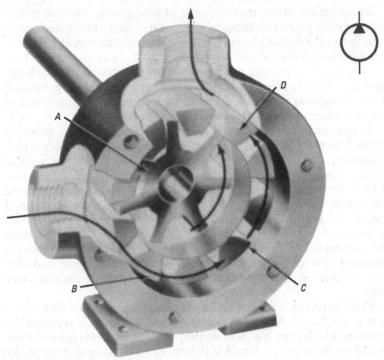

**Figure 4-3** Basic parts and principle of operation of the internal type of spur-gear rotary pump. *(Courtesy Viking Pump Division)*

### Helical-Gear
Figure 4-7 shows a rotary pump with helical gears. This pump is adaptable to jobs such as pressure lubrication, hydraulic service, fuel supply, or general transfer work pumping clean liquids. This pump is self-priming and operates in either direction. The rotary pump (see Figure 4-8) is adapted to applications where a quiet and compact unit is required (such as a hydraulic lift or elevator application).

### Herringbone-Gear
A rotary pump with herringbone gears is shown in Figure 4-9. This is also a quiet and compact unit.

### Vane-Type Pumps
In the rotary vane-type pump (see Figure 4-10), operation is also based on the principle of increasing the size of the cavity to form a

**Figure 4-4** Cross section of an internal-type spur-gear rotary pump.
*(Courtesy Viking Pump Division)*

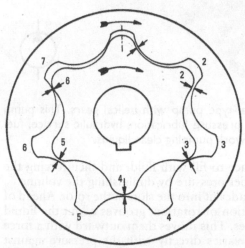

**Figure 4-5** Schematic of the Gerotor mechanism.
*(Courtesy Double A Products Co.)*

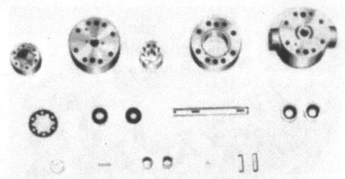

**Figure 4-6**  Basic parts of a Gerotor pump. *(Courtesy Double A Products Co.)*

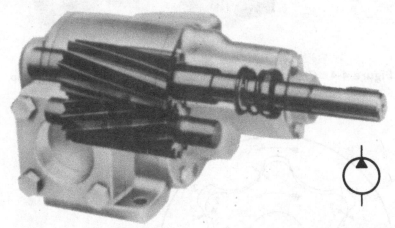

**Figure 4-7**  A rotary gear-type pump with helical gears. This pump is adapted to jobs such as pressure lubrication, hydraulic service, fuel supply, or general transfer work pumping clean liquids.

vacuum, allowing the space to fill with fluid, and then forcing the fluid out of the pump under pressure by diminishing the volume.

The sliding vanes or blades fit into the slots in the rotor. Ahead of the slots and in the direction of rotation, grooves admit the liquid being pumped by the vanes. This moves them outward with a force or locking pressure that varies directly with the pressure against which the pump is operating. The grooves also serve to break the vacuum on the admission side. The operating cycle and the alternate action of centrifugal force and hydraulic pressure hold the vanes in contact with the casing, as shown in Figure 4-11.

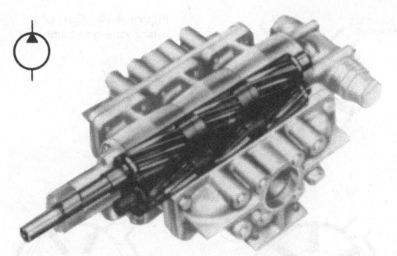

**Figure 4-8** A rotary gear-type pump with helical gears adapted to applications requiring a quiet and compact unit, such as a hydraulic lift or elevator application. *(Courtesy Roper Pump Company)*

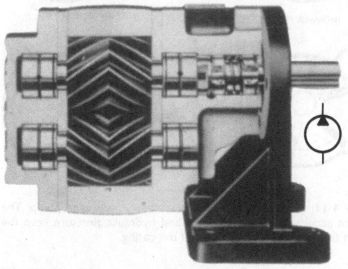

**Figure 4-9** Sectional view of a rotary gear-type pump with herring-bone gears. *(Courtesy Brown & Sharpe Mfg. Co.)*

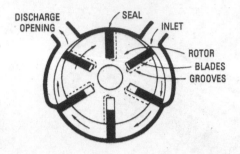

**Figure 4-10** Parts of a rotary vane-type pump.

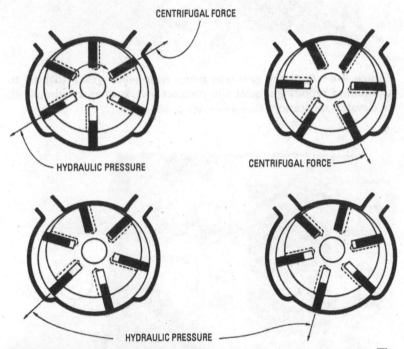

**Figure 4-11** Operating principle of a rotary vane-type pump. The alternate actions of centrifugal force and hydraulic pressure keep the vanes in firm contact with the walls of the casing.

Vane-type pumps are available as:

- *Single-stage* (see Figure 4-12) pumps
- *Double-stage* (see Figure 4-13) pumps

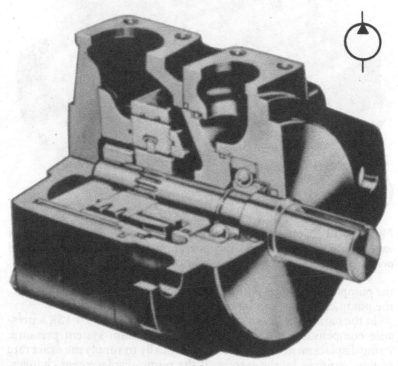

**Figure 4-12** Cutaway view of a high-speed high-pressure single-stage vane-type pump. This pump can operate at speeds to 2700 rpm and pressure to 2500 psi. *(Courtesy Sperry Vickers, Div. of Sperry Rand Corp.)*

The double-stage vane-type pump may consist of two single-stage pumps. They would be mounted end to end on a single shaft. A *combination* vane-type pump is also available. It may be two pumps mounted on a common shaft. The larger pump may be pumping at low pressure, and the smaller pump may be delivering at high pressure. This type of pump may be called a "hi-low" pump.

In the pumps shown in Figure 4-12 and Figure 4-13, the wearing parts are contained in replaceable cartridges (see Figure 4-14). Since

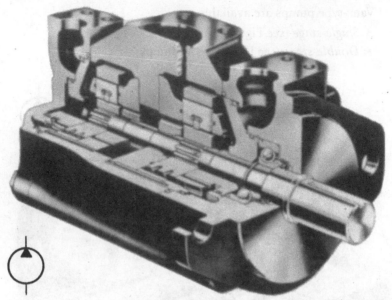

**Figure 4-13**  Cutaway view of a high-speed high-pressure double-stage vane-type pump. This pump can operate at speeds to 2700 rpm and pressures at 2500 psl. *(Courtesy Sperry Vickers, Div. of Sperry Rand Corp.)*

the pumping cartridges within each pump series are interchangeable, the pump capacities can be modified quickly in the field.

In the *variable-volume* vane-type pump (see Figure 4-15), a pressure compensator is used to control maximum system pressure. Pump displacement is changed automatically to supply the exact rate of flow required by the system. If the pump displacement changes, system pressure remains nearly constant at the value selected by the compensator setting.

If the hydraulic system does not require flow, the pressure ring of the pump is at a nearly neutral position, supplying only leakage losses at the set pressure. If full-capacity pump delivery is required, the pressure drops sufficiently to cause the compensator spring to stroke the pressure ring to full-flow position. Any flow rate from zero to maximum is automatically delivered to the system to match the circuit demands precisely, by the balance of reaction pressure and compensator spring force. This reduces horsepower consumption as flow rate is reduced. Bypassing of pressure oil does not occur and excess heat is not generated, which are important factors in the efficiency of the circuit.

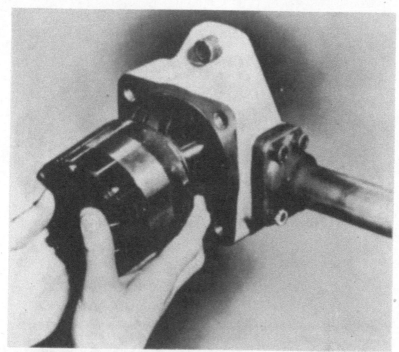

**Figure 4-14**   A replaceable pump cartridge.
*(Courtesy Sperry Vickers, Div. of Sperry Rand Corp.)*

## Piston-Type Pumps

Rotary piston-type pumps are either radial or axial in design. Each of these pumps may be designed for either constant displacement or variable displacement.

The pistons are arranged around a rotor hub in the radial pump (see Figure 4-16). In the Figure 4-16, the side block is at the right-hand side of the centerline of the cylinder barrel. Reciprocating motion is imparted to the pistons, so that those pistons passing over the lower port of the pintle deliver oil to that port, while the pistons passing over the upper port are filling with oil. The delivery of the pump can be controlled accurately from zero to maximum capacity, because the piston and movement of the slide block can be controlled accurately.

In the axial piston-type pump, the pistons are arranged parallel to the shaft of the pump rotor. The driving means of the pump rotates the cylinder barrel. The axial reciprocation of the pumping pistons

**Figure 4-15**   A variable-volume vane-type pump with spring-type pressure compensator. *(Courtesy Continental Hydraulics)*

that are confined in the cylinder is caused by the shoe retainer, which is spring-loaded toward the cam plate. The piston stroke and the quantity of oil delivered are limited by the angle of the cam plate (see Figure 4-17).

A mechanism is used to change the angle of the cam plate in the variable-volume pump. The mechanism may be a hand wheel, a pressure-compensating control, or a stem control that actuates a free-swing hanger attached to the cam plate for changing the angle of the cam plate.

## Construction

To ensure dependable, long-life operation, rotary pumps are of heavy-duty construction throughout. Fluid power components that

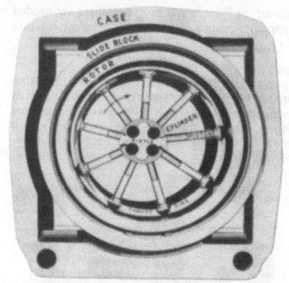

**Figure 4-16**  Sectional view of a variable-delivery radial piston-type pump. *(Courtesy Oilgear Company)*

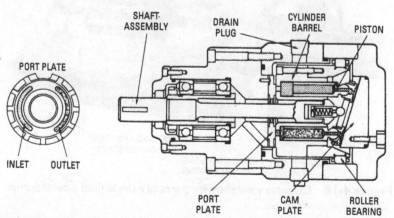

**Figure 4-17**  Cross-sectional view of a constant-displacement axial piston-type pump. *(Courtesy Abex Corp., Denison Division)*

can perform satisfactorily at higher operating pressures are needed to satisfy the ever-increasing demand for faster, more positive-acting original equipment.

## Gear-Type Pumps

Heavy-duty gear-type pumps are able to withstand rugged operating conditions, are simple in construction, and are economical in cost and maintenance (see Figure 4-18). High volumetric efficiency of gear-type pumps depends on maintaining complete sealing of all gear tooth contact surfaces. All gear surfaces are precision finished, and each pair is matched carefully.

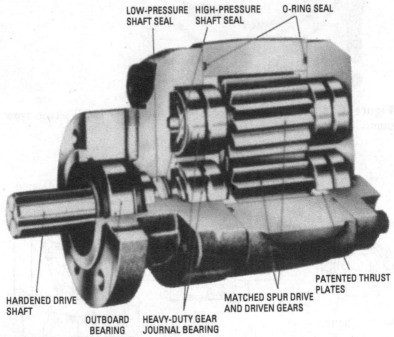

**Figure 4-18** Cutaway view showing parts of a single fluid-power pump.
(Courtesy Commercial Shearing, Inc.)

Gear-type pumps are made with fewer working parts than many other types of pumps. Gear housings are available with tapered thread, SAE split flange or straight thread fittings, and with no porting, or left- and/or right-hand side porting (see Figure 4-19). The pumps are used to handle all kinds of liquids over a wide range of capacities and pressures. Viscosities cover the range from gasoline,

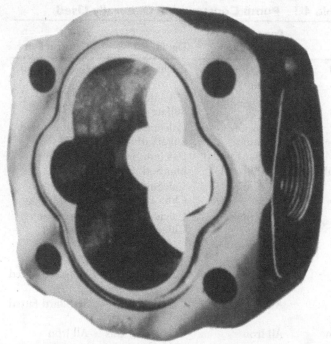

**Figure 4-19**   Housing for a gear-type pump. *(Courtesy Commercial Shearing, Inc.)*

water, all petroleum products, paint, white lead, and molasses. They are generally made of the following materials:

- *Standard fitted*—Case is cast iron, internal working parts to suit individual manufacturer's design.
- *All iron*—All parts of the pump in direct contact with the liquid are made of iron or ferrous metal.
- *Bronze fitted*—Iron casing with bronze pumping elements.
- *All bronze*—All parts in direct contact with liquid are made of bronze.
- *Corrosive-resisting*—All parts in contact with liquid are of materials offering maximum resistance to corrosion.

Table 4-1 indicates the kind of construction generally used, based on the liquid to be pumped. If there were a hot liquid to be pumped, it would be best to consult with the engineering department of the particular company before making a selection.

### Table 4-1    Pump Construction Generally Used

| Liquid | Pump Construction | Liquid | Pump Construction |
|---|---|---|---|
| Acetic Acid | Corrosion Resisting | Cachaza | Bronze Fitted |
| Acetone | Standard Fitted | Calcium Acid Sulfate (conc.) | All Bronze |
| Acid Mine Water | Corrosion Resisting | Calcium Acid Sulfate (dil.) | Bronze Fitted |
| Alcohol (commercial) | Standard Fitted | Calcium Brine plus Sod. Chl. | All Bronze |
| Alcohol (medicinal) | All Bronze | Calcium Chlorate | Corrosion Resisting |
| Alum | All Bronze | Calcium Chloride | Corrosion Resisting |
| Aluminum Chloride | Corrosion Resisting | Calcium Magn. So. Chl. | All Bronze |
| Aluminum Sulfate | Corrosion Resisting | Cane Juice | Standard Fitted |
| Ammonia | All Iron | Carbolic Acid (dil.) | Standard Fitted |
| Ammonium Bicarbonate | All Iron | Carb. Acid Gas in $H_2O$ | All Iron |
| Ammonium Chloride | Corrosion Resisting | Carbonate of Soda | All Bronze |
| Ammonium Nitrate | All Iron | Carbon Bisulfide | All Iron |
| Ammonium Sulfate | All Iron | Carbon Tetrachloride | Standard Fitted |
| Aniline Water | Standard Fitted | Caustic Cl. of Magn. | Corrosion Resisting |
| Asphaltum | Standard Fitted | Caustic Cyanogen | All Iron |
| | | Caustic Potash | All Iron |
| Barium Chloride | All Iron | Caustic Soda | All Iron |
| Barium Nitrate | All Iron | Caustic Strontia | All Iron |
| Beer | All Bronze | Caustic Zinc Chloride | Corrosion Resisting |
| Beer Wort | All Bronze | Cellulose | Corrosion Resisting |

(continued)

## Table 4-1 (continued)

| Liquid | Pump Construction | Liquid | Pump Construction |
|---|---|---|---|
| Beet Juice | All Bronze | Cellulose Acetate | Corrosion Resisting |
| Benzene (coal tar) | Standard Fitted | Chloride of Zinc | Corrosion Resisting |
| Benzine (oil Dist.) | Standard Fitted | Cider | All Bronze |
| Bichloride of Mercury | Corrosion Resisting | Citric Acid | Corrosion Resisting |
| Blood | Standard Fitted | Coal Tar Oil | Standard Fitted |
| Body Deadener | Special Construction | Copperas (Green Vit.) | Corrosion Resisting |
| Brine (Calcium) | All Iron | Copper Sulphate (Blue Vit.) | Corrosion Resisting |
| Brine (Sodium) | All Bronze | Creosote | Standard Fitted |
| Brine (Sod. Chl.) | All Bronze | Creosote Oils | Standard Fitted |
| Butane | Standard Fitted | Cyanic Acids | All Iron |
| Cyanic Liquors | All Iron | Iron Pyritic Acid | All Bronze |
| Cyanide | All Iron | | |
| Cyanide Potassium | All Iron | Kerosene | Standard Fitted |
| Cyanogen | All Iron | | |
| | | Lard | All Iron |
| Distillery Wort | All Bronze | Lime Water | All Iron |
| Dog Food | Special Construction | Linseed Oil | Standard Fitted |
| Duco (hot) | Special Construction | Lye (Caustic) | All Iron |
| Dye Wood Liquors | Bronze Fitted | Lye (Salty) | Corrosion Resisting |
| Ethyl Acetate | Standard Fitted | Magnesium Sulfate | All Iron |
| Ethyl Chloride | All Bronze | Mash | Bronze Fitted |
| Ethylene Chloride | All Bronze | Methanol | Standard Fitted |

(continued)

### Table 4-1   (continued)

| Liquid | Pump Construction | Liquid | Pump Construction |
|---|---|---|---|
| Ethylene Glycol | Standard Fitted | Milk of Lime | All Iron |
|  |  | Molasses | Standard Fitted |
| Fatty Acids | Corrosion Resisting | Mustard | Corrosion Resisting |
| Ferrous Chloride | Corrosion Resisting |  |  |
| Ferrous Sulfate | Corrosion Resisting | Naptha | Standard Fitted |
| Fuel Oil(See Petr. Oils) | Standard Fitted | Nitric Acid (diluted) | Corrosion Resisting |
| Furfural | Standard Fitted |  |  |
|  |  | Paraffin (hot) | Standard Fitted |
| Gasoline | Standard Fitted | Peroxide of Hydrogen | Corrosion Resisting |
| Glue | Standard Fitted | Petroleum Ether | Bronze Fitted |
| Glycerine | Standard Fitted | Petroleum Oil | Standard Fitted |
| Grape Juice | All Bronze | Potash | Corrosion Resisting |
| Gun Cotton Brine | All Bronze | Potassium Carbonate | All Iron |
|  |  | Potassium Chloride | Corrosion Resisting |
| Heptane | Standard Fitted | Potassium Cyanide | All Iron |
| Hydrocyanic Acid | Bronze Fitted | Potassium Nitrate | All Iron |
| Hydroflusilic Acid | All Bronze | Potassium Sulfate | All Bronze |
| Propane | Standard Fitted | Sweet Water | All Bronze |
|  |  | Syrup | All Bronze |
| Rapeseed Oil | All Bronze |  |  |
| Rhigolene (oil dust) | Standard Fitted | Tan Liquor | All Bronze |
|  |  | Tar | Standard Fitted |

(continued)

## Table 4-1 (continued)

| Liquid | Pump Construction | Liquid | Pump Construction |
|---|---|---|---|
| Sal Ammoniac | Corrosion Resisting | Tar and Ammonia in Water | Standard Fitted |
| Salt Brine (3 percent salt) | Standard Fitted | Tomato Pulp | Corrosion Resisting |
| Salt Brine (over 3 percent salt) | Corrosion Resisting | Trisodium Phosphate | All Iron |
| Sea Water | Standard Fitted | Turpentine Oil | Standard Fitted |
| Soap Water | All Iron | | |
| Soda Ash (cold) | All Iron | Urine | All Bronze |
| Sodium Bicarbonate | All Iron | | |
| Sodium Hydroxide | Corrosion Resisting | Vegetable Oil (general) | Standard Fitted |
| Sodium Hyposulfite | Corrosion Resisting | Vinegar | All Bronze |
| Sodium Nitrate | All Iron | Vitriol, Blue | Corrosion Resisting |
| Sodium Sulfate | Corrosion Resisting | Vitriol, Green | Corrosion Resisting |
| Sodium Sulfide | Corrosion Resisting | | |
| Starch | Standard Fitted | Water (constant duty) | Standard Fitted |
| Strontium Nitrate | All Iron | Water (intermittent duty) | All Bronze |
| Sugar | All Bronze | Whiskey | All Bronze |
| Sulfide of Hydrogen | Corrosion Resisting | Wine | All Bronze |
| Sulfide of Sodium (hot) | Corrosion Resisting | Wood Pulp | Bronze Fitted |
| Sulfur Dioxide | Corrosion Resisting | | |
| Sulfur in Water | All Bronze | Yeast | All Bronze |
| Sulfuric Acid (conc.) | Corrosion Resisting | | |

(continued)

**Table 4-1 (continued)**

| Liquid | Pump Construction | Liquid | Pump Construction |
|---|---|---|---|
| Sulfuric Acid (diluted) | Corrosion Resisting | Zinc Chloride | Corrosion Resisting |
| Sulfuric Acid (fuming) | Corrosion Resisting | Zinc Nitrate | All Bronze |
| Sulfurous Acid (conc.) | Corrosion Resisting | Zinc Sulfate | All Bronze |
| Sulfurous Acid (diluted) | Corrosion Resisting | | |

*End covers* for fluid power pumps are usually specified and coded. These details are important in specifying and ordering parts. The *shaft-end* cover (see Figure 4-20) may be furnished in either flange or pad mounting, and the *port-end* cover (see Figure 4-21) may be provided either with no porting or with end porting arrangements.

**Figure 4-20** A four-bolt shaft-end cover for a gear-type pump.
*(Courtesy Commercial Shearing, Inc.)*

**Figure 4-21** A port-end cover for a gear-type pump.
(Courtesy Commercial Shearing, Inc.)

The *drive shafts* (see Figure 4-22) are also specified and ordered by code number. They may be either splined or straight-keyed shafts.

**Figure 4-22** Straight-splined drive shaft for a gear-type pump.
(Courtesy Commercial Shearing, Inc.)

A *bearing carrier* (see Figure 4-23) is used on tandem pumps and motors. It is positioned between adjacent pumps or motors. The bearing carrier is also available with tapered thread, *SAE* thread,

**Figure 4-23**   A bearing carrier which is placed between the two adjacent pumps in a tandem gear-type pump. *(Courtesy Commercial Shearing, Inc.)*

and straight-thread fittings, and either with no porting or with left-and/or right-hand side porting.

Some rotary gear-type pumps for general-purpose applications use a packed box for the shaft seal (see Figure 4-24). The packing gland should be adjusted to permit slight seepage for best performance. A *mechanical seal* (see Figure 4-25) uses less power than the packed box, has longer service life under proper conditions, and does not require adjustment. Special mechanical seals, such as *Teflon* (500°F and corrosion resistant), can be supplied for special conditions.

In the rotary pump, a *steam chest* (see Figure 4-26), located between the casing and the outboard bearing, effectively transfers heat to both the pump and the packing. It can be used with hot water, steam, and heat transfer oil; or it can be used as a cooling chamber. The steam chest is ideal for transferring thick, viscous liquids (such as asphalt mixes, creosote, refined sugars, corn starch, and so on).

An *adjustable relief valve* (see Figure 4-27) in the pump faceplate eliminates outside piping and protects the pump from excessive

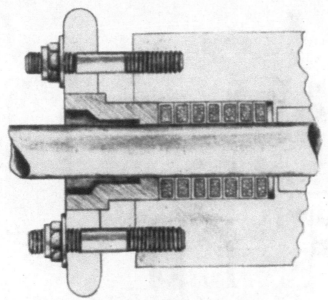

**Figure 4-24**  Packing for a rotary gear-type pump.
*(Courtesy Roper Pump Company)*

outlet line pressure; it also permits the operator to close the discharge line without stopping the pump, under standard operating conditions. Various spring sizes are available to provide adjustments over the full operating range of the pump from 30 to 100 psi.

Many mounting styles are available for convenience in mounting rotary pumps (see Figure 4-28). Bedplates and mounting brackets are available to permit the coupling of pumps and motors for complete motor-driven units (see Figure 4-29). The rotary gear-type pumps also may be mounted in various other styles, including the *foot-mounted* (see Figure 4-30) and the *end-bell mounted* (see Figure 4-31) pumps.

As mentioned, the rotary gear-type pumps are simple in design, and have fewer working parts, than the rotary vane-type and piston-type pumps. Figure 4-32 shows the names of the various parts of a typical gear-type pump and parts list.

## Vane-Type Pumps
Figure 4-33 shows the design principle of hydraulic balance. Bearing loads resulting from pressure are eliminated, and the only radial

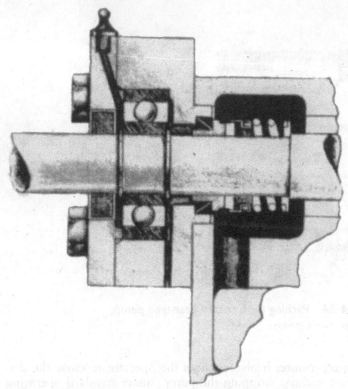

**Figure 4-25** Mechanical seal for a rotary gear-type pump.
(Courtesy Roper Pump Company)

loads are imposed by the drive itself. Communication holes in the
rotor direct pressure from spaces behind the vanes to their lower
edges. The outside edges of the vanes are machined to a bevel, holding them in continuous hydraulic balance, except for the pre-selected
area of the intra-vane ends.

## Piston-Type Pumps

The *radial piston-type pump* (see Figure 4-34) is a rugged and compact high-pressure pump. The walls of the pistons are tapered to a
thin-edged section capable of expanding against the cylinder wall as
pressure is applied. The higher the pressure, the tighter the seal becomes, to increase the efficiency of high-pressure systems. Positive-
acting check valves properly port the suction and discharge oil for

**Figure 4-26** A steam chest used on a rotary pump for steam and hot-water applications. *(Courtesy Roper Pump Company)*

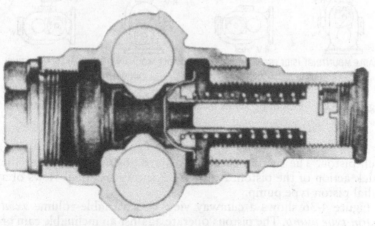

**Figure 4-27** An adjustable relief valve protects the rotary gear-type pump from excessive outlet line pressure. *(Courtesy Roper Pump Company)*

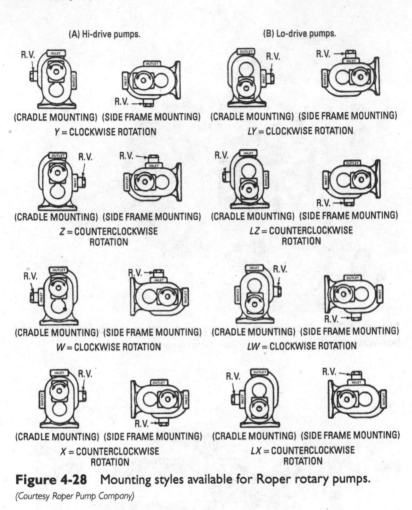

**Figure 4-28** Mounting styles available for Roper rotary pumps.
(Courtesy Roper Pump Company)

each piston. The suction check valves are positively seated by the quick action of the piston. Figure 4-35 shows sectional views of a radial piston-type pump.

Figure 4-36 shows a cutaway view of a variable-volume *axial piston-type pump*. The pistons operate against an inclinable cam or hanger (see inset). The pump delivery is in direct proportion to the tilt of the hanger that is tilted either manually or automatically by means of various types of controls

**Figure 4-29**   Bedplates and mounting brackets permit coupling of pumps and motors for complete motor-driven units.
(Courtesy Roper Pump Company)

**Figure 4-30**   Foot-mounted rotary gear-type pump.
(Courtesy Roper Pump Company)

**Figure 4-31** End-bell mounted rotary gear-type pump.

*(Courtesy Roper Pump Company)*

## Installation and Operation

Many of the installation, operation, and maintenance principles that apply to centrifugal pumps can also be applied to rotary pumps. Since rotary pumps are commonly much smaller than centrifugal pumps, the foundation that is required is usually smaller, but the requirements are similar.

### Alignment

Correct alignment is required for successful operation of the pump. A flexible coupling cannot compensate for incorrect alignment. If the rotary pumping unit is aligned accurately, the flexible coupling can then serve its purpose—to prevent the transmission of end thrust from one machine to another and to compensate for slight changes in alignment that may occur during normal operation.

Each pumping unit should be aligned accurately at the factory before it is shipped. After the unit is assembled, it is aligned by placing the base plate on a surface plate and then leveling the machined pads. Shims are inserted beneath the feet of both the pump and the driver to obtain correct alignment.

In many instances, the manufacturer cannot assume full responsibility for proper mechanical operation, because the base plates are not rigid. This means that the unit must be aligned correctly after it is erected on the foundation. The unit is usually supported on the foundation by wedges placed near the foundation bolts. The wedges

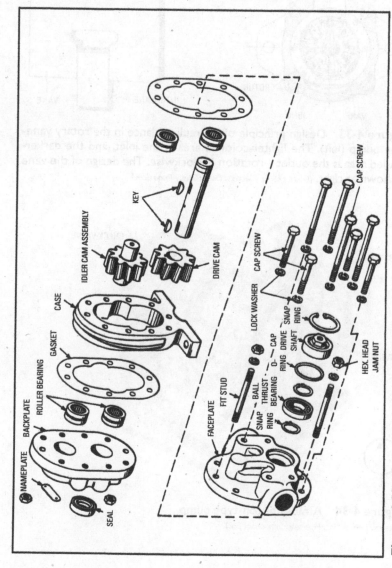

**Figure 4-32** Parts of a gear-type pump. *(Courtesy Roper Pump Company)*

NAMEPLATE
BACKPLATE
ROLLER BEARING
GASKET
CASE
IDLER CAM ASSEMBLY
KEY
DRIVE CAM
CAP SCREW
SEAL
FACEPLATE
FIT STUD
SNAP RING
BALL THRUST BEARING
D-RING
CAP
DRIVE SHAFT
SNAP RING
LOCK WASHER
CAP SCREW
HEX. HEAD
JAM NUT

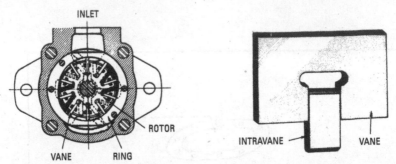

**Figure 4-33** Design principle of hydraulic balance in the rotary vane-type pump (left). The lighter-colored area is the inlet, and the darker-shaded area is the outlet—rotation is clockwise. The design of the vane is shown at right. *(Courtesy Sperry Vickers, Div. of Sperry Rand Corp.)*

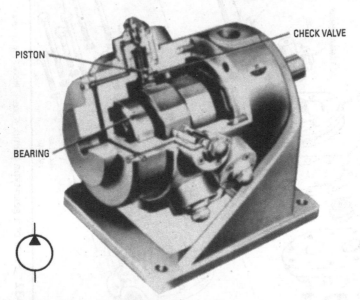

**Figure 4-34** A radial piston-type pump.

*(Courtesy Rexnard, Inc., Hydraulic Component Div.)*

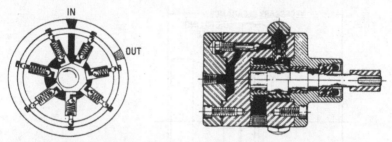

**Figure 4-35** Sectional views of a radial piston-type pump.
*(Courtesy Rexnard, Inc., Hydraulic Component Div.)*

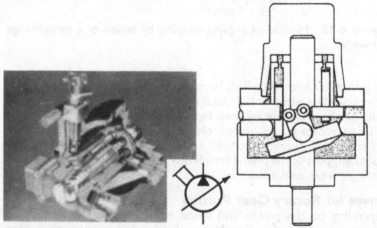

**Figure 4-36** Cutaway view of a variable-volume axial piston-type pump (left) and Inclinable cam or hanger (right) that is tilted to control pump's delivery. *(Courtesy Roper Pump Company)*

underneath the base plate are adjusted, until a spirit level placed on the pads indicates that the pump shaft is level.

The alignment should be checked and corrected to align the coupling half of the driver with the coupling half of the pump. The shafts can be aligned by means of a straightedge and thickness gage (see Figure 4-37). The clearances between the coupling halves should be set so that they cannot strike, rub, or exert end thrust on either the pump or the driver.

Before placing the unit in operation, oil should be added to the coupling. The alignment should be checked again after the piping to

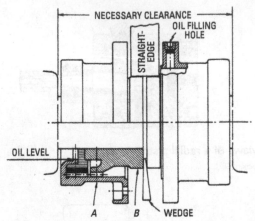

**Figure 4-37** Method of aligning coupling by means of a straightedge and wedge.

the pump has been installed, because the units are often sprung or moved out of position when the flange bolts are tightened, especially if the flanges were not squared before tightening.

To prevent a strain or pull on the pump, extreme care should be exercised to support the inlet and discharge piping properly. Improper support of the piping frequently causes misalignment, heated bearings, wear, and vibration.

### Drives for Rotary Gear Pumps

Depending on the power and speed requirements, the pump and power source may be directly coupled or driven through a gear reduction. Speed reduction can also be achieved with V belts or and chain-and-sprocket. The latter puts radial load on the pump drive shaft. Therefore, precautions should be taken to ensure that the pump is built to withstand the conditions set up by the drive. Improper alignment and excess radial loads can cause a pump to wear out rapidly.

### Power for Driving Pumps

The manufacturer of the pump can provide complete information on power sources for rotary gear pumps. Following, though, are some tips.

*Electric motors* are probably the most common drive for rotary pumps. Manufacturers' catalogs show proper horsepower requirements for a given set of conditions. If the power requirements are somewhere between a smaller and a larger motor, opt for the larger. An electric motor can be momentarily overloaded without serious consequences, but continuous duty under overload conditions will burn it out.

Installation conditions will indicate the type of motor to select. An open motor is fine where conditions are dry and clean. Where there is a moisture possibility, select a motor that guards against this. If flammable liquids are used, then an enclosed, explosion-proof motor must be selected.

Most communities have laws as to conditions of installation. It would be prudent to check these out before selecting a motor.

*Internal combustion engines* are used where there is no electricity, or where portability is required.

In estimating the size of an engine for a job, it is necessary to allow a reasonable safety factor. A gasoline engine should not be operated continuously at its maximum capacity, but should have a 20 percent to 25 percent allowance for losses caused by atmospheric conditions and wear, to ensure a reasonable engine life. Added torque will be needed for starting and for abnormal application conditions. This means that, according to present practice of power ratings, a gas engine should be selected with twice the rating of an electric motor for the same job. For best performance, an engine should not be operated at less than 30 percent of its top-rated speed.

Power take-offs are commonly used on equipment where the pump is part of the unit, and the required speed, power, and control can be made available.

## Piping

The general requirements for installation of piping are similar to those for centrifugal pumps. Sufficient static negative lift (static head) should be provided on the inlet line, in addition to the vapor pressure, to prevent vaporization of the liquid inside the pump when highly volatile liquids (butane, propane, hot oils, and so on) are being pumped. The discharge piping should be extended upward through a riser that is approximately five times the pipe diameter (see Figure 4-38). This prevents gas or air pockets in the pump and acts as a seal in high-vacuum service. A valve at the top of the riser can be used as a vent when starting the pump. A bypass line with a relief valve can be installed to protect the pump from excessive pressure caused by increased pipe friction in cold weather, and from accidental closing of the valve in the discharge line. The relief valve should be set not more than 10 percent higher than the maximum pump discharge pressure.

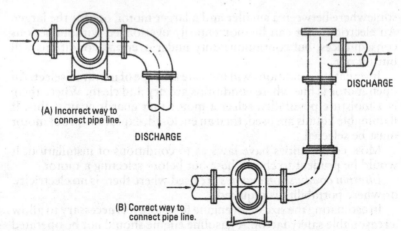

(A) Incorrect way to
connect pipe line.

DISCHARGE

DISCHARGE

(B) Correct way to
connect pipe line.

**Figure 4-38**  Incorrect (left) and correct (right) methods of installing discharge piping.

If steam jackets are necessary, the inlet is located at the top and the outlet at the bottom. On water jackets, the inlet is at the bottom and the outlet is at the top. Valves should be installed in the inlet lines to regulate the quantity of fluid to the jackets.

## Direction of Rotation
The direction of rotation of a pump is usually indicated by an arrow on the body of the pump. This varies with the type of pump. For example, in a double-helix gear-type pump, the direction of rotation is counterclockwise when standing at and facing the shaft extension end.

Rotation of internal-gear roller bearing pumps can be reversed by removing the outside bearing cover and stuffing box. Then the small plug in the side plate casting is transferred to the opposite side. These small plugs (one plug in each side plate) should be on the discharge side, to induce circulation through the bearings to the inlet, and to maintain inlet pressure on the stuffing box and ends of the drive shafts.

For another example, a pump operating on the internal-gear principle has the following instructions for determining direction of rotation:

- In determining the desired direction of rotation, the observer should stand at the shaft end of the pump.

- Note that the balancing groove in the shoe should be located on the inlet side.
- If a change in direction of rotation is desired, it is necessary only to remove the cover. Then, remove both the upper and lower shoes, turn them end for end to place the grooves on the new inlet side, and reassemble the pump.

Another model built by the same pump manufacturer is listed as an "automatic-reversing" pump. Regardless of the direction of shaft rotation and without the use of check valves, a unidirectional flow is maintained.

The instructions for determining the direction of rotation for a helical-gear type of rotary pump are as follows. To determine the direction of rotation, stand at the driving end, facing the pump. If the shaft revolves in a left-hand to right-hand direction, its rotation is *clockwise;* if the shaft revolves in a right-hand to left-hand direction, its rotation is *counterclockwise.* As shown in the Figure 4-39, a change in direction of rotation of the pump drive shaft reverses the direction of flow of the liquid, causing the inlet and discharge openings to be reversed.

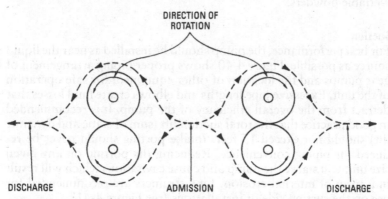

**Figure 4-39** Flow of liquids and direction of rotation in rotary gear-type pumps.

Motors usually rotate in a counterclockwise direction. The direction of rotation for a motor is determined from a position at the end of the motor that couples to the pump.

As indicated in the foregoing examples, the direction of rotation for the various types of pumps should be obtained from the manufacturers' instructions. Some types of rotary pumps are reversible and some types are nonreversible.

## Starting and Operating the Pump

Before starting the pump, it should be primed. Then, the prime mover should be checked for proper direction of rotation. Pressure or vacuum should be checked on the inlet side, and pressure should be checked on the outlet side, to determine whether they conform to specifications and whether the pump can deliver full capacity without overloading the driver.

Operation should be started at a reduced load, gradually increasing to maximum service conditions. Pumps with external bearings may require occasional lubrication or soft grease for the bearings. If grease fittings are not furnished on pumps with internal bearings, lubrication is not necessary.

## Practical Installation

Gear pumps are simple and rugged in construction. A minimum number of moving parts ensures longer service life and lower maintenance cost. Their design criteria are for high flow at moderate pressures. Gear pumps are self-priming and excellent for viscous liquids. All fluids must be free of abrasives such as sand, silt, and wettable powders.

### Suction

For best performance, the pump should be installed as near the liquid source as possible. Figure 4-40 shows proper piping arrangement of gear pumps and the location of other equipment for safe operation of the unit. Excessive pipe lengths and elbows create fluid losses that detract from the overall efficiency of the pump. It is recommended in good practice that the total suction lift (sum of static and dynamic lift) should not exceed 15 feet. Intake porting should never be reduced for piping convenience. Reducing the porting on any given size unit will starve the pump and cause cavitations which will result in a form of internal erosion. Line strainers are recommended for use on the suction side of installations (see Figure 4-41).

### Discharge

Since gear pumps are positive displacement, care must be taken with the use of restrictive devices (such as gate valves) in the discharge line for purpose of throttling. Accidental closing of the discharge line can cause extreme overpressure, which may result in serious damage to the pump or motor. If the discharge line throttling is necessary, selection of the pump with a built-in relief valve should be considered.

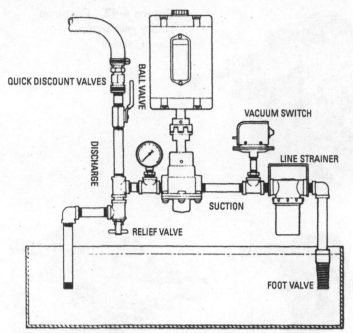

**Figure 4-40** Installation of a gear pump. *(Courtesy Sherwood)*

**Figure 4-41** Line strainer. *(Courtesy Sherwood)*

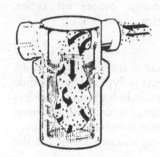

## Types of Gear Pumps

Cast-iron rotary gear pumps are ideal for pumping oils and other self-lubricating liquids. Figure 4-42 shows one of a number of models available.

Figure 4-43 shows bronze rotary gear pumps with self-lubricating carbon bearings and relief valve. High-grade brass castings machined to close tolerances are used for maximum pumping efficiency. A stainless steel shaft is provided for proper corrosion resistance.

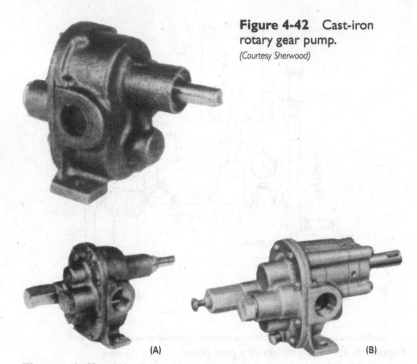

**Figure 4-42** Cast-iron rotary gear pump.
*(Courtesy Sherwood)*

(A)　　　　　　　　　　　　　　　(B)

**Figure 4-43** (A) Bronze rotary gear pump, model BB series. (B) Bronze rotary gear pump, model S&V series. *(Courtesy Sherwood)*

The stainless steel rotary gear pump has a stainless steel body, cover, and shaft (see Figure 4-44). It also has polyphenylene sulfide rotary gears with a Viton dripless mechanical shaft seal. Oil-less carbon graphite bushings and self-lubricating, sealed ball bearings minimize maintenance and provide long time trouble free service.

Gear pumps are well suited for pumping viscous liquids if the following rules are observed:

- Pump speed (rpm) must be reduced. Use Table 4-2 as a guide.
- Suction and discharge lines must be increased by at least one pipe size, or better, two pipe sizes, over the size of the pump parts.
- Horsepower of the motor must be increased over whatever power rating would be required for pumping water under the same pressure and flow.

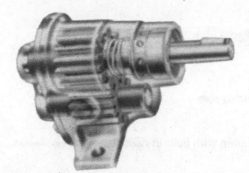

**Figure 4-44** Stainless steel rotary gear pump.
*(Courtesy Sherwood)*

#### Table 4-2 Speed Reduction

| Viscosity in SSU | Recommended Speed (rpm) |
|---|---|
| 50 | 1725 |
| 500 | 1500 |
| 1000 | 1300 |
| 5000 | 1000 |
| 10,000 | 600 |
| 50,000 | 400 |
| 100,000 | 200 |

Table 4-3 shows the percent in increase in horsepower for pumping viscous liquids.

#### Table 4-3 Percent Increases in Horsepower

| | Viscosity in SSU | | | | | | |
|---|---|---|---|---|---|---|---|
| Pressure psi | 30 | 500 | 1000 | 5000 | 10,000 | 50,000 | 100,000 |
| 2 | — | 30 | 60 | 120 | 200 | 300 | 400 |
| 20 | — | 25 | 50 | 100 | 160 | 260 | 350 |
| 40 | — | 20 | 40 | 80 | 120 | 220 | 300 |
| 60 | — | 15 | 30 | 60 | 105 | 180 | 250 |
| 80 | — | 12 | 25 | 50 | 90 | 150 | 200 |
| 100 | — | 10 | 20 | 40 | 80 | 120 | 150 |

#### Pressure Relief Valve

On pumps that are supplied with built-in relief valves such as shown in Figure 4-45, the valve may be adjusted and used to temporarily

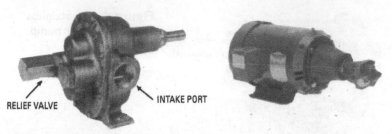

RELIEF VALVE                          INTAKE PORT

**Figure 4-45**   Rotary gear pump with built-in relief valve. *(Courtesy Sherwood)*

prevent pump and motor damage that can occur when the discharge line is closed off.

This relief valve is not set at the factory. Extended operation under shut-off conditions could cause overheating and damage the pump.

To regulate the relief valve, remove the cap covering the valve adjustment screw. Turning the adjustment screw in will increase the pressure setting. Turning the adjustment screw out will reduce the pressure setting.

The relief valve must always be on the discharge side of the pump. The valve assembly can be removed and reversed on the pump cover to accommodate this. If operation calls for prolonged closing-off of the discharge line while the pump is running, then a full-line bypass relief valve should be permanently placed in the discharge line with a return line to the supply screw.

## Rotary Pump Troubleshooting

Rotary pumps, like centrifugal pumps, normally require little attention while they are running. However, most troubles can be avoided if they are given only a small amount of care, rather than no attention at all. Some of the more frequent causes of trouble are indicated here.

### No Liquid Delivered
The following steps should be taken if no liquid is delivered:

1. Stop pump immediately.
2. If pump is not primed, prime according to instructions.
3. Lift may be too high. Check this factor with a vacuum gage on the inlet. If the lift is too high, lower the position of the pump and increase the size of the inlet pipe; check the inlet line for air leaks.
4. Check for wrong direction of rotation.

## Insufficient Liquid Delivered

One of the following causes may result in delivery of insufficient liquid:

- Air leak in the inlet line or through the stuffing box. Oil and tighten the stuffing-box gland. Paint the inlet pipe joints with shellac.
- Speed too slow. The rpm should be checked. The driver may be overloaded, or the cause may be low voltage or low steam pressure.
- Lift may be too high. Check with a vacuum gage. Small fractions in some liquids vaporize easily and occupy a portion of the pump displacement.
- Excess lift for hot liquids.
- Pump may be worn.
- Foot valve may not be deep enough (not required on many pumps).
- Foot valve may be either too small or obstructed.
- Piping is improperly installed, permitting air or gas to pocket inside the pump.
- Mechanical defects, such as defective packing or damaged pump.

## Pump Delivers for a Short Period, Then Quits

This problem may be a result of one of the following causes:

- Leak in the inlet.
- End of the inlet valve is not deep enough.
- Air or gas in the liquid.
- Supply is exhausted.
- Vaporization of the liquid in the inlet line. Check this with the vacuum gage to be sure that the pressure in the pump is greater than the vapor pressure of the liquid.
- Air or gas pockets in the inlet line.
- Pump is cut by presence of sand or other abrasives in the liquid.

## Rapid Wear

Some of the causes of rapid wear in a pump are the following:

- Grit or dirt in the liquid that is being pumped. A fine-mesh strainer or filter can be installed in the inlet line.

- Pipe strain on the pump casing causes the working parts to bind. The pipe connections can be released and the alignment checked to determine whether this factor is a cause of rapid wear.
- Pump operating against excessive pressure.
- Corrosion roughens surfaces.
- Pump runs dry or with insufficient liquid.

## Pump Requires Too Much Power

Too much power required to operate the pump may be caused by the following:

- Speed too fast.
- Liquid either heavier or more viscous than water.
- Mechanical defects, such as a bent shaft, binding of the rotating element, stuffing boxes too tight, misalignment caused by improper connections to the pipelines, or installation on the foundation in such a way that the base is sprung.
- Misalignment of the coupling (direct-connected units).

## Noisy Operation

The causes of noisy operation may be the following:

- Insufficient supply, which may be because of liquid vaporizing in the pump. This may be corrected by lowering the pump and by increasing the size of the inlet pipe.
- Air leaks in the inlet pipe, causing a crackling noise in the pump.
- Air or gas pockets in the inlet.
- A pump that is out of alignment, causing metallic contact between the rotors and the casing.
- Operating against excessive pressure.
- Coupling out of balance.

## Calculations

In nearly all installations, it is important to calculate the size of pump, horsepower, lift, head, total load, and so on. This is important in determining whether the system is operating with maximum efficiency.

## Correct Size of Pump
It is always important to determine whether the pump is too large or too small for the job—either before or after it has been installed.

### Problem
In a given installation, a pump is required that can fill an 8000-gallon tank in two hours. What size pump is required?

### Solution
Since the capacity of a pump is rated in gallons per minute (gpm), the 8000 gallons in two hours that is required can be reduced to gpm as follows:

$$\text{gpm} = \frac{8000}{2 \times 60} = 66^{2}/_{3}, \text{ or approximately 70 gpm}$$

If the rated capacity of a given pump indicates that the pump can deliver 70 gpm at 450 rpm, and since the capacity of a pump is nearly proportional to its speed, the rpm required to deliver $66^{2}/_{3}$ gallons of water per minute can be determined as follows:

$$\text{required rpm} = 450 \times \frac{66^{2}/_{3}}{70} = 428.6 \text{ rpm}$$

Therefore, if the pump is rated to deliver 70 gpm at 450 rpm, it should be capable of delivering $66^{2}/_{3}$ gpm at 428.6 rpm, which is within the rated capacity of the pump.

## Friction of Water in Pipes
The values for loss of head caused by friction can be obtained from Chapter 3. The values in Table 3-3 are based on 15-year-old wrought-iron or cast-iron pipe when pumping clear soft water. The following coefficients can be used to determine the friction in pipes for various lengths of service:

- New, smooth pipe—0.71
- 10-year-old pipe—0.84
- 15-year-old pipe—1.00
- 20-year-old pipe—1.22

## Friction Loss in Rubber Hose
The friction loss in smooth bore rubber hose is about the same as corresponding sizes of steel pipe. Table 4-4 shows the loss in pounds per 100 feet of hose.

## Table 4-4   Loss in Pounds per 100 Feet of Smooth Bore Rubber Hose

| U.S. Gal. Per Min. | Actual Inside Diameter in Inches | | | | | | | |
|---|---|---|---|---|---|---|---|---|
| | 3/4 | 1 | 1 1/4 | 1 1/2 | 2 | 2 1/2 | 3 | 4 |
| 15 | 30.0 | 8.9 | 2.5 | 1.1 | 0.4 | 0.1 | | |
| 20 | 53.0 | 14.0 | 4.3 | 1.8 | 0.7 | 0.2 | | |
| 25 | 79.0 | 22.0 | 6.5 | 2.9 | 1.0 | 0.3 | | |
| 30 | 112.0 | 31.0 | 9.2 | 4.0 | 1.4 | 0.4 | 0.1 | |
| 40 | | 53.0 | 15.0 | 6.7 | 2.4 | 0.6 | 0.3 | |
| 50 | | 80.0 | 24.0 | 10.0 | 3.6 | 1.0 | 0.5 | |
| 60 | | 101.0 | 35.0 | 14.0 | 5.1 | 1.4 | 0.6 | |
| 70 | | | 45.0 | 19.0 | 6.6 | 1.8 | 0.8 | |
| 80 | | | 58.0 | 24.0 | 8.6 | 2.3 | 1.1 | |
| 90 | | | 71.0 | 30.0 | 11.0 | 3.0 | 1.4 | 0.3 |
| 100 | | | 88.0 | 37.0 | 12.5 | 3.5 | 1.7 | 0.4 |
| 125 | | | 132.0 | 55.0 | 20.0 | 5.3 | 2.5 | 0.6 |
| 150 | | | 183.0 | 78.0 | 27.0 | 7.5 | 3.5 | 0.7 |
| 175 | | | | 100.0 | 37.0 | 10.0 | 4.6 | 1.1 |
| 200 | | | | 133.0 | 46.0 | 13.0 | 5.9 | 1.4 |
| 250 | | | | | 70.0 | 19.0 | 9.1 | 2.1 |
| 300 | | | | | 95.0 | 27.0 | 12.0 | 2.9 |
| 350 | | | | | 126.0 | 36.0 | 17.0 | 4.0 |
| 400 | | | | | | 46.0 | 21.0 | 5.1 |
| 500 | | | | | | 70.0 | 32.0 | 7.4 |
| 600 | | | | | | 105.0 | 46.0 | 10.0 |
| 700 | | | | | | 148.0 | 62.0 | 13.0 |
| 800 | | | | | | 190.0 | 79.0 | 17.0 |
| 900 | | | | | | | 97.0 | 22.0 |
| 1000 | | | | | | | 116.0 | 27.0 |
| 1250 | | | | | | | 170.0 | 43.0 |
| 1500 | | | | | | | 226.0 | 61.0 |
| 1750 | | | | | | | | 80.0 |
| 2000 | | | | | | | | 100.0 |

Data shown is for liquid having specific gravity of 1 and a viscosity of 30 SSU

## Dynamic Column or Total Load

The *dynamic column* or *total load* must be calculated before the horsepower required to drive the pump can be calculated. The dynamic column (often referred to as *total load*) consists of the *dynamic lift* plus the *dynamic head* (see Figure 4-46).

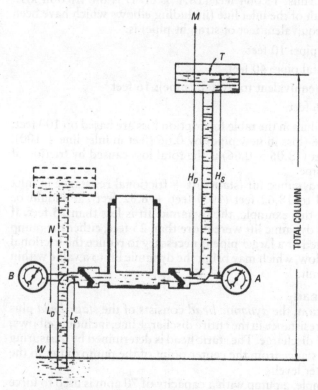

**Figure 4-46** The dynamic or total column for a pumping unit.

## Dynamic Lift

The dynamic lift (in an installation) consists of *static lift* and frictional resistance in the entire length of inlet piping from the water level to the intake opening of the pump. To determine the static lift, measure the vertical distance from the water level in the well to the center point of the inlet opening of the pump. The frictional loss is then added to the static lift, which gives the dynamic lift.

As an example, a delivery of 70 gpm is required from a pump located 10 feet (static lift) above the level of the water in the well.

The horizontal distance is 40 feet, and new 2-inch pipe with two 2-inch elbows is to be used.

The friction loss for 70 gpm through 100 feet of 15-year-old pipe is 18.4 feet (from Table 3-3). Since new pipe is to be used, multiply 18.4 feet (loss in 15-year-old pipe) by the coefficient 0.71 for new, smooth pipe. Thus, 13.064 feet (18.4 × 0.71) is the friction loss. The total length of the inlet line (including elbows which have been converted to equivalent feet of straight pipe) is:

Vertical pipe: 10 feet

Horizontal pipe: 40 feet

Elbows (equivalent to straight pipe): 16 feet

Total: 66 feet

Since the values in the table for friction loss are based on 100 feet, multiply 13.06 (loss in new pipe) by 0.66 (feet in inlet line ÷ 100). Thus, 8.62 feet (13.06 × 0.66) is the total loss caused by friction of water in the pipe.

Therefore, dynamic lift (static lift + frictional resistance in inlet pipe) is equal to 18.62 feet (10.0 feet + 8.62 feet). It should be noted that, in this example, the dynamic lift is less than 25 feet. If the calculated dynamic lift were more than 25 feet, either the pump must be lowered or a larger pipe is necessary to reduce the frictional resistance to flow, which may bring the dynamic lift to a value within the 25-foot limit.

## Dynamic Head

In an installation, the *dynamic head* consists of the *static head* plus the frictional resistance in the entire discharge line, including elbows, to the point of discharge. The static head is determined by measuring the vertical distance from the center point of the pump outlet to the discharge water level.

As an example, a pump with a capacity of 70 gpm is used to force water through a vertical pipe that is 30 feet (static head) in length and a horizontal pipe that is 108 feet in length (2-inch new pipe with three elbows).

The friction loss at 70 gpm through 100 feet of 15-year-old 2-inch pipe is 18.4 feet (from Table 3-3). Since the coefficient for new pipe is 0.71, the friction loss is 13.064 feet (18.4 × 0.71). The friction loss in the total length of the discharge line is:

Total discharge line (30 + 108): 138 feet

Equivalent for three elbows: 24 feet

Total: 162 feet

Since values in Table 3-3 are based on 100 feet of pipe, multiply 13.06 (loss in new pipe) by 1.62(162 ÷ 100). Thus, 21.16 feet (13.06 × 1.62), is the total loss due to friction in the pipe.

Therefore, dynamic head (static head + friction resistance in the discharge pipe) is equal to 51.16 feet (30 feet + 21.16 feet).

After the dynamic lift (13 feet) and the dynamic head (51 feet) for the pump used in the two previous examples have been calculated, the dynamic column (total column) can be calculated as follows:

dynamic column = dynamic lift + dynamic head

Substituting,

dynamic or total column = 13 + 51 = 64 feet

Since the pressure, in pounds per square inch, of a column of water is equal to head (in feet) times 0.433, the pressure in psi for the previous example is equivalent to the dynamic column (or total column) times 0.433. Thus, 27.7 psi (64 feet × 0.433) is the pressure of the column of water (Table 4-5 and Table 4-6).

**Table 4-5  Converting Head of Water (feet) to Pressure (psi)**

| Feet Head | Pounds Per Square Inch | Feet Head | Pounds Per Square Inch | Feet Head | Pounds Per Square Inch |
|---|---|---|---|---|---|
| 1 | 0.43 | 60 | 25.99 | 200 | 86.62 |
| 2 | 0.87 | 70 | 30.32 | 225 | 97.45 |
| 3 | 1.30 | 80 | 34.65 | 250 | 108.27 |
| 4 | 1.73 | 90 | 38.98 | 275 | 119.10 |
| 5 | 2.17 | 100 | 43.31 | 300 | 129.93 |
| 6 | 2.60 | 110 | 47.64 | 325 | 140.75 |
| 7 | 3.03 | 120 | 51.97 | 350 | 151.58 |
| 8 | 3.40 | 130 | 56.30 | 400 | 173.24 |
| 9 | 3.90 | 140 | 60.63 | 500 | 216.55 |
| 10 | 4.33 | 150 | 64.96 | 600 | 259.85 |
| 20 | 8.66 | 160 | 69.29 | 700 | 303.16 |
| 30 | 12.99 | 170 | 73.63 | 800 | 346.47 |
| 40 | 17.32 | 180 | 77.96 | 900 | 389.78 |
| 50 | 21.65 | 190 | 83.29 | 1000 | 433.09 |

Also, the head, in feet, of a column of water is equivalent to the pressure, in psi, times 2.31. Therefore 27.7 psi (64 feet ÷ 2.31) is the pressure of the column of water. This value is identical to the result in the previous paragraph.

### Table 4-6 Converting Pressure (psi) to Head of Water (feet)

| Pounds Per Square Inch | Feet Head | Pounds Per Square Inch | Feet Head | Pounds Per Square Inch | Feet Head |
|---|---|---|---|---|---|
| 1 | 2.31 | 40 | 90.36 | 170 | 392.52 |
| 2 | 4.62 | 50 | 115.45 | 180 | 415.61 |
| 3 | 6.93 | 60 | 138.54 | 190 | 438.90 |
| 4 | 9.24 | 70 | 161.63 | 200 | 461.78 |
| 5 | 11.54 | 80 | 184.72 | 225 | 519.51 |
| 6 | 13.85 | 90 | 207.81 | 250 | 577.24 |
| 7 | 16.16 | 100 | 230.90 | 275 | 643.03 |
| 8 | 18.47 | 110 | 253.98 | 300 | 692.69 |
| 9 | 20.78 | 120 | 277.07 | 325 | 750.41 |
| 10 | 23.09 | 125 | 288.62 | 350 | 808.13 |
| 15 | 34.63 | 130 | 300.16 | 375 | 865.89 |
| 20 | 46.18 | 140 | 323.25 | 400 | 922.58 |
| 25 | 57.72 | 150 | 346.34 | 500 | 1154.48 |
| 30 | 69.27 | 160 | 369.43 | 1000 | 2308.00 |

## Horsepower Required

The power required to raise a given quantity of water to a given elevation is the *actual horsepower*, rather than *theoretical horsepower*. That is, the actual horsepower is equivalent to the theoretical horsepower, plus the additional power required for overcoming frictional resistance and inefficiency of the pump. Theoretical horsepower (*hp*) can be determined by the following formulas:

$$\text{theoretical hp} = \frac{\text{gpm} \times 8^{1/3} \times \text{d.c.}}{33,000}$$

or

$$\text{theoretical hp} = \frac{\text{cu ft} \times 62.4 \times \text{d.c.}}{33,000}$$

in which:

gpm is gallons per minute

$8^{1/3}$ is the approximate weight of one gallon of water in pounds

62.4 is the weight of 1 cubic foot of water at room temperature

d.c. is dynamic column

To obtain the actual horsepower, the theoretical horsepower can be divided by the efficiency $E$ of the pumping unit, expressed as a decimal. Thus, the formula can be changed to:

$$\text{actual hp} = \frac{\text{gpm} \times 8^{1}/_{3} \times \text{d.c.}}{33{,}000 \times E}$$

**Problem**

What is the actual horsepower required to drive a pump that is required to pump 200 gpm against a combined static lift and static head (static column) of 50 feet, if the pump efficiency is 57 percent and the 4-inch pipeline consisting of three 90-degree elbows is 200 feet in length?

**Solution**

The friction loss per 100 feet of 4-inch pipe discharging 200 gpm is 4.4 feet; therefore, for 200 feet of pipe, the loss is 8.8 feet $(2 \times 4.4)$. The friction loss in one 90-degree elbow is 16 feet; or 48 feet $(3 \times 16 \text{ feet})$ for three 90-degree elbows. Therefore, the dynamic column is:

$$\text{d.c.} = 50 + 8.8 + 48 + 106.8 \text{ feet}$$

Substituting in the formula for actual horsepower:

$$\text{actual hp} = \frac{200 \times 8^{1}/_{3} \times 106.8}{33{,}000 \times 0.57} = 9.46, \text{ or a 10-hp pump}$$

## Summary

The rotary pump is used widely in machine-tool, aircraft, automotive, press, transmission, and mobile-equipment applications. It is a primary source of fluid power in hydraulic systems (see Figure 4-47).

The rotary pump continuously scoops the fluid from the pump chamber. It is a positive-displacement pump with a rotary motion.

Rotary pumps are classified with respect to the impelling element as gear-type, vane-type, and piston-type pumps. The gear-type pump is classified as to the type of gears used as: spur-gear, helical-gear, and herringbone-gear types of pumps. The two types of spur-gear rotary pumps are the external and internal types.

Rotary vane-type pumps are available as single-stage and double-stage pumps and as constant-volume and variable-volume pumps. In the variable-volume vane-type pump, a pressure compensator is used to control maximum system pressure.

Rotary piston-type pumps are either radial or axial in design. Each pump may be designed for either constant displacement or variable displacement.

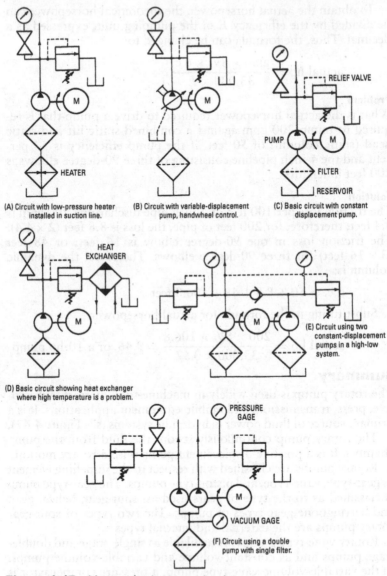

(A) Circuit with low-pressure heater installed in suction line.

(B) Circuit with variable-displacement pump, handwheel control.

(C) Basic circuit with constant displacement pump.

RELIEF VALVE

PUMP

FILTER

RESERVOIR

HEATER

HEAT EXCHANGER

(D) Basic circuit showing heat exchanger where high temperature is a problem.

(E) Circuit using two constant-displacement pumps in a high-low system.

PRESSURE GAGE

VACUUM GAGE

(F) Circuit using a double pump with single filter.

**Figure 4-47** Various basic pump circuits.

Rotary pumps are of heavy-duty construction throughout. These pumps and components can perform satisfactorily at high operating pressures.

The direction of rotation for a rotary pump is usually indicated by an arrow on the body of the pump. The manufacturer's directions should be followed in determining the direction of rotation for a rotary pump.

A newly installed pump should be operated at a reduced load, gradually increasing to maximum service conditions. Rotary pumps normally require little attention while they are operating. However, most troubles can be avoided if pumps are given only a minimum amount of care, rather than no attention at all.

The size of pump, lift, head, total load, horsepower, and so on, should be calculated for most pump installations. This is important in determining whether the pump is too large or too small—before or after it has been installed.

The dynamic column (total column) must be calculated before the horsepower required to drive the pump can be calculated. The dynamic column is the total of the dynamic lift (static lift + frictional resistance in the inlet piping) and the dynamic head (static head + frictional resistance in the outlet piping) or:

$$\text{dynamic column} = \text{dynamic lift} + \text{dynamic head}$$

The pressure, in psi, of a column of water is equal to the dynamic column times 0.433. Also, the head, in feet, of a column of water is equal to pressure (in psi) times 2.31.

The theoretical horsepower, plus the additional horsepower required to overcome frictional resistance and inefficiency, is equal to the actual horsepower required to raise a given quantity of water to a given elevation. The formulas for theoretical horsepower are as follows:

$$\text{theoretical hp} = \frac{\text{gpm} \times 8^{1/3} \times \text{dynamic column}}{33,000}$$

or

$$\text{theoretical hp} = \frac{\text{gpm} \times 62.3 \times \text{dynamic column}}{33,000}$$

Actual horsepower can be determined by dividing the theoretical horsepower by the efficiency $E$ of the pumping unit as follows:

$$\text{actual} = \frac{\text{gpm} \times 8^{1/3} \times \text{dynamic column}}{33,000 \times E}$$

## Review Questions

1. How does the basic principle of the rotary pump differ from that of the centrifugal pump?
2. What are the three types of rotary pumps?
3. List three types of gears used in gear-type rotary pumps.
4. What two designs are used for rotary piston-type pumps?
5. How is the direction of rotation indicated on a rotary pump?
6. What is meant by *dynamic column* or total load with respect to delivery capacity of a pump?
7. Where is the rotary pump used?
8. What does *ANS* represent?
9. What type of displacement does the rotary pump utilize?
10. Describe the Gerotor's operation.
11. What is the main favorable feature of a herringbone gear pump?
12. How is the maximum system pressure controlled in a variable-volume vane-type pump?
13. Where is the pintle located in a variable delivery radial piston-type pump?
14. What type of construction is utilized in making a pump to move barium chloride?
15. Why is a steam chest used on a rotary pump utilized in steam and hot water pumping?
16. Are rotary pumps smaller or larger than centrifugal pumps?
17. Can a flexible coupling compensate for incorrect alignment between the pump and its driver?
18. What is the most common drive for rotary pumps?
19. What causes noisy operation of a rotary pump?
20. How do you obtain the actual horsepower needed to raise a given quantity of water to a given elevation?

# Chapter 5

## Reciprocating Pumps

A reciprocating pump is described as a pump having a to-and-fro motion. Its motion is backward and forward or upward and downward—as distinguished from the circular motion of centrifugal and rotary pumps. A piston or plunger differentiates the reciprocating pump from a centrifugal or rotary pump. In a reciprocating pump, the reciprocating motion of the wrist pin is converted to a circular motion by means of a connecting link or connecting rod (see Figure 5-1). Following are three moving elements necessary for operation of a reciprocating pump:

- Piston or plunger
- Inlet or admission valve
- Outlet or discharge valve

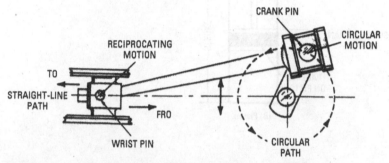

**Figure 5-1** The basic difference between reciprocating motion (left) and circular motion (right).

The piston or plunger works within a watertight cylinder. The basic difference between a piston and a plunger should be noted (see Figure 5-2). A piston is shorter than the stroke of the cylinder; the plunger is longer than the stroke. Another distinguishing feature is that the packing is inlaid on the rim of the piston for a tight seal. When a plunger is used, the packing is moved in a stuffing box located at the end of the cylinder to provide a tight seal.

### Principles of Operation

In general (and with respect to the way that the water is handled), reciprocating pumps may be classified as *lift pumps* or *force pumps*, which in turn, are either single-acting or double-acting pumps.

229

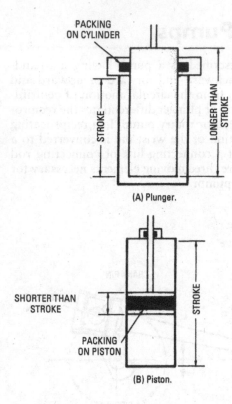

**Figure 5-2** Basic differences between a piston and a plunger.

(A) Plunger.

(B) Piston.

## Lift Pumps

A lift pump is a single-acting pump; it consists of an open cylinder and a discharge or bucket-type valve (see Figure 5-3). An open cylinder and a discharge or bucket-type valve in combination are the basic parts of the lift pump—it *lifts* the water, rather than forces it. In the lift pump, the bucket valve is built into the piston and moves upward and downward with the piston.

A four-stroke cycle is necessary to start the lift pump in operation (see Figure 5-4). The strokes are as follows:

- *Air exhaust*—The piston descends to the bottom of the cylinder, forcing out the air.
- *Water inlet*—On this upward stroke, a vacuum is created. Atmospheric pressure causes the water to flow into the cylinder.

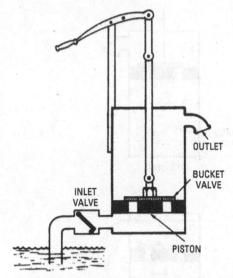

**Figure 5-3**   Basic construction of a single-acting lift pump.

- *Water transfer*—During this downward stroke, the water flows through the bucket valve (that is, it is transferred to the upper side of the piston).
- *Water discharge*—As the piston rises, the water is discharged (that is, it runs from the pump).

After the pump has been primed and is in operation, the working cycle is completed in two strokes of the piston—a downward stroke and an upward stroke (see Figure 5-5). The downward stroke of the piston is called the *transfer* stroke, and the upward stroke is called the *intake and discharge* stroke, because water enters the cylinder as the preceding charge of water is being discharged.

## Force Pumps
The force pump is actually an extension of a lift pump, in that it both *lifts and forces* the water against an external pressure. The basic operating principle of the force pump is that it forces water above the atmospheric pressure range, as distinguished from the lift pump, which elevates the water to flow from a spout.

### Single-Action
In a force pump, the water is forced out of the cylinder by means of a piston or plunger working against a pressure that corresponds

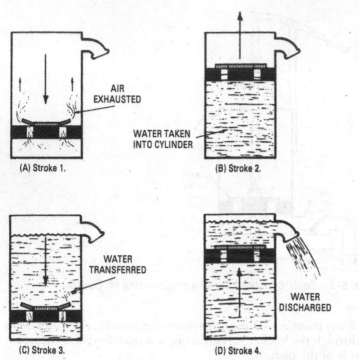

**Figure 5-4** Four-stroke starting cycle for a single-acting lift pump.

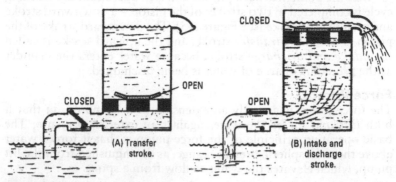

**Figure 5-5** Two-stroke working cycle (after priming) for a single-acting lift pump.

to the head or elevation above the inlet valve to which the water is pumped. In its simplest form, the force pump consists of an inlet valve, a discharge valve, and a single-acting plunger (see Figure 5-6).

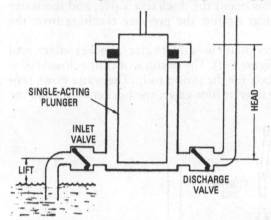

**Figure 5-6**  Basic construction of a single-acting plunger-type force pump.

In a single-acting force pump, the working cycle is completed in two strokes—an upward (intake) stroke and a downward (discharge) stroke (see Figure 5-7). During the upward stroke, the

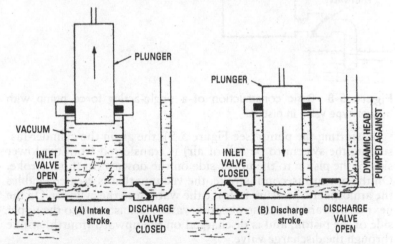

**Figure 5-7**  Two-stroke working cycle of a single-acting plunger-type force pump.

vacuum that is created enables atmospheric pressure to force the water into the cylinder. During this stroke the inlet valve is open and the discharge valve is closed. During the downward stroke, the plunger displaces (forces open) the discharge valve, and the water flows out of the cylinder against the pressure resulting from the dynamic head.

Another type of force pump uses inlet valves, bucket valves, and discharge valves (see Figure 5-8). The piston works in a closed cylinder (note the stuffing box for the piston rod). The water flows progressively through the inlet or foot valve, the bucket valve, and the discharge valve.

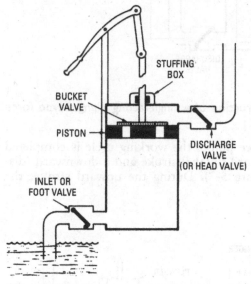

**Figure 5-8** Basic construction of a single-acting force pump with bucket-type valve in piston.

In starting the pump (see Figure 5-9), the air in the cylinder (assuming the system to be full of air) is transferred from the lower side of the piston to the upper side on the downward (first) stroke. On the upward (second) stroke, the vacuum that is created enables the atmospheric pressure to force the water into the cylinder. On the next downward (third) stroke, the water is transferred to the upper side of the piston, and is discharged on the upward (fourth) stroke through the discharge valve.

When the system is cleared of air and in operation, the working cycle is completed in two strokes of the piston—a downward (transfer) stroke and an upward (discharge) stroke. On the downward

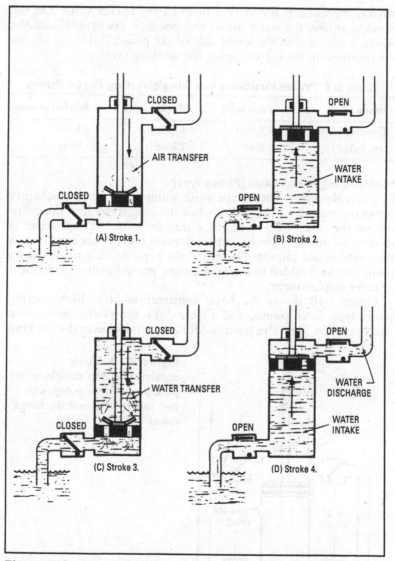

**Figure 5-9** Four-stroke starting cycle of a single-acting force pump with a bucket-type valve in the piston.

stroke, the water is transferred through the bucket valve. On the upward stroke, the water above the piston is discharged, and the water is admitted to the lower side of the piston. Table 5-1 shows the positions of the valves during the working cycle.

**Table 5-1  Valve Positions in a Single-Acting Force Pump**

| Stroke | Foot valve | Bucket valve | Discharge valve |
|---|---|---|---|
| Transfer (down) | Closed | Open | Closed |
| Discharge (up) | Open | Closed | Open |

### Double-Acting Force Pumps (Piston-Type)

In a double-acting piston-type force pump, the piston discharges water on one side of the piston while drawing water into the cylinder on the other side—without a transfer stroke. Thus, water is discharged on every stroke, rather than on every other stroke as in the single-acting pumps. Therefore, the capacity of a single-acting pump can be doubled in a double-acting pump having an identical cylinder displacement.

Figure 5-10 shows the basic construction of a double-acting piston-type force pump, and Figure 5-11 shows the two-stroke working cycle. The valve positions for the two strokes in the working

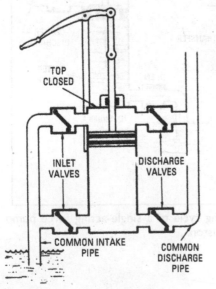

TOP
CLOSED

INLET
VALVES

DISCHARGE
VALVES

COMMON INTAKE
PIPE

COMMON
DISCHARGE
PIPE

**Figure 5-10**  Basic construction of a double-acting piston-type force pump with two sets of inlet and discharge valves.

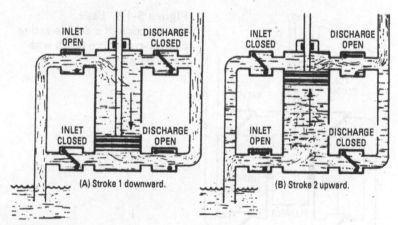

**Figure 5-11** Two-stroke working cycle of a double-acting piston-type force pump.

cycle should be noted. The diagonally-opposite inlet and discharge valves work in unison (that is, they are either open or closed at a given time).

### Double-Acting Force Pump (Plunger-Type)

The operation of this pump is identical to the operation of the piston-type double-acting force pump, except for the replacement of the piston with a plunger. These pumps are of two different types with respect to the location of the packing:

- Inside packed
- Outside packed

Figure 5-12 shows the basic construction of a pump with inside packing.

In the pump with inside packing (see Figure 5-12), the long cylinder is virtually divided into two separate chambers by the packing. Figure 5-13 shows the basic two-stroke working cycle of this pump. In its upward and downward movements, the plunger alternately displaces water in the two chambers. A disadvantage of this type of pump is that it is necessary to remove the cylinder head to adjust or to renew the packing. Leakage through the packing cannot be determined while the pump is in operation.

These disadvantages are overcome in the plunger-type pump with outside packing (see Figure 5-14). This design requires two plungers.

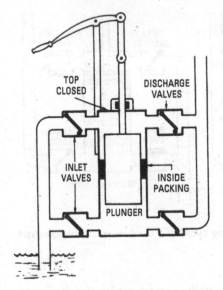

**Figure 5-12** Basic construction of a double-acting plunger-type force pump with inside packing.

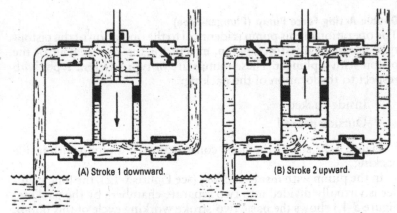

**Figure 5-13** The two-stroke working cycle of a double-acting plunger-type pump with inside packing.

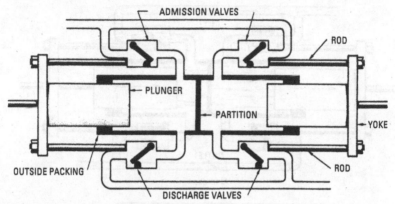

**Figure 5-14** Basic construction of double-acting plunger-type force pump with outside packing.

that are connected rigidly by yokes and rods. The packing is on the outside. It is serviced easily and its condition can be checked easily.

Figure 5-15 shows the two-stroke working cycle of the outside-packed plunger-type pump. Note that the plungers move in unison. Water is discharged at one end, while the opposite plunger is receding to fill the other end. A disadvantage of the outside-packed pump is that its construction is more complicated, and, therefore, more expensive than the inside-packed pump.

### Self-Priming or Siphon Pumps
Since there is a considerable clearance or enclosed space between the inlet and discharge valves (which becomes filled with air), most pumps require priming. Priming is necessary because the piston or plunger attempts to remove the air, creating a vacuum. Then, atmospheric pressure can force the water into the pump chamber.

Priming is an operation in which the pump chamber is filled with water to increase the vacuum. Thus, water is drawn in from the source. Unless the pump is provided with a vent and inlet opening, it may be necessary to remove a pump part to get water into the cylinder.

Figure 5-16 shows the basic construction of a self-priming or siphon pump. The pump casting consists of a pump barrel and a concentric outer chamber—the lower end of the barrel opening into the outer chamber. As shown in Figure 5-16, the piston contains a bucket-type valve with a discharge valve at the top. The inlet is also at the top, so that the water is trapped in the outer chamber.

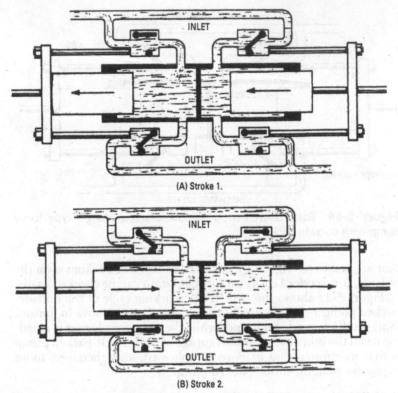

(A) Stroke 1.

(B) Stroke 2.

**Figure 5-15** Two-stroke working cycle of a double-acting plunger-type pump with outside packing.

Figure 5-17 shows the two-stroke working cycle of the self-priming pump. Initially, the outer chamber is filled with water. On the downward stroke, water is transferred through the bucket-type valve in the piston. This does not change the water level in the outer barrel. On the upward or discharge stroke, water is drawn into the barrel, which causes the water level in the outer chamber to recede to a low point. This, in turn, creates a vacuum in the outer chamber, which causes the water to flow in from the source and fill the outer chamber.

## Construction

In industrial hydraulic applications, the reciprocating piston pumps are usually of a large capacity. These pumps are often used to supply

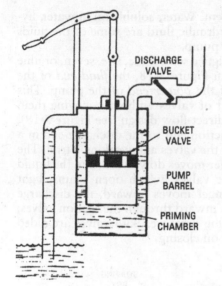

**Figure 5-16** Basic construction of a siphon (self-priming) pump.

DISCHARGE VALVE

BUCKET VALVE

PUMP BARREL

PRIMING CHAMBER

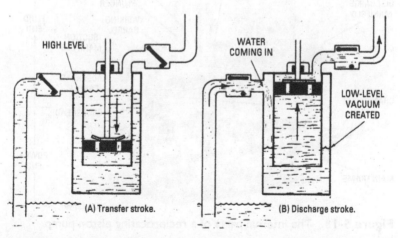

HIGH LEVEL

(A) Transfer stroke.

WATER COMING IN

LOW-LEVEL VACUUM CREATED

(B) Discharge stroke.

**Figure 5-17** Two-stroke working cycle of a siphon (self-priming) pump.

fluid to a central hydraulic system. Water, soluble oil in water, hydraulic oil, and fire-resistant hydraulic fluid are some of the fluids that are handled by this type of pump.

Reciprocating pumps are designed with three, five, seven, or nine plungers. In the pump shown in Figure 5-18, the *fluid end* of the pump is mounted at the top of the *power end* of the pump. This permits removal or replacement of valves without removing them from the pump entirely. In the direct-flow design (see Figure 5-19), the liquid is passed from the suction end to the discharge end in a straight horizontal line through the valves and working barrel. The suction valve closes as the plunger moves downward, and the liquid is forced through the discharge valves, which open against light spring pressure. When the plunger moves upward, the discharge valves close and liquid is drawn inward through the suction valves, which open against similar spring pressure. Both valves are aided and sealed by the fluid pressure on closing.

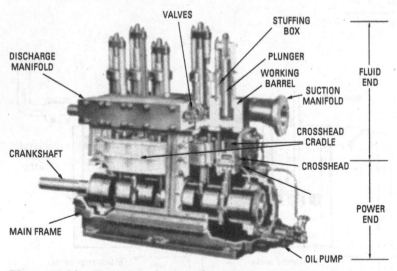

**Figure 5-18** The internal parts of a reciprocating piston pump.

The plunger-type pumps are widely used in industry to accomplish medium- to high-pressure chemical feeding. For example, a plunger-type pump can be used to meter chemicals and general industrial fluids, ranging from 0.2 to 1200 gallons per hour in pressures up to 1600 psi. This pump is available in either simplex

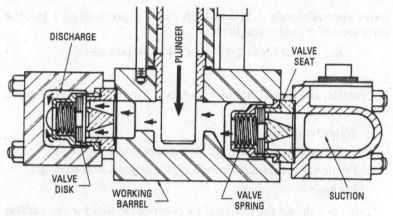

**Figure 5-19**  Direct-flow principle in the reciprocating piston pump.
*(Courtesy The Aldrich Pump Company, Standard Pump Div.)*

or duplex design, with an optional stroke-adjustment-in-motion attachment. All the common water-treatment chemicals and liquids (such as acids, caustics, solvents, and many other industrial processing liquids) can be handled by this pump. The pump is readily adapted to applications requiring continuous, intermittent, or flow-responsive feeding.

## Calculations

The numerous calculations, data, and tables relative to the various types of pumps can be useful to both operators and installation personnel. The calculations are diversified, and are intended to cover situations commonly encountered.

### Lift

Theoretical lift and actual lift are quite different. A number of factors cause the actual lift of a pump to be considerably less than its theoretical lift. Numerous installation failures have resulted from calculating the actual lift incorrectly.

#### Theoretical Lift

Although Torricelli demonstrated by experiment that atmospheric pressure of 14.7 pounds per square foot will support a 33.83-foot column of water at its maximum density, a pump cannot perform this feat. A barometer reading of 30 inches of mercury corresponds to the atmospheric pressure reading of 14.74 psi. Since a column of water approximately 2.31 feet high exerts a pressure of 1 psi, the corresponding theoretical lift is:

$$\text{theoretical lift} = 2.31 \times 14.74 = 34.049 \text{ feet}$$

**Actual Lift**

In practice, the actual lift of a pump is limited by the following factors:

- Water temperature
- Air pressure decreasing at higher elevations
- Frictional resistance through pipes, fittings, and passages
- Leakage

Table 5-2 shows the practical or permissible lifts for the various temperatures and elevations. Values that are preceded by a minus (−) sign indicate *lift*, and values preceded by a plus (+) sign indicate *head*. The table applies only for water, not for any other liquids that may be pumped. The actual lift depends on the specific gravity of the liquid being pumped. The thicker liquids (such as tar and molasses) should be moved toward the pump by means of gravity. Then, the inlet head is often called *negative lift*.

The inlet pipe is often called the *suction pipe*. To minimize frictional resistance, avoid elbows, or a number of closely spaced elbows; the inlet pipe should be proportioned for a flow rate of 250 feet per minute, for example. The following examples may be used to determine the nominal pipe size.

**Problem**

Calculate the diameter of the inlet pipe (at 250 feet per minute) required for a double-acting duplex pump with 10-inch cylinders by 12-inch stroke, operating at 50 rpm.

**Solution**

Since each pump makes two discharge strokes per revolution, the piston in each pump travels:

$$\frac{12 \times 2}{12} \times 50 = 100 \text{ feet}$$

At 100 feet per minute per pump, the total distance traveled by the two pistons, or total piston speed, is 200 feet per minute (2 × 100). The area of the two pistons is:

$$\text{area} = 2 \times 10^2 \times 0.7854 = 157.1 \text{ sq. in.}$$

## Table 5-2  Practical Lifts at Various Temperatures and Altitudes

| Altitude | Temperature of Water in Degrees F | | | | | | | | | | | | | | | |
|---|---|---|---|---|---|---|---|---|---|---|---|---|---|---|---|---|
| | 60 | 70 | 80 | 90 | 100 | 110 | 120 | 130 | 140 | 150 | 160 | 170 | 180 | 190 | 200 | 210 |
| sea level | −22 | −20 | −17 | −15 | −13 | −11 | −8 | −6 | −4 | −2 | 0 | +3 | +5 | +7 | +10 | +12 |
| 2000 ft. | −19 | −17 | −15 | −13 | −11 | −8 | −6 | −4 | −2 | +1 | +3 | +5 | +7 | +10 | +12 | +15 |
| 4000 ft | −17 | −15 | −13 | −10 | −8 | −6 | −4 | −1 | +1 | +3 | +5 | +7 | +10 | +12 | +14 | – |
| 6000 ft. | −15 | −13 | −11 | −8 | −6 | −4 | −2 | +1 | +3 | +5 | +7 | +10 | +12 | +14 | +16 | – |
| 8000 ft. | −13 | −11 | −9 | −6 | −4 | −2 | 0 | +3 | +5 | +7 | +9 | +12 | +14 | +16 | – | – |
| 10,000 ft. | −11 | −9 | −7 | −4 | −2 | 0 | +2 | +4 | +7 | +9 | +11 | +14 | +16 | +18 | – | – |

Therefore, the area of the inlet pipe is as much smaller than the cylinder area, as 200 is to 250, that is:

$$\text{area inlet pipe} = 157.1 \times \frac{200}{250} = 125.7$$

$$\text{diameter of inlet pipe} = \sqrt{\frac{125.7}{0.7854}}$$

$$= 12^{5}/_{8} \text{ inches (approximately)}$$

**Problem**
What diameter is required for the inlet pipe for a 1,000,000-gallon (per 24 hours) capacity pump for a flow through the pipe of 250 feet per minute?

**Solution**
For 1,000,000 gallons per 24-hour period, the gallons per minute (gpm) rating is as follows:

$$\text{gpm} = \frac{1,000,000}{(24 \times 60)} = 694$$

Since 1 gallon of water has a volume of 231 cubic inches, the volume of flow per minute is 160,314 cubic inches (694 × 231). For a flow of 250 feet per minute, the area of the inlet pipe must be as follows:

$$\text{area inlet pipe} = \frac{160,314}{(250 \times 12)} = \frac{160,314}{3000} = 53.4 \text{ sq. in.}$$

$$\text{diameter of inlet pipe} = \sqrt{\frac{53.4}{0.7854}}$$

$$= 8^{1}/_{4} \text{ inches (approximately)}$$

Therefore, the next larger nominal pipe size (9 inches) is required. This is especially important for the inlet side, because increased flow with a smaller pipe adds to the frictional resistance of the pipe. Many pump manufacturers make the outlet opening one nominal pipe size smaller than the inlet opening. Nominal pipe size is smaller than the inlet opening. Table 5-3 shows nominal pipe sizes for the water end and the steam end.

## Table 5-3 Pipe Sizes for Pumps (for Simplex Boiler Feed and General Service Piston-type Pumps)

| Size (inches) | Boiler feed capacity Gal. per min. | Boiler horsepower | Water and liquids up to 250 ssu visc. Gal. per min. | Piston speed, ft. per min. | Liquids 250 to 500 ssu visc. Gal. per min. | Piston speed, ft. per min. | Liquids 500 to 1000 ssu visc. Gal. per min. | Piston speed, ft. per min. | Liquids 1000 to 2500 ssu visc. Gal. per min. | Piston speed, ft. per min. | Liquids 2500 to 5000 ssu visc. Gal. per min. | Piston speed, ft. per min. | Pipe Diameters (inches) Steam | Exhaust | Suction | Discharge |
|---|---|---|---|---|---|---|---|---|---|---|---|---|---|---|---|---|
| 4½ × 2¾ × 6 | 8 | 110 | 13 | 45 | 11 | 39 | 10 | 35 | 8 | 27 | 6 | 22 | ¾ | 1 | 1¼ | 1 |
| 5¼ × 3¼ × 7 | 13 | 180 | 21 | 50 | 18 | 43 | 16 | 39 | 13 | 30 | 10 | 25 | ¾ | 1 | 1½ | 1¼ |
| 6 × 4½ × 8 | 23 | 325 | 38 | 55 | 33 | 48 | 29 | 43 | 23 | 33 | 19 | 27 | ¾ | 1 | 2½ | 2 |
| 7½ × 4½ × 10 | 31 | 450 | 52 | 63 | 45 | 55 | 40 | 49 | 31 | 38 | 25 | 31 | 1¼ | 1½ | 3 | 2½ |
| 8 × 5 × 12 | 42 | 600 | 71 | 70 | 61 | 60 | 55 | 54 | 42 | 42 | 35 | 35 | 1¼ | 1½ | 3 | 3 |
| 10 × 6 × 12 | 62 | 875 | 103 | 70 | 88 | 60 | 79 | 54 | 62 | 42 | 51 | 35 | 1¼ | 1½ | 4 | 3 |
| 12 × 7 × 12 | 84 | 1200 | 140 | 70 | 120 | 60 | 108 | 54 | 84 | 42 | 70 | 35 | 1¼ | 1½ | 5 | 4 |
| 14 × 8 × 12 | 109 | 1575 | 182 | 70 | 156 | 60 | 140 | 54 | 109 | 42 | 91 | 35 | 2½ | 3 | 5 | 4 |
| 16 × 10 × 18 | 219 | 3150 | 365 | 90 | 316 | 80 | 285 | 70 | 220 | 54 | 183 | 45 | 2½ | 3 | 8 | 6 |

* These normal operating capacities are for continuous duty. For emergency service increase about 15 percent.

## Size of Discharge Pipe

The nominal size of the discharge pipe is usually smaller than the nominal size of the inlet pipe. The required size of the discharge line is dependent on its length, number of elbows in the line, and other conditions that tend to resist the flow of water. In most calculations, a flow of 400 feet per minute can be used for building installations or tank service.

## Head

As shown in Figure 5-20, the static head ($H_s$) or dynamic head ($H_d$) may be determined by means of a test gage $A$ placed on the discharge side of a pump. If the pump is at rest, the gage reading can be used to determine the static head $H_s$. If the pump is in operation, the dynamic head $H_d$ can be determined.

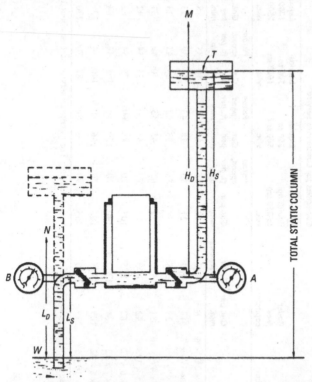

**Figure 5-20**  Gage method of obtaining static and dynamic head, static and dynamic lift, and static and total dynamic column.

**Problem**

When the gage $A$ indicates 35 pounds, what is the static head $H_s$ with the pump at rest?

**Solution**

Since a column of water 2.31 feet in height exerts a pressure of 1 psi, the corresponding head $H_s$ for a gage reading of 35 pounds is as follows:

$$H_s = 2.31 \times 35 = 80.9 \text{ feet}$$

If the pump were in motion, the gage $A$ would indicate a higher reading because of frictional resistance to flow in the pipe, thus indicating the dynamic head $H_d$.

As indicated in Figure 5-20, the head is not the total load on the pump, because lift must also be considered. The pump must raise the water from the level $W$ in the well to the level $T$ in the tank; this is referred to as *total static column* ($C_s$) or:

$$C_s = L_s + H_s$$

Also, the total dynamic column ($C_d$) is equal to:

$$C_d = L_d + H_d$$

The static lift ($L_s$) and the dynamic lift ($L_d$) can be determined by means of the vacuum gage $B$ (see Figure 5-20); the gage reading can be converted to the elevation in feet as shown in the following problem.

**Example**

What is the dynamic lift $L_d$ if the vacuum gage $B$ reads 18 inches?

lift (in feet) = vacuum reading $\times$ 0.49116 $\times$ 2.31

In the equation, 0.49116 psi corresponds to 1 inch of mercury, and a 2.31-foot column of water exerts a pressure of 1 psi. Substituting the vacuum gage $B$ reading into the equation:

lift, in feet = 18 $\times$ 0.49116 $\times$ 2.31 = 20.4 feet (approximately)

To calculate the total column when the water flows toward the pump (from point $N$ in Figure 5-20), the inlet head should be subtracted from the discharge head. Therefore, the total column is $(H - L_n)$, in which the symbol $n$ represents the negative lift. In direct-acting pumps, the allowance for velocity head is negligible if the velocities are low.

## Displacement

The volume of fluid that is displaced by the piston or plunger in a single stroke is called *displacement* in a reciprocating pump. Displacement is expressed as cubic inches per stroke, cubic inches per minute, and gallons per minute. Displacement is more commonly stated in terms of cubic inches per stroke. *To determine piston displacement, in cubic inches per stroke, multiply the effective area of the piston or plunger by the length of the stroke.*

### Problem

Calculate the displacement, in cubic inches per stroke, in a double-acting simplex reciprocating pump with a water cylinder that is 5 inches by 12 inches, with a 1-inch piston rod.

### Solution

The effective piston area is:

$$\text{piston area} = (5)^2 \times 0.7854 = 19.635 \text{ sq. in.}$$

$$\tfrac{1}{2} \text{ area piston rod} = \frac{(1)^2 \times 0.7854}{2} = 0.393$$

$$\text{total effective area of piston} = 19.242 \text{ sq. in.}$$

(the rod reduces displacement on only one side of piston; therefore, one-half the rod area is an average)

$$\text{displacement} = \text{effective piston area} \times \text{stroke}$$

$$= 19.242 \times 12$$

$$= 230.9 \text{ cu in./stroke}$$

*To determine the displacement, in cubic inches per minute, multiply the cylinder displacement per stroke by the number of discharging strokes per minute.* Thus, in the preceding problem, each stroke is a discharging stroke, because the pump is a double-acting pump. Therefore, if the pump is operating at 92 strokes per minute:

$$\text{displacement} = 230.9 \times 92$$

$$= 21,242.8 \text{ cubic inches per minute}$$

*To determine the displacement, in gallons per minute, divide the displacement per minute by 231 (volume of 1 gallon of water).* Therefore, displacement is:

$$\text{displacement} = \frac{21,242.8}{231} = 91.96 \text{ gal/min}$$

## Piston Speed

The total distance (in feet) traveled by a piston (or plunger) in 1 minute is referred to as the piston speed.

### Example

If a piston having a 16-inch stroke operates at 60 strokes per minute, its piston speed is:

$$\text{piston speed} = \frac{16 \times 60}{12} = 80 \text{ ft/min}$$

In calculating displacement, piston speed can be used as a factor. Depending on the type of pump, a coefficient stated in terms of the "number of discharging strokes per revolution" must be used. For example, in the simplex *single-acting* pump (see Figure 5-21), only *one* discharging stroke per revolution occurs, and in the *double-acting* pump, *two* discharging strokes per revolution occur.

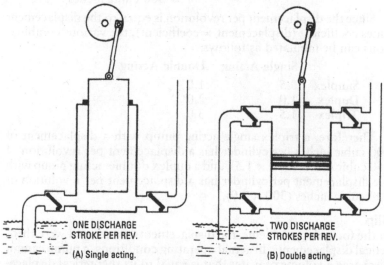

ONE DISCHARGE STROKE PER REV.

(A) Single acting.

TWO DISCHARGE STROKES PER REV.

(B) Double acting.

**Figure 5-21**  Discharge strokes per revolution as a factor in determining piston speed for calculating displacement in a simplex single-acting pump (left) and double-acting pump (right).

Since piston speed is based on both the upward (charging) stroke and the downward (discharging) stroke, and since only one discharging stroke occurs per two-stroke revolution in a single-acting pump, the *coefficient* is one-half (0.5). That is, the piston speed must be multiplied by $\frac{1}{2}$, or 0.5, as in the following problem.

**Problem**

What is the displacement per revolution in a single-acting pump with a 300-cubic inch displacement, operating at 100 rpm?

**Solution**

Since, in a single-acting pump, only one discharging stroke per revolution occurs, the displacement per revolution is:

displacement per revolution = 300 × 0.5

= 150 cubic inches

Since two discharging strokes per revolution occur in a double-acting pump, the coefficient is one-half (0.5)—that is, if the piston in the preceding problem were a *double-acting* pump, the displacement per revolution would be:

displacement per revolution = 300 × 1

= 300 cubic inches

Since the displacement per revolution is equal to the displacement times coefficient (displacement × coefficient), the various combinations can be tabulated as follows:

|         | Single-Acting | Double-Acting |
|---------|---------------|---------------|
| Simplex | 0.5           | 1.0           |
| Duplex  | 1.0           | 2.0           |
| Triplex | 1.5           | 3.0           |

Therefore, a triplex single-acting pump with a displacement of 300 cubic inches per cylinder has a displacement per revolution of 450 cubic inches (300 × 1.5), and a duplex double-acting pump with the displacement per cylinder has a displacement per revolution of 600 cubic inches (300 × 2.0).

**Slip**

In the foregoing problems, the displacements mentioned were theoretical displacements. In actual operating conditions, a pump cannot discharge a volume of water that is equal to its theoretical displacement because of slip through the valves and leakage. The *slip* is generally expressed as a percentage of the displacement. It is that amount by which the volume of water delivered per stroke falls short of the pump's displacement.

**Problem**

What is the slip (in percentage of the displacement) in a pump with a displacement of 300 cubic inches per stroke, but which discharges only 285 cubic inches of water per stroke?

**Solution**

The slip is 15 cubic inches (300 − 285); Therefore:

$$slip = \frac{15}{300} \times 100 = 5\%$$

That means the percentage is determined by:

- The percentage of slip. It varies from 2 to 10 percent.
- The type of pump. It is piston or plunger.
- The condition of the pump.
- The pressure that the pump is working against.

In actual practice, a slip of 2 percent is a common value for a plunger-type pump. For light-service piston pumps, the slip is approximately 5 percent. A slip of 10 percent is common for pressure-type piston pumps. Another factor that may reduce the output of a high-speed reciprocating pump occurs when the speed is too high for the water to flow through the inlet valves fast enough to completely fill the cylinder.

In pumps with bucket valves, operating on low lift where the column of water has sufficient dynamic inertia to continue in motion during a portion (or all) of the return stroke, the discharge volume may be greater than the displacement. This condition is referred to as *negative slip*.

## Capacity

The *capacity* of a pump is the actual volume of water or fluid delivered. It is usually stated in terms of gallons per stroke or gallons per minute when discharging at a given speed.

### Problem

A single-acting pump with a displacement of 300 cubic inches is operating at 100 strokes per minute. What is its capacity in gallons per minute with a 5 percent slip?

### Solution

For a single-acting pump, there are 50 discharging strokes per minute (100 ÷ 2). Hence:

$$displacement/min = 300 \times 50 = 15,000 \text{ cu in.}$$

$$5\% \text{ slip} = 15,000 \times 0.05 = 750 \text{ cu in.}$$

$$capacity/min = 15,000 - 750 = 14,250 \text{ cu in.}$$

$$capacity/min = \frac{14,250}{231} = 61.69 \text{ gal}$$

To calculate the capacity of a pump, *multiply* the area of the piston (in square inches) by the length of the stroke (in inches) and by the number of delivery strokes per minute. *Divide* the product by 1728 (to obtain theoretical capacity in cubic feet) or by 231 (to obtain theoretical capacity in gallons). *Multiply* the result by the efficiency factor for the pump to obtain the approximate net capacity. Expressed as a formula:

$$\text{approximate net capacity}$$

$$= \frac{0.7854 \times D^2 \times L \times N}{1728} \times (f) \text{ cu ft}$$

or

$$= \frac{0.7854 \times D^2 \times L \times N}{231} \times (f) \text{ gal}$$

in which the following is true

$D$ is diameter of piston or plunger (in inches)

$L$ is length of stroke (in inches)

$N$ is number of delivery strokes per minute

$f$ is slip, in percentage of displacement

1728 cubic inches in 1 cubic foot

231 cubic inches in 1 U.S. gallon

**Problem**
Calculate the approximate net capacity of a 3-inch diameter × 5-inch piston stroke double-acting pump, operating at 75 rpm, assuming a 5 percent slip.

**Solution**
Using the formula:

$$\text{approximate net capacity} = \frac{0.7854 \times (3)^2 \times 5 \times 150}{1728}$$

$$\times (1.0 - 0.05) = 2.914 \text{ cu ft}$$

$$= \frac{0.7854 \times (3)^2 \times 5 \times 150}{231}$$

$$\times (1.0 - 0.05) = 21.80 \text{ gal}$$

**Efficiency**
In general, *efficiency* is the ratio of the useful work performed by a prime mover to the energy expended in producing it. In regard

to pumps, there are several types of efficiency (such as hydraulic, volumetric, thermal, mechanical, and overall efficiency).

### Hydraulic Efficiency

*Hydraulic efficiency* is the ratio of the total column (dynamic head + dynamic lift) pumped against, to the total column, plus hydraulic losses, including all losses (such as velocity head) from the source of supply through the water-end cylinders to the point of attachment of the discharge gage.

### Volumetric Efficiency

The ratio of the capacity to the displacement, as:

$$\text{volumetric efficiency} = \frac{\text{capacity}}{\text{displacement}}$$

### Thermal Efficiency

The ratio of the heat utilized by the pump in doing useful work to the heat supplied. The formula is:

$$E_t = 2\left[\frac{42.44 \times P \times 60}{S(H - h)}\right]$$

in which the following is true:

> $P$ is horsepower
>
> $S$ is steam consumed (in pounds per hour)
>
> $H$ is total heat in 1 pound of steam at initial pressure
>
> $h$ is total heat in 1 pound of water supplied
>
> 42.44 is heat equivalent of 1 horsepower (in Btu per minute)

### Mechanical Efficiency

The ratio of the indicated horsepower of the water end to the indicated horsepower of the steam end:

$$\text{mechanical efficiency} = \frac{\text{indicated horsepower (water end)}}{\text{indicated horsepower (steam end)}}$$

This factor can be determined only by actual testing. A manufacturer states that the mechanical efficiency of direct-acting pumps varies with the size and type of pump from 50 to 90 percent (see Table 5-4).

### Table 5-4  Mechanical Efficiency of Pumps

| Stroke of Pump (Inches) | Piston-Type (percent) | Outside-Packed Plunger-Type (percent) |
|---|---|---|
| 3 | 55 | 50 |
| 5 | 60 | 56 |
| 6 | 65 | 61 |
| 7 | 68 | 64 |
| 8 | 72 | 68 |
| 10 | 76 | 72 |
| 12 | 78 | 75 |
| 16 | 80 | 77 |
| 20 | 83 | 80 |
| 24 | 85 | 82 |

### Horsepower at Water End of Pump

The horsepower required for a given pump capacity can be calculated by the following formulas:

$$thp = \frac{\text{cu ft} \times W(L_s + H_s)}{33,000}$$

$$ihp = \frac{\text{cu ft} \times W(L_d + H_d)}{33,000}$$

in which the following is true:

   $thp$ is theoretical horsepower

   $ihp$ is indicated horsepower

   $W$ is weight of one cubic foot of water (in pounds)

   $L_s$ is static lift (in feet)

   $H_s$ is static head (in feet)

   $H_d$ is dynamic head (in feet)

### Problem

What theoretical horsepower ($thp$) is required to raise 100 cubic feet of water to 200 feet? The lift is 10 feet and the water temperature is 75°F.

### Solution

At 75°F, 1 cubic foot of water weighs 62.28 pounds. Substituting in the formula:

$$thp \text{ (at 75°F)} = \frac{100 \times 62.28 \times (10 + 200)}{33,000} = 39.63$$

At 35°F (cold weather), the weight of 1 cubic foot of water increases to 62.42. The horsepower increases in proportion to the ratio of the two weights, as follows:

$$thp \text{ (at 35°F)} = \frac{39.63 \times 62.42}{62.28} = 39.7$$

In comparing the preceding equations, it may be noted that temperature causes only a slight difference in the results. Thus, the temperature factor can be disregarded for most calculations, and the common weight value of 62.4 pounds per cubic foot of water can be used. Table 5-5 shows the theoretical horsepower required to raise water to various heights.

### Water Supply Pumps

Careful consideration must be given to the selection of pumps for water supply purposes, for each pump, application will differ, not only in capacity requirements, but also in the pressure against which the pump will have to operate. For example, a ½-inch hose with nozzle is to be used for sprinkling water, will be consumed at the rate of 200 gallons per hour. To permit use of water for other purposes at the same time, it is necessary to have a pump capacity in excess of 200 gallons per hour. Where ½-inch hose with nozzle is to be used, it is necessary to have a pump with a capacity of at least 220 gallons per hour. In determining the desired pump capacity, even for ordinary requirements, it is advisable to select a size large enough so that the pump will not run more than a few hours per day.

### Typical Water-Supply Pumps

Automatic and semiautomatic pump installations for water supply purposes generally employ the following four types of pumps (all conforming to the same general principles):

- Reciprocating or plunger
- Rotary
- Centrifugal
- Jet or ejector

The ejector pump gets its name from the introduction of a jet stream attached to the centrifugal or reciprocating type of pump. Table 5-6 provides a brief tabulation of the characteristics of the various types of pumps. For convenience, the pumps are classified according to speed, suction lift, and practical pressure head.

Reciprocating pumps will deliver water in quantities proportional to the number of strokes and the length and size of the cylinder. They are adapted to a wide range of speeds, and to practically any depth

## Table 5-5 Table of Theoretical Horsepower Required to Elevate Water to Various Heights

| Gallons per Minute | 5 feet | 10 feet | 15 feet | 20 feet | 25 feet | 30 feet | 35 feet | 40 feet | 45 feet | 50 feet | 60 feet | 75 feet | 90 feet | 100 feet | 125 feet | 150 feet | 175 feet | 200 feet | 250 feet | 300 feet | 350 feet | 400 feet | Gallons per Minute |
|---|---|---|---|---|---|---|---|---|---|---|---|---|---|---|---|---|---|---|---|---|---|---|---|
| 5 | 0.006 | 0.012 | 0.019 | 0.025 | 0.031 | 0.037 | 0.044 | 0.05 | 0.06 | 0.06 | 0.07 | 0.09 | 0.11 | 0.12 | 0.16 | 0.19 | 0.22 | 0.25 | 0.31 | 0.37 | 0.44 | 0.50 | 5 |
| 10 | 0.012 | 0.025 | 0.037 | 0.050 | 0.062 | 0.075 | 0.087 | 0.10 | 0.11 | 0.12 | 0.15 | 0.18 | 0.22 | 0.25 | 0.31 | 0.37 | 0.44 | 0.50 | 0.62 | 0.75 | 0.87 | 1.00 | 10 |
| 15 | 0.019 | 0.037 | 0.056 | 0.075 | 0.094 | 0.112 | 0.131 | 0.15 | 0.17 | 0.19 | 0.22 | 0.28 | 0.34 | 0.37 | 0.47 | 0.56 | 0.66 | 0.75 | 0.94 | 1.12 | 1.31 | 1.50 | 15 |
| 20 | 0.025 | 0.050 | 0.075 | 0.100 | 0.125 | 0.150 | 0.175 | 0.20 | 0.22 | 0.25 | 0.30 | 0.37 | 0.45 | 0.50 | 0.62 | 0.75 | 0.87 | 1.00 | 1.25 | 1.50 | 1.75 | 2.00 | 20 |
| 25 | 0.031 | 0.062 | 0.093 | 0.125 | 0.156 | 0.187 | 0.219 | 0.25 | 0.28 | 0.31 | 0.37 | 0.47 | 0.56 | 0.62 | 0.78 | 0.94 | 1.09 | 1.25 | 1.56 | 1.87 | 2.19 | 2.50 | 25 |
| 30 | 0.037 | 0.075 | 0.112 | 0.150 | 0.187 | 0.225 | 0.262 | 0.30 | 0.34 | 0.37 | 0.45 | 0.56 | 0.67 | 0.75 | 0.94 | 1.12 | 1.31 | 1.50 | 1.87 | 2.25 | 2.62 | 3.00 | 30 |
| 35 | 0.043 | 0.087 | 0.131 | 0.175 | 0.219 | 0.262 | 0.306 | 0.35 | 0.39 | 0.44 | 0.52 | 0.66 | 0.79 | 0.87 | 1.09 | 1.31 | 1.53 | 1.75 | 2.19 | 2.62 | 3.06 | 3.50 | 35 |
| 40 | 0.050 | 0.100 | 0.150 | 0.200 | 0.250 | 0.300 | 0.350 | 0.40 | 0.45 | 0.50 | 0.60 | 0.75 | 0.90 | 1.00 | 1.25 | 1.50 | 1.75 | 2.00 | 2.50 | 3.00 | 3.50 | 4.00 | 40 |
| 45 | 0.056 | 0.112 | 0.168 | 0.225 | 0.281 | 0.337 | 0.394 | 0.45 | 0.51 | 0.56 | 0.67 | 0.84 | 1.01 | 1.12 | 1.41 | 1.69 | 1.97 | 2.25 | 2.81 | 3.37 | 3.94 | 4.50 | 45 |
| 50 | 0.062 | 0.125 | 0.187 | 0.250 | 0.312 | 0.375 | 0.437 | 0.50 | 0.56 | 0.62 | 0.75 | 0.94 | 1.12 | 1.25 | 1.56 | 1.87 | 2.19 | 2.50 | 3.12 | 3.75 | 4.37 | 5.00 | 50 |
| 60 | 0.075 | 0.150 | 0.225 | 0.300 | 0.375 | 0.450 | 0.525 | 0.60 | 0.67 | 0.75 | 0.90 | 1.12 | 1.35 | 1.50 | 1.87 | 2.25 | 2.62 | 3.00 | 3.75 | 4.50 | 5.25 | 6.00 | 60 |
| 75 | 0.093 | 0.187 | 0.281 | 0.375 | 0.469 | 0.562 | 0.656 | 0.75 | 0.84 | 0.94 | 1.12 | 1.40 | 1.68 | 1.87 | 2.34 | 2.81 | 3.28 | 3.75 | 4.69 | 5.62 | 6.56 | 7.50 | 75 |
| 90 | 0.112 | 0.225 | 0.337 | 0.450 | 0.562 | 0.675 | 0.787 | 0.90 | 1.01 | 1.12 | 1.35 | 1.68 | 2.02 | 2.25 | 2.81 | 3.37 | 3.94 | 4.50 | 5.62 | 6.75 | 7.87 | 9.00 | 90 |
| 100 | 0.125 | 0.250 | 0.375 | 0.500 | 0.625 | 0.750 | 0.875 | 1.00 | 1.12 | 1.25 | 1.50 | 1.87 | 2.25 | 2.50 | 3.12 | 3.75 | 4.37 | 5.00 | 6.25 | 7.50 | 8.75 | 10.00 | 100 |
| 125 | 0.156 | 0.312 | 0.469 | 0.625 | 0.781 | 0.937 | 1.094 | 1.25 | 1.41 | 1.56 | 1.87 | 2.34 | 2.81 | 3.12 | 3.91 | 4.69 | 5.47 | 6.25 | 7.81 | 9.37 | 10.94 | 12.50 | 125 |
| 150 | 0.187 | 0.375 | 0.562 | 0.750 | 0.937 | 1.125 | 1.312 | 1.50 | 1.69 | 1.87 | 2.25 | 2.81 | 3.37 | 3.75 | 4.69 | 5.62 | 6.56 | 7.50 | 9.37 | 11.25 | 13.12 | 15.00 | 150 |
| 175 | 0.219 | 0.437 | 0.656 | 0.875 | 1.093 | 1.312 | 1.531 | 1.75 | 1.97 | 2.19 | 2.62 | 3.28 | 3.94 | 4.37 | 5.47 | 6.56 | 7.66 | 8.75 | 10.94 | 13.12 | 15.31 | 17.50 | 175 |
| 200 | 0.250 | 0.500 | 0.750 | 1.000 | 1.250 | 1.500 | 1.750 | 2.00 | 2.25 | 2.50 | 3.00 | 3.75 | 4.50 | 5.00 | 6.25 | 7.50 | 8.75 | 10.00 | 12.50 | 15.00 | 17.50 | 20.00 | 200 |
| 250 | 0.312 | 0.625 | 0.937 | 1.250 | 1.562 | 1.875 | 2.187 | 2.50 | 2.81 | 3.12 | 3.75 | 4.69 | 5.62 | 6.25 | 7.81 | 9.37 | 10.94 | 12.50 | 15.62 | 18.75 | 21.87 | 25.00 | 250 |
| 300 | 0.375 | 0.750 | 1.125 | 1.500 | 1.875 | 2.250 | 2.625 | 3.00 | 3.37 | 3.75 | 4.50 | 5.62 | 6.75 | 7.50 | 9.37 | 11.25 | 13.12 | 15.00 | 18.75 | 22.50 | 26.25 | 30.00 | 300 |
| 350 | 0.437 | 0.875 | 1.312 | 1.750 | 2.187 | 2.625 | 3.062 | 3.50 | 3.94 | 4.37 | 5.25 | 6.56 | 7.87 | 8.75 | 10.94 | 13.12 | 15.31 | 17.50 | 21.87 | 26.25 | 30.62 | 35.00 | 350 |
| 400 | 0.500 | 1.000 | 1.500 | 2.000 | 2.500 | 3.000 | 3.500 | 4.00 | 4.50 | 5.00 | 6.00 | 7.50 | 9.00 | 10.00 | 12.50 | 15.00 | 17.50 | 20.00 | 25.00 | 30.00 | 35.00 | 40.00 | 400 |
| 500 | 0.625 | 1.250 | 1.875 | 2.500 | 3.125 | 3.750 | 4.375 | 5.00 | 5.62 | 6.25 | 7.50 | 9.37 | 11.25 | 12.50 | 15.62 | 18.75 | 21.87 | 25.00 | 31.25 | 37.50 | 43.75 | 50.00 | 500 |

This table gives the actual water horsepower. When selecting motors and turbines, allowance must be made for pipe friction and loss in the pump, gears, belts, etc. One foot of head equals 0.43 pounds pressure per square inch.

## Table 5-6  Pump Characteristics

| Type Pump | Speed | Practical Suction Lift | Pressure Head | Delivery Characteristics |
|---|---|---|---|---|
| **Reciprocating:** | | | | |
| Shallow Well (Low Pressure) | Slow | 22 to 25 ft. | 40 to 43 lbs. | Pulsating (air chamber evens pulsations) |
| (Medium Pressure) | 250 to 550 strokes per min. | | Up to 100 lbs. | Pulsating (air chamber evens pulsations) |
| (High Pressure) | | | Up to 350 lbs. | Pulsating (air chamber evens pulsations) |
| Deep Well | Slow 30 to 52 strokes per min. | Available for lifts up to 875 ft. Suction lift below cylinder 22 ft. | Normal 40 lbs. | Pulsating (air chamber evens pulsations) |
| **Rotary Pump:** | | | | |
| Shallow Well | 400 to 1725 rpm | 22 ft. | About 100 lbs. | Positive (slightly pulsating) |
| **Ejector Pump:** | | | | |
| (Shallow Well and limited Deep Wells) | Used with Centrifugal-turbine or Shallow Well reciprocating pump. | Max. around 120 ft. Practical at lifts of 80 ft. or less | 40 lbs. (normal) Available at up to 70 lbs pressure head | Continuous non-pulsating, high capacity with low-pressure head |
| **Centrifugal:** | | | | |
| Shallow Well (Single Stage) | High, 1750 and 3600 rpm | 15 ft. maximum | 40 lbs. (normal) 70 lbs (maximum) | Continuous non-pulsating, high capacity with low-pressure head |
| **Turbine Type:** | | | | |
| (Single Impeller) | High, 1750 rpm | 28 ft. maximum at sea level | 40 lbs. (normal) Available up to 100 lbs. pressure head | Continuous non-pulsating, high capacity with low-pressure head |

of well. Since reciprocating pumps are positive in operation, they should be fitted with automatic relief valves to prevent rupture of pipes or other damage should power be applied against abnormal pressure.

Centrifugal pumps are somewhat critical as far as speed is concerned. They should be used only where power can be applied at a reasonably constant speed. The vertical type centrifugal pump is used in deep wells. They are usually driven through shafting by vertical motors mounted at the top of the well.

Turbine pumps (as used in domestic water systems) are self-priming. Their smooth operation makes them suitable for applications where noise and vibrations must be kept to a minimum.

Ejector pumps operate quietly, and neither the deep well nor the shallow well type need be mounted over the well.

## Summary

A reciprocating pump has a to-and-fro motion. Its motion is backward and forward (or upward and downward) as distinguished from the circular motion of centrifugal and rotary pumps. A piston or plunger differentiates the reciprocating pump from a centrifugal or rotary pump. In the reciprocating pump, the reciprocating motion of the wrist pin is converted to circular motion by means of a connecting link or connecting rod.

A lift pump is a single-acting pump. It consists of an open cylinder and a discharge or bucket-type valve. It lifts the water, rather than forces it. A force pump is a single-acting or a double-acting pump.

After the lift pump has been primed, its working cycle is completed in two strokes of the piston: a downward or transfer stroke, and the upward stroke (which is an intake and discharge stroke, because water enters the cylinder as the preceding charge of water is being discharged).

The force pump is actually an extension of the lift pump. It lifts and forces the water against an external pressure. The basic operating principle of the force pump is that it forces water above the level attained by the atmospheric pressure range, as distinguished from the lift pump, which elevates the water to flow from a spout.

In a single-acting force pump, the water is forced from the cylinder by means of a piston or plunger working against a pressure that corresponds to the head or elevation above the inlet valve to which the water is pumped. The working cycle is completed in two strokes: an upward (intake) stroke and a downward (discharge) stroke. During the intake stroke, the vacuum that is created enables atmospheric pressure to force the water into the cylinder. During the discharge stroke, the plunger displaces (forces open) the discharge valve, and the water flows from the cylinder against the pressure resulting from the dynamic head.

In a double-acting force pump, the piston discharges water from one side of the piston while drawing water into the cylinder on the other side—without a transfer stroke. Thus, water is discharged on each stroke, rather than on alternate strokes, as in the single-acting pumps. Therefore, the capacity of a double-acting pump can be twice that of a single-acting pump having an identical cylinder displacement.

Reciprocating piston pumps having a large capacity are used in many industrial applications. They are designed with three, five, seven, or nine plungers. The plunger-type pumps are widely used in industry to accomplish medium- to high-pressure chemical feeding. These pumps are readily adapted to applications requiring continuous, intermittent, or flow-responsive feeding.

Theoretical lift and actual lift are quite different. Actual lift is a lesser value than theoretical lift; it is limited by factors, such as water temperature, decreased air pressure at higher elevations, frictional resistance (through pipes, fittings, and passages), and leakage.

The nominal size of the discharge pipe is usually smaller than that of the inlet pipe. The required size of the discharge line is dependent on its length, number of elbows in the line, and other conditions that tend to resist the flow of water.

The total load on a pump is not equal to the head alone. The lift must also be considered. The pump must raise the water from the surface level in the wall to the surface level in the tank. This is termed total static column. The formula for total static column is:

$$C_s = L_s + H_s$$

When all resistances to flow are considered, the total load becomes total dynamic column, and the formula is:

$$C_d = L_d + H_d$$

The volume of fluid that is displaced by a piston or plunger in a single stroke is called displacement. Displacement is expressed in cubic inches per stroke, cubic inches per minute, or gallons per minute. To determine piston displacement in cubic inches per stroke, multiply the effective area of the piston or plunger by the length of the stroke.

The total distance, in feet, traveled by a piston (or plunger) in 1 minute is referred to as the piston speed. This is a factor in calculating displacement. In a simplex single-acting pump, only one stroke per revolution occurs. In the double-acting pump, two discharging strokes per revolution occur.

In a pump, slip is generally expressed as a percentage of displacement. Pumps may fall short of the water previously designed to handle. The amount or the volume of water (delivered per stroke) that falls short of the pump's displacement that is being used in the problem. The percentage of slip varies from 2 to 10 percent. It depends on the type of pump (piston or plunger). The condition of the pump also figures into the slip variation. So does the pressure against which the pump is working. When the discharge volume is greater than the displacement, the condition is called *negative slip*.

The capacity of a pump is the actual volume of water or fluid delivered. It is usually stated in terms of gallons per stroke or gallons per minute when discharging at a given speed. To calculate capacity, multiply the area of the piston (square inches) by the length of the stroke (inches) and by the number of delivery strokes per minute. Divide the product by 1728 (to obtain theoretical capacity, in cubic feet) or by 231 (to obtain theoretical capacity, in gallons).

Efficiency is the ratio of the useful work performed by a prime mover to the energy expended in producing it. There are several types of efficiency in regard to pumps (such as hydraulic, volumetric, thermal, mechanical, and overall efficiency).

## Review Questions

1. How does a reciprocating pump differ from a centrifugal or rotary pump?
2. How does the basic operation of a lift pump differ from that of a force pump?
3. What is the chief difference in the working cycles of the single-acting force pump and the double-acting force pump?
4. Compare the capacities of single-acting and double-acting force pumps having identical cylinder displacements.
5. What are the disadvantages of the inside-packed double-acting plunger-type force pump?
6. Describe the working cycle of the self-priming reciprocating pump.
7. What types of industrial applications use reciprocating piston pumps?
8. Compare theoretical lift and actual lift.

9. Why is the nominal size of the discharge pipe usually smaller than the nominal size of the intake line?

10. Why is the total load on a pump not equal to just the head alone?

11. What is meant by *displacement*, and how is it expressed?

12. What is meant by *slip*, and how is it expressed?

13. How is capacity of a pump calculated?

14. What are three moving elements necessary for operation of a reciprocating pump?

15. What is the difference between a piston and a plunger?

16. List the four strokes of a single-acting lift pump.

17. List the four strokes of a single-acting force pump with a bucket-type valve in the piston.

18. How many plungers can reciprocating pumps have?

19. What is the lift at sea level of water at 120°F?

20. What is the term used to indicate the volume of fluid that is displaced by the piston or plunger in a single stroke in a reciprocating pump?

21. Define *hydraulic efficiency*.

22. Define *volumetric efficiency*.

23. Define *thermal efficiency*.

24. Define *mechanical efficiency*.

25. How much horsepower would it take to lift 40 gallons of water to a height of 10 feet?

# Chapter 6

## Special-Service Pumps

Pump selection should be based on a thorough understanding of the characteristics and fundamental principles of the basic types of pumps (such as centrifugal, rotary, reciprocating, and so on), which have been discussed in preceding chapters. The operating principles and various design characteristics or features adapt these pumps to specialized or unusual service conditions. In some instances, pumps have been designed especially for a given type of job or service condition.

Many of the pumps that have been introduced are specifically designed for special types of service or operating conditions. Some of these special types of pumps are used by fire departments, railroads, automobiles, diesel engines, contractors, mines, drainage, hydrostatic transmissions, machine tools, and numerous other applications throughout industry. Liquids requiring a special type of pump are pure water, sewage, paper stock, oil, milk, chemicals, magma, and various types of thick liquids.

### Service Pumps

Simplex or duplex reciprocating piston-type pumps are well-adapted for the service requirements found in tanneries, sugar refineries, bleacheries, and so on. The duplex high-pressure reciprocating piston pump shown in Figure 6-1 is adapted for high-pressure service on water supply systems in country clubs, dairies, industrial plants, and so on. Two double-acting pistons actually provide four strokes per revolution, which produces a constant and even flow of liquid through the pump. This pump can handle either cold or hot water at temperatures up to 200°F. This type of pump is capable of capacities ranging from 295 to 1080 gallons per hour, and working pressures ranging from 150 to 300 psi.

The general-purpose rotary gear-type pump (see Figure 6-2) is designed to handle either thick or thin liquids. It can operate smoothly in either direction of rotation with equal efficiency. These pumps can handle heavy, viscous materials (such as roofing materials or printing inks), as well as fuel oils, gasoline, and similar thin liquids. The pump shown is capable of capacities ranging from 40 to 600 gallons per minute and pressures to 100 psi.

The rotary gear-type pump shown in Figure 6-3 is adaptable to a wide range of jobs, such as pressure lubrication, hydraulic service, fuel supply, or general transfer work, which includes pumping clean liquids. The bearings are lubricated by the liquid being pumped.

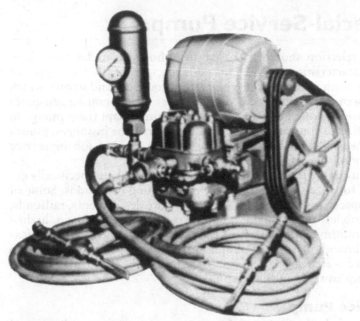

**Figure 6-1** Double-acting, high-pressure, horizontal duplex reciprocating piston pump. This pump has two cylinders and is capable of capacities ranging from 5 to 9 gallons per minute at working pressures up to 350 psi.

These pumps are self-priming and operate in either direction of rotation with equal efficiency. This type of pump is available in various sizes ranging from 3/4 to 110 gallons per minute at various pressure settings to 150 psi.

The self-priming motor-mounted centrifugal pump (see Figure 6-4) has several applications. It can be used for swimming pools, lawn sprinklers, booster service, recirculation, irrigation, dewatering, sump and bilge, liquid fertilizers, and chemical solutions. Its capacity is 10 to 130 gallons per minute against heads ranging to 120 feet.

The horizontally split casing double-suction single-stage centrifugal pump (see Figure 6-5) is designed for high efficiency with heavy casing walls for 175 psi working pressure. This pump is used for general water supply, booster service, municipal waterworks, air washing, condenser cooling, water circulation, industrial service, building service, and chemical plants.

**Figure 6-2** General-purpose rotary gear-type pump designed to handle either thick or thin liquids. It is capable of capacities ranging from 40 to 600 gallons per minute and pressures up to 100 psi.
*(Courtesy Roper Pump Company)*

The dry-pit non-clog centrifugal pump (see Figure 6-6) is adapted to solids-handling applications (such as sanitary waste, sewage lift stations, treatment plants, industrial waste, general drainage, sump service, dewatering, industrial process service, food processing, and chemical plants). The impeller is of a non-clog fully enclosed type with extra-smooth passageways. The pump capacity ranges to 3000 gallons per minute against heads to 150 feet. This pump is available in both right- and left-hand rotation and in five styles of mechanical assemblies for ease of installation.

The close-coupled single-suction single-stage centrifugal pump (see Figure 6-7) is available in various sizes and designs for handling capacities to 2200 gallons per minute and heads to 500 feet. It is available in a wide variety of constructions for handling most liquids ranging from clear water to highly corrosive chemicals.

**Figure 6-3** Rotary gear-type pump. *(Courtesy Roper Pump Company)*

**Figure 6-4** Self-priming centrifugal pump.

**Figure 6-5**  Single-stage double-suction centrifugal pump with horizontally split casing. *(Courtesy Deming Division, Crane Co.)*

**Figure 6-6**  Dry-pit non-clog centrifugal pump, which is generally used for solids-handling applications.

**Figure 6-7** A close-coupled single-suction single-stage centrifugal pump.

The double-suction single-stage pump shown in Figure 6-8 is used extensively in nearly all types of service. The horizontally split casing provides easy access for maintenance and inspection of rotating parts without disturbing the piping. This pump is designed for capacities ranging from 10 to 20,000 gallons per minute.

For handling clear water at any temperature, the turbine-driven single-suction four-stage pump (see Figure 6-9) is capable of capacities ranging from 20 to 900 gallons per minute at heads up to 1500 feet. These pumps are both mechanically and hydraulically efficient.

## Chemical and Process Pumps

Centrifugal pumps are extensively used in handling the wide variety of corrosive and abrasive liquids in the chemical and paper industries. Formerly, pump life was exceedingly short, because the pump casings, impellers, shafts, and other parts were reduced rapidly by the liquid. The application of vulcanized rubber to the parts contacted by the liquid has extended the service-life of these parts. Liquids handled by rubber-lined pumps are chlorinated paper stock, hypochlorous acid, chlorinated brines, sodium chloride, potassium chloride, brine slurries, hydrofluoric slurry, sulfurous water, caustic soda, and many other acids and caustics of various concentrations and temperatures.

In the operation of a paper mill, pumps are required to move the material from point to point throughout the mill—starting with the raw water and then the paper pulp—and finally to the head box on the papermaking machine. The various chemicals that are necessary in the papermaking process are handled by pumps, and the

**Figure 6-8**    A double-suction single-stage pump with horizontally split casing. *(Courtesy Buffalo Forge Company)*

various types of waste liquids are disposed of by means of pumps. The horizontal end-suction centrifugal pump shown in Figure 6-10 is designed specifically for a wide range of applications in the chemical, petrochemical, and other types of material-processing industries that handle various corrosive and non-corrosive liquids, ranging from water and light hydrocarbons to heavy slurries.

Centrifugal pumps are widely used in the paper industry, with the most important feature being the impeller. In general, the characteristics of the impellers are few vanes, large inlets, and heavy rigid

**Figure 6-9** A turbine-driven single-suction four-stage pump.
*(Courtesy Buffalo Forge Company)*

**Figure 6-10** A horizontal end-suction centrifugal pump used as a chemical and process pump. *(Courtesy Deming Division, Crane Co.)*

block plates and flanges. The impellers are shrouded and do not require wearing plates. The shape and number of vanes are governed by service conditions.

Centrifugal pumps made of Pyrex or glass is used to pump acids, milk, fruit juices, and other acid solutions. The glass is resistant to acid solutions. Therefore, the liquid being pumped is not contaminated by the chemical reactions between the liquid and the material in the pump parts. Glass is resistant to all the acids, except hydrofluoric acid and glacial phosphoric acid. Pumps made of glass should not be used with alkaline solutions.

## Pumps for Medical Use

Vacuum pumps for medical use have been around for many years. However, there have been some improvements in recent years.

Historically, almost all types of pumps that were used for medical-surgical vacuum systems used either water or oil for operation. Water-sealed liquid-ring pumps gained acceptance because they could safely handle the flammable anesthetic gases that previously were used, without any risk of fire or explosion. Oil-lubricated pumps were popular because they were generally more efficient than the water-sealed liquid-ring pumps and were less costly. Both types worked well, and were dependable. Both water-sealed liquid-ring and oil-lubricated pumps were not without their faults, however.

Water-ring vacuum pumps were first invented in the late 1800s. They required dependency upon another utility—water. These pumps could transfer bacteria from the air stream to the water. That meant the water could be considered contaminated. The facility had to pay for the water as well as the sewage.

Oil-lubricated pumps came in several design variations: reciprocating, rotary vane, rotary screw, and oil-sealed liquid-ring.

Reciprocating vacuum pumps are the oldest known type of pump widely used until the early 1980s. They became less popular when quieter and smoother-running pumps appeared on the market. These pumps were lubricated by the splash lubrication method. Today, the reciprocating pump is almost extinct in hospitals.

Rotary-screw pumps are a variation of the rotary-screw compressor. They were oil-flooded and became available in 1980s and were used in larger facilities. They used a large oil reservoir and recirculated the oil once it was filtered. Rotary-screw pumps are more complex than flooded vane pumps. Because the rotors are connected, they require high starting power, and rotor friction is such that it was best to keep them running continuously.

Oil-sealed liquid-ring pumps have been used for decades. They were used in hospital vacuum-type pumps only recently. The design is almost identical to the rotary-screw pump. It is sometimes called the *air end pump*.

Rotary sliding-vane pumps have been used for many years. They are mostly lubricated by the drip-style of oiler and are difficult to set properly. These pumps tend to run very hot. They are limited to how long they can run on a continuous basis before damage occurs. Today, they are only used for intermittent duty. The oil-flooded variety is now the most popular vacuum pump.

The typical oil-less pump used for surgical centers and waste anesthetic gas evacuation systems is shown in Figure 6-11.

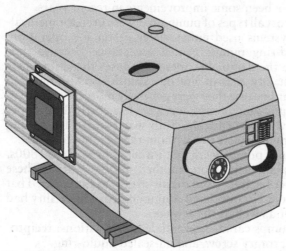

**Figure 6-11** Typical oil-less pump used for surgical centers.
*(Courtesy Becker Pumps)*

## Pumps for Handling of Sewage

Pumping stations are built when sewage must be raised to a point of higher elevation, or where the topography prevents downhill gravity flow. Special non-clogging pumps are available to handle raw sewage. They are insulated in structures called *lift stations*. Two basic types of lift station are dry well and wet well. A *wet-well installation* has only one chamber or tank to receive and hold the sewage until it is pumped out. Specially designed submersible pumps and motors can be located at the bottom of the chamber, completely below the water level. *Dry-well installations* have two separate chambers, one to receive the wastewater, and one to enclose and protect the pumps and controls. The protective dry chamber allows easy access for inspection and maintenance.

All sewage lift stations, whether wet well or dry well, should include at least two pumps. One pump can operate while the other is removed for repair. In this type of service, centrifugal pumps are required to pump either raw sewage or sludge. The solid precipitant that remains after the raw sewage has been treated chemically or bacterially is called *sludge*.

The chief difference between centrifugal pumps used for raw sewage and those used for sludge is the design of the impeller that is used. To avoid clogging, the impeller used for raw sewage is of an enclosed type, and it is usually wider, with two to four vanes, depending on the size of the pump. The inlet portion of the vanes is usually rounded to reduce resistance to flow; it is shaped to prevent clogging by strings, rags, and paper, which tend to form a wad or ball of material.

The handling of sludge from a sewage treatment plant is more difficult than the handling of raw sewage, because larger quantities of the various solids are present. In addition to a properly designed impeller, the sludge pump is designed with a double-threaded screw in the inlet connection to force the sludge into the impeller. Each thread of the screw connects to an impeller vane. Solids or stringy materials that extend beyond the edges of the screw are cut up as they pass between the edges of the screw and the flues of the screw housing.

When sewage pumps are installed they are usually relatively permanent, therefore the stationary pump parts (casing and base) should be substantial. The vertical type of sewage pump (see Figure 6-12) is designed for convenience in inspection and maintenance. The cutaway view in Figure 6-13 shows how the motor is placed in a watertight enclosure and the ball bearings are lubricated for life. The semi-axial impellers with large free passage ensure blockage-free operation at high efficiencies (see Figure 6-14).

A horizontal-type sewage pump used on lift-station service is shown in Figure 6-15.

## Other Special-Service Pumps

In addition to the chemical, process, and sewage pumps, various other special services require specifically designed pumps designed specifically for that type of service. The sugar-making industry is one example.

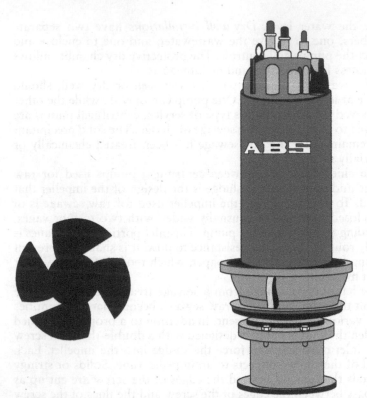

**Figure 6-12** Submersible propeller pump for large volumes of water or sewage. *(Courtesy ABS)*

## Magma Pumps

The term *magma* includes any crude mixture (especially of organic matter) in the form of a thin paste. Therefore, a magma pump is a reciprocating pump designed to move the various heavy confectionery mixtures and non-liquids involved in the sugar-making process. These pumps are designed without inlet valves—the liquid flows by gravity to the pump (negative lift), so the function of the inlet valves is performed by the piston of the reciprocating pump.

## Sump Pumps

A sump pump (see Figure 6-16) is not a sewage pump. The liquid that is pumped is not as thick as sewage, and it is relatively free from

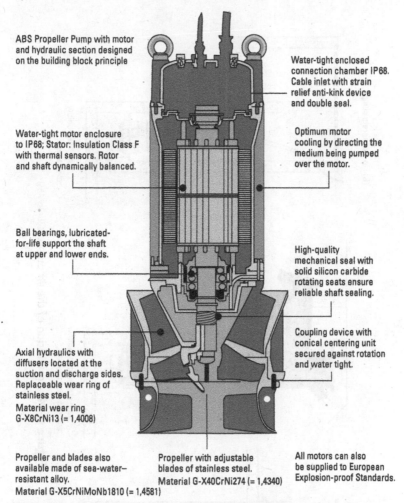

ABS Propeller Pump with motor and hydraulic section designed on the building block principle

Water-tight enclosed connection chamber IP68. Cable inlet with strain relief anti-kink device and double seal.

Water-tight motor enclosure to IP68; Stator: Insulation Class F with thermal sensors. Rotor and shaft dynamically balanced.

Optimum motor cooling by directing the medium being pumped over the motor.

Ball bearings, lubricated-for-life support the shaft at upper and lower ends.

High-quality mechanical seal with solid silicon carbide rotating seats ensure reliable shaft sealing.

Coupling device with conical centering unit secured against rotation and water tight.

Axial hydraulics with diffusers located at the suction and discharge sides. Replaceable wear ring of stainless steel.
Material wear ring G-X8CrNi13 (= 1,4008)

Propeller and blades also available made of sea-water-resistant alloy.
Material G-X5CrNiMoNb1810 (= 1,4581)

Propeller with adjustable blades of stainless steel.
Material G-X40CrNi274 (= 1,4340)

All motors can also be supplied to European Explosion-proof Standards.

**Figure 6-13**  Cutaway view of the submersible lift pump. *(Courtesy ABS)*

foreign matter. A *sump* is a cistern or reservoir constructed at a low point. The water that accumulates is drainage water.

These pumps are often used to remove excess drainage water from non-waterproof basements that have become flooded during periods of heavy rainfall. The sump pump is entirely automatic in its action. Since the pump is submerged, it does not require priming. The motor

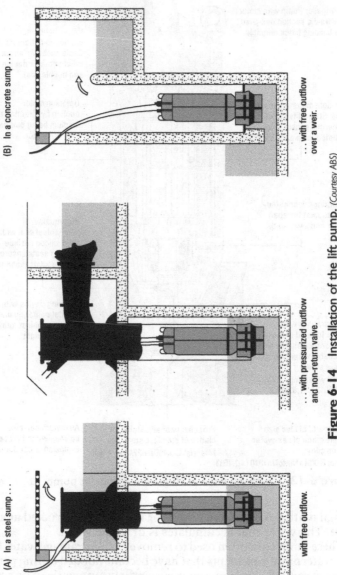

**Figure 6-14** Installation of the lift pump. *(Courtesy ABS)*

(A) In a steel sump...
with free outflow.

...with pressurized outflow
and non-return valve.

(B) In a concrete sump...
...with free outflow
over a weir.

**Figure 6-15** A horizontal type of centrifugal pump for sewage.
*(Courtesy Buffalo Forge Company)*

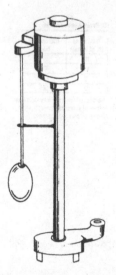

**Figure 6-16** Sump pump used to remove water from a sump or pit. It starts automatically when the water level reaches a certain point, and it stops automatically when the sump is emptied.
*(Courtesy Gould Pumps)*

is controlled by a float that is arranged to start the motor automatically when the water level in the pit or sump reaches a given point. The motor is stopped automatically when the sump is emptied.

Sump pumps are designed to operate intermittently and usually seasonally. They should be tested before the rainy season starts. The pump can be placed directly on the bottom of a poly or fiberglass sump basin, or a concrete sump bottom. If the bottom is packed gravel, the stones must be larger than ¹/₂ inch (13 mm) in diameter and the pump should be placed on bricks for support (see Figures 6-17 and 6-18).

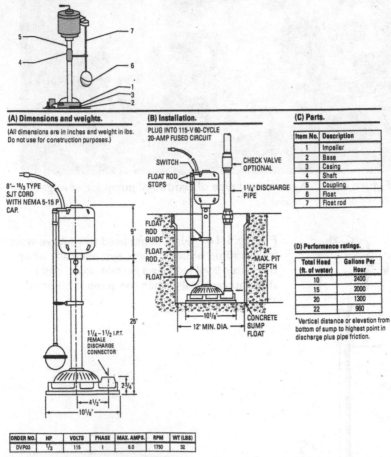

**Figure 6-17** General specifications for sump pumps. *(Courtesy Gould Pumps)*

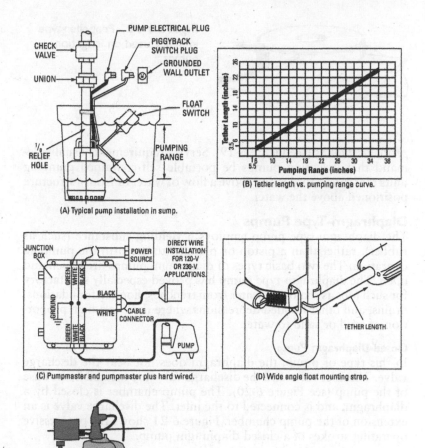

**Figure 6-18** Installation of a sump pump. *(Courtesy Gould Pumps)*

## Irrigation Pumps

The pumps used in irrigation service are designed for large capacities and low heads. These pumps are also used for land drainage, flood control, storm water disposal, and so on. Since the impeller resembles a marine type of impeller, the pumps are sometimes called

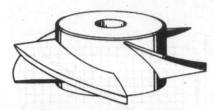

**Figure 6-19** Propeller-type impeller used on irrigation pumps.

*propeller pumps* (see Figure 6-19). Service requirements usually demand that irrigation pumps be portable self-contained pumping units that can be suspended above a flow of water or from a structure positioned above the water.

## Diaphragm-Type Pumps

The diaphragm-type pump employs a yielding substance (such as rubber), rather than a piston or plunger, to perform the pumping operation. The two basic types of diaphragm pumps are *closed* and *open*. The diaphragm-type pump has proved especially satisfactory for such jobs as removing water from trenches, flooded foundations, drains, and other flooded depressions where there is a high proportion of mud or sand to water.

### Closed-Diaphragm Pump

In this type of pump, the diaphragm does not bear the discharge valve. The inlet valve and the discharge valve are located in the base of the pump (see Figure 6-20). The pump chamber is closed by a diaphragm, and is connected to the inlet. The discharge valve is an extension of the pump chamber. Figure 6-21 shows the progressive operating strokes of a closed-diaphragm pump.

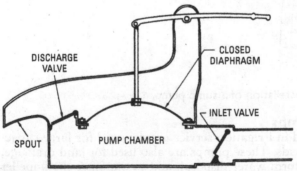

**Figure 6-20** Basic parts of a closed-diaphragm pump.

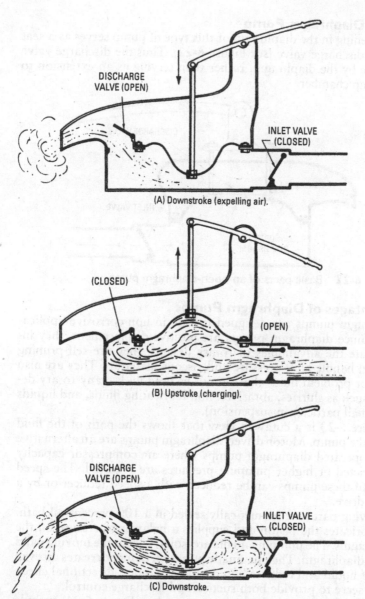

(A) Downstroke (expelling air).

(B) Upstroke (charging).

(C) Downstroke.

**Figure 6-21** Progressive operating strokes of a closed-diaphragm pump.

## Open-Diaphragm Pump

The opening in the diaphragm of this type of pump serves as a seat for the discharge valve (see Figure 6-22). Thus the discharge valve is borne by the diaphragm, rather than serving as an extension to the pump chamber.

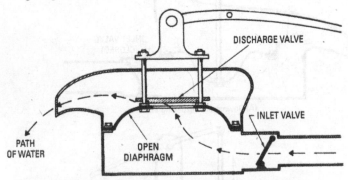

**Figure 6-22**  Basic parts of an open-diaphragm pump.

## Advantages of Diaphragm Pumps

Diaphragm pumps are designed for use in non-corrosive applications. Since diaphragm pumps are positive displacement, they incorporate the advantages of rotary designs: they are self-priming and can handle high-pressure and viscous materials. They are also ideal for problem fluids that cannot be used with many rotary designs (such as slurries, abrasives, non-lubricating fluids, and liquids with small particles in suspension).

Figure 6-23 is a cutaway view that shows the path of the fluid inside the pump. Motor-driven diaphragm pumps are an alternative to air-operated diaphragm pumps where air compressor capacity is restricted or higher pumping pressures are required. The speed (rpm) of these pumps can be reduced with a speed reducer or by a pulley drive.

Moving parts are hermetically sealed in a 100-percent oil bath. Oil lubricates the piston and supplies a balancing pressure on the diaphragms. The pumping media are isolated from the moving parts by the diaphragm. The pneumatic pulsation damper creates an even flow of liquid and reduces liquid surge. Two non-directional check valves serve to provide both suction and discharge control.

Design features of diaphragm pumps (see Figure 6-24) include the following:

- No metal-to-metal contact of pumping mechanism
- Elimination of shaft seal

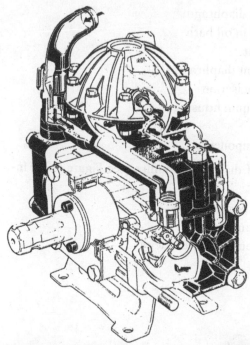

**Figure 6-23**  Diaphragm pump cutaway view showing flow path.
(Courtesy Sherwood)

**Figure 6-24**  Diaphragm pump that delivers 5.3 gallons per minute at 275 psi. (Courtesy Sherwood)

- Mechanically driven diaphragm
- Moving parts sealed in oil bath
- Positive displacement
- Rugged wear-resistant diaphragms
- Built-in pneumatic pulsation
- Epoxy-coated aluminum housing
- Semi-hydraulic
- Commonality of components

Operational benefits of diaphragm pumps (see Figure 6-25) include the following:

- No deterioration of performance
- Runs dry without damage
- Not limited to air-compressor capacity
- Long reliable life
- Self-priming

**Figure 6-25** Diaphragm pump that delivers 59.4 gallons per minute at 250 psi. *(Courtesy Sherwood)*

- Low repair cost, minimum maintenance
- Quiet operation
- Lightweight
- Higher pressures
- Reduced spare-parts stocking

## Shallow-Well and Deep-Well Pumps

Figure 6-26 shows a *shallow-well* pump installation. The working level of the water in the well should be less than 22 feet below the inlet opening of the pump.

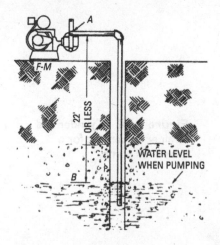

**Figure 6-26**   Shallow-well pump installation.

Figure 6-27 shows *deep-well* construction details. In successful deep-well pump installations, the cylinder is placed at least 5 feet below the working level of the water in the well. A strainer should be fitted to the bottom of the cylinder (see Figure 6-28). The deep-well jet-type pump is illustrated in Figure 6-29.

In the submersible deep-well pump illustrated in Figure 6-30, the pump-motor unit is completely submerged in the well. The pump cannot freeze or lose its prime. This type of pump does not require a pit or well house, and it can easily be lowered if the water level in the well should drop.

Figure 6-31 is a chart on ten-gallon-per-minute submersible pumps showing the total head in feet and gallons per minute. Note the most efficient operating range for each pump. Table 6-1 is a selection table for 10-gpm pumps showing the depth of water level in feet and the capacities in gallons per hour. Figures 6-32(a) and 6-32(b)

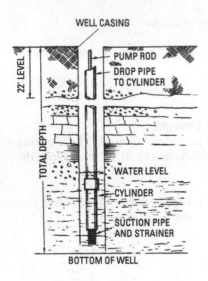

**Figure 6-27** Deep-well construction details.

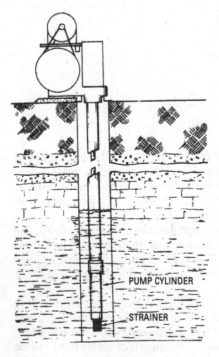

**Figure 6-28** Cylinder and strainer placement for a deep-well pump.

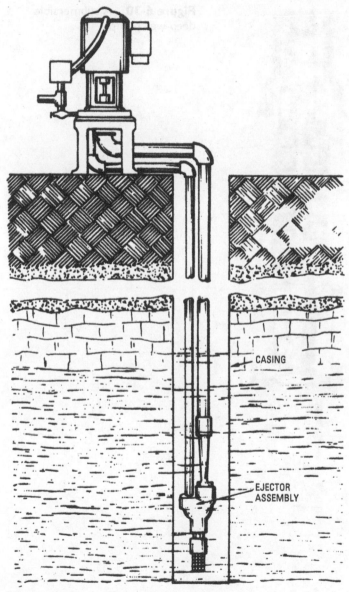

CASING

EJECTOR
ASSEMBLY

**Figure 6-29**   A deep-well jet-type pump.

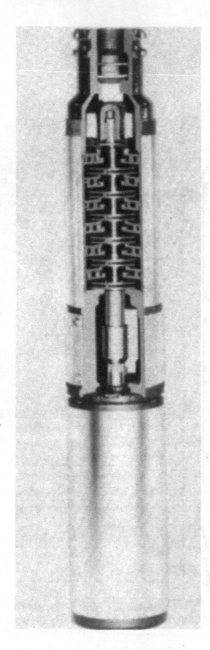

**Figure 6-30**    A submersible deep-well pump.

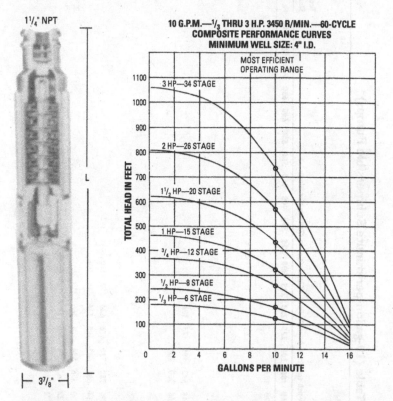

**1¼" NPT**

L

3⅞"

10 G.P.M.—⅓ THRU 3 H.P. 3450 R/MIN.—60-CYCLE
COMPOSITE PERFORMANCE CURVES
MINIMUM WELL SIZE: 4" I.D.

MOST EFFICIENT
OPERATING RANGE

3 HP—34 STAGE

2 HP—26 STAGE

1½ HP—20 STAGE

1 HP—15 STAGE

¾ HP—12 STAGE

½ HP—8 STAGE

⅓ HP—6 STAGE

TOTAL HEAD IN FEET

GALLONS PER MINUTE

| PUMP TYPE | 3BA6 | | | 5BA8 | | | 7BA12 | |
|---|---|---|---|---|---|---|---|---|
| MTR. TYPE | 1 PH 3 WIRE | 1 PH-C/S 2 WIRE | 1 PH-S/P 2 WIRE | 1 PH 3 WIRE | 1 PH-C/S 2 WIRE | 1 PH-S/P 2 WIRE | 1 PH 3 WIRE | 1 PH-C/S 2 WIRE |
| LENGTH | 23 | 35 | 32 | 25 | 37 | 33 | 29 | 42 |
| WEIGHT | 26 | 43 | 40 | 29 | 44 | 41 | 33 | 50 |

| PUMP TYPE | 10BA15 | | 15BA20 | | 20BA26 | | 30BA34 | |
|---|---|---|---|---|---|---|---|---|
| MTR. TYPE | 1 PH 3 WIRE | 1 PH-C/S 2 WIRE | 1 PH 3 WIRE | 3 PH 3 WIRE | 1 PH 3 WIRE | 3 PH 3 WIRE | 1 PH 3 WIRE | 3 PH 3 WIRE |
| LENGTH | 33 | 45 | 43 | 41 | 55 | 53 | 68 | 62 |
| WEIGHT | 37 | 52 | 60 | 60 | 78 | 71 | 103 | 92 |

**Figure 6-31**  Composite performance curves—minimum well size 4" ID. *(Courtesy Flint & Walling, Inc.)*

## Table 6-1 Selection Table (10-gallon-per-minute Submersible Pumps)

| Pump Type | H.P | Stages | Discharge Pressure PSI | \*Depth to Water Level in Feet—Capacities in Gallons per Hour | | | | | | | | | | | | | | | | | | | | | | | | | Total Shut Off Ft | Total Shut Off PSI |
|---|---|---|---|---|---|---|---|---|---|---|---|---|---|---|---|---|---|---|---|---|---|---|---|---|---|---|---|---|---|---|
| | | | | 20 | 40 | 60 | 80 | 100 | 120 | 140 | 160 | 180 | 200 | 225 | 250 | 275 | 300 | 325 | 350 | 400 | 450 | 500 | 550 | 600 | 650 | 700 | 800 | 900 | | |
| 3BA6 | 1/3 | 6 | 0 | 900 | 840 | 786 | 720 | 636 | 540 | 420 | 192 | | | | | | | | | | | | | | | | | | | 190 | 82 |
| | | | 20 | 828 | 765 | 708 | 606 | 510 | 396 | 290 | | | | | | | | | | | | | | | | | | | | | |
| | | | 40 | 702 | 588 | 480 | 330 | | | | | | | | | | | | | | | | | | | | | | | | |
| | | | 60 | 450 | 240 | | | | | | | | | | | | | | | | | | | | | | | | | | |
| | | | 80 | | | | | | | | | | | | | | | | | | | | | | | | | | | | |
| 5BA8 | 1/2 | 8 | 0 | 836 | 846 | 882 | 852 | 804 | 756 | 702 | 648 | 582 | 495 | 375 | 126 | | | | | | | | | | | | | | 250 | 108 |
| | | | 20 | 876 | 846 | 798 | 750 | 696 | 630 | 570 | 468 | 360 | 204 | | | | | | | | | | | | | | | | | | |
| | | | 40 | 768 | 720 | 660 | 600 | 522 | 414 | 288 | | | | | | | | | | | | | | | | | | | | | |
| | | | 60 | 654 | 594 | 516 | 402 | 270 | | | | | | | | | | | | | | | | | | | | | | | | |
| | | | 80 | 468 | 294 | 216 | 270 | | | | | | | | | | | | | | | | | | | | | | | | | |
| 7BA12 | 3/4 | 12 | 0 | 924 | 894 | 930 | 900 | 876 | 846 | 816 | 786 | 750 | 720 | 678 | 624 | 582 | 510 | 444 | 342 | | | | | | | | | | 375 | 162 |
| | | | 20 | 858 | 828 | 870 | 834 | 804 | 780 | 744 | 714 | 684 | 642 | 588 | 534 | 444 | 386 | | | | | | | | | | | | | | |
| | | | 40 | 792 | 758 | 798 | 768 | 744 | 684 | 660 | 630 | 576 | 534 | 474 | 366 | 210 | | | | | | | | | | | | | | | |
| | | | 60 | 714 | 684 | 726 | 690 | 654 | 618 | 576 | 522 | 462 | 396 | 228 | | | | | | | | | | | | | | | | | |
| | | | 80 | | | 636 | 600 | 558 | 504 | 444 | 372 | 228 | | | | | | | | | | | | | | | | | | | |

Submersible pump performance data table:

| Model | HP | | Setting | Performance values (head vs. capacity) | | Spec |
|---|---|---|---|---|---|---|
| 10BA15 | 1 | 15 | 0 | 924 906 882 858 834 816 786 756 720 684 648 612 558 444 228 | | 470 203 |
| | | | 20 | 918 900 870 846 828 804 780 756 726 696 654 612 576 510 456 276 | | |
| | | | 40 | 894 870 846 822 798 774 738 708 672 654 618 570 516 474 390 270 | | |
| | | | 60 | 834 810 792 762 738 708 672 654 618 582 534 468 396 300 | | |
| | | | 80 | 786 756 726 702 672 642 606 582 540 492 408 318 156 | | |
| 15BA20 | 1½ | 20 | 0 | 924 912 900 882 864 846 828 804 774 756 732 708 648 588 510 288 234 | 510 438 162 | 630 272 |
| | | | 20 | 906 894 876 858 846 828 804 780 756 732 708 684 654 594 516 420 246 | | |
| | | | 40 | 900 888 879 852 834 816 804 762 744 690 660 630 600 516 426 264 | | |
| | | | 60 | 846 834 816 792 774 762 738 720 690 666 635 606 570 528 438 288 | | |
| | | | 80 | 810 774 756 732 708 684 660 630 606 576 540 492 444 294 | | |
| 20BA26 | 2 | 26 | 0 | 924 912 906 900 888 876 858 852 840 828 816 804 786 768 744 708 666 624 570 438 162 | | 815 352 |
| | | | 20 | 936 924 906 900 882 870 858 834 822 804 786 768 750 708 672 624 576 | | |
| | | | 40 | 930 918 906 875 864 852 840 828 804 786 768 756 732 714 672 630 576 | | |
| | | | 60 | 918 900 876 864 852 834 816 804 774 758 738 714 696 678 650 630 582 | | |
| | | | 80 | 882 870 858 846 828 816 804 792 780 756 738 720 702 684 650 630 582 | | |
| 30BA34 | 3 | 34 | 0 | 930 918 900 888 876 864 852 846 840 834 816 810 786 756 726 702 666 636 558 450 | | 1070 482 |
| | | | 20 | 930 918 900 888 876 864 852 840 828 816 804 780 762 738 708 678 630 600 510 408 | | |
| | | | 40 | 924 912 900 888 876 864 852 846 840 834 828 816 804 792 756 732 702 678 642 606 576 456 336 216 | | |
| | | | 60 | 924 912 900 888 876 864 852 846 840 834 828 816 804 786 768 750 726 708 678 636 612 570 522 408 | | |
| | | | 80 | 918 900 888 876 864 852 846 840 834 822 816 804 786 768 750 732 714 684 648 612 570 516 450 342 | | |

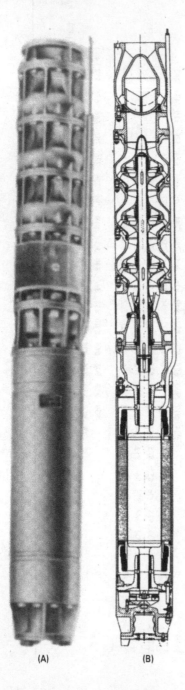

**Figure 6-32** Typical submersible pump motor with cross-sectional view. *(Courtesy Pleuger Submersible Pumps, Inc.)*

(A)          (B)

show another submersible pump. Figure 6-32(b) shows the cross section of the pump in Figure 6-32(a). This pump has a flange-type connection whereas the one in Figure 6-30 has a threaded connection. Figure 6-33 shows a cross-section and names the parts of the electric motor that is used in Figure 6-32.

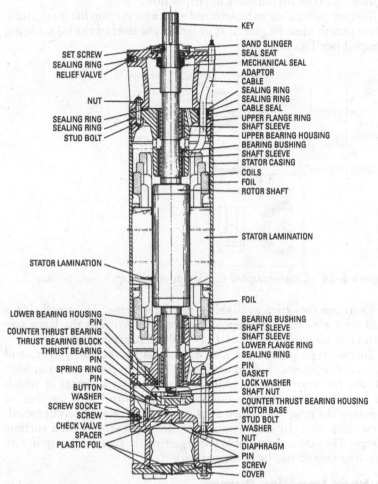

**Figure 6-33** Names of various parts of submersible pump motor shown in Figure 6-32.

Turbine pumps are designed for low flow at medium pressures. Since turbines operate without metal-to-metal contact, they are an excellent selection for pumping non-lubricating fluids at higher pressures, especially those applications requiring continuous operation of the pump for extended periods.

Corrosive or abrasive liquids should not be pumped with a turbine. Nor is the turbine design suitable for viscous fluids or for liquids that contain particles in suspension.

Turbine pumps are to be selected where no suction lift is required. These pumps must be placed at or below the level of the liquid being pumped (see Figure 6-34).

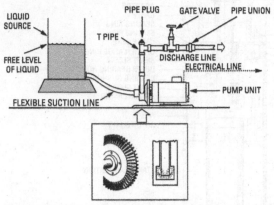

**Figure 6-34** Close-coupled turbine pump set-up. *(Courtesy Sherwood)*

There are two sizes available—small and large. The small size is used with a $1/2$ to $3/4$ horsepower electric motor and the large size pump is used with motors of 1, $1^1/2$, and 2 horsepower.

Turbine-type pumps are also known as *vortex*, *periphery*, and *regenerative pumps*. Liquid is whirled by the impeller vanes at high velocity for nearly one revolution in an annular channel in which the impeller turns. Energy is added to the liquid in a number of impulses; the insert in Figure 6-34 shows how this is accomplished. Some deep-well diffuser-type pumps are sometimes called turbine pumps. They do not resemble the regenerative turbine pump in any way, and should not be confused with it.

## Rubber Impeller Pumps

Rubber impeller pumps are designed for low flow at low pressure (see Figures 6-35 and 6-36). One part rotation without metal-to-metal contacts makes rubber impeller pumps suitable for non-lubricating liquid, mild abrasives, and fluids with small particles

**Figure 6-35** A rubber impeller pump with a bronze body, cover, and housing. Note the stainless-steel shaft. *(Courtesy Sherwood)*

**Figure 6-36** A rubber impeller pump for general low pressure transfer mounted on a motor. *(Courtesy Sherwood)*

in suspension. These pumps are self-priming, but *should never be run dry.*

## Principles of Operation

Pumping action is developed through a flexing or bending of a rubber impeller by an offset cam (see Figure 6-37). As each compressed impeller vane rotates from the cam, it draws liquid from the inlet and displaces it in a squeezing action out the discharge port. This cycle produces a continuous uniform flow of liquid.

Impeller size, cam thickness, rotation speed, and pressure are all variables in determining both pump flow characteristics and pump life span.

## Marine Applications

The raw water pump is the heart of the marine engine, supplying a continuous flow of seawater to maintain temperatures consistent

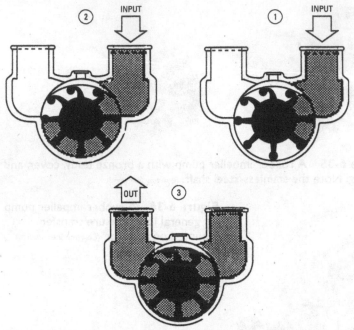

**Figure 6-37**   Pumping action of a rubber impeller pump. *(Courtesy Sherwood)*

with engine cooling requirements. The high demanding service of marine engine coolant applications is best accomplished by the flexible rubber impeller-type pump. The following unique combination of features makes a flexible rubber impeller pump the dominant type:

- Self-priming
- One part rotation without metal-to-metal contact
- Reversibility
- The ability to handle abrasive or corrosive liquid (see Figure 6-38)

## Tubing Pumps

A tubing pump is self-priming, pumps abrasives, runs dry without harm, handles viscous material, pumps without shear, and pumps sterile solutions. It can operate with a blocked suction, but cannot operate against a blocked discharge.

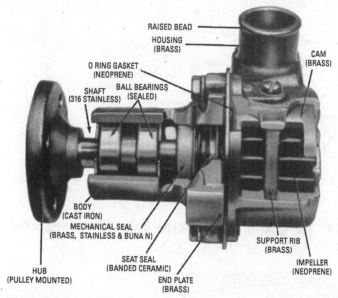

RAISED BEAD
HOUSING
(BRASS)
CAM
(BRASS)
O RING GASKET
(NEOPRENE)
SHAFT
(316 STAINLESS)
BALL BEARINGS
(SEALED)
BODY
(CAST IRON)
MECHANICAL SEAL
(BRASS, STAINLESS & BUNA N)
SUPPORT RIB
(BRASS)
HUB
(PULLEY MOUNTED)
SEAT SEAL
(BANDED CERAMIC)
END PLATE
(BRASS)
IMPELLER
(NEOPRENE)

**Figure 6-38**  A cutaway view of a typical brass fitted marine engine coolant pump. It shows design features and component materials.
*(Courtesy Sherwood)*

## Basic Principle of Operation

The basic principle of operation for a tubing pump can be demonstrated by placing a soft rubber or plastic tube on a hard, flat surface (see Figure 6-39). With a rolling pin, pinch the tube closed at line *A-A*, then, roll the rolling pin toward the line *B-B*. As the rolling pin is pushed forward, any fluid in the tube (liquid or gas) is pushed along in front of it. Behind the rolling pin, the tube springs back into its original shape, that creates a vacuum. Atmospheric pressure pushes the fluid in to fill the vacuum. As the rolling pin approaches line *B-B*, start a second rolling pin at *A-A*. This will allow the pump to stay in operation when the first rolling pin is removed.

By placing the tubing in a U shape slightly larger than 180 degrees, you can use two rollers to seal the tube against a rigid, concentric housing of slightly more than 180 degrees (see Figure 6-40). As the first roller goes around, the fluid in front of it gets pushed out. As the tube springs back, atmospheric pressure pushes more fluid into the tube. Before the first roller leaves the sealed condition, the second roller starts to squeeze the tubing tight. Now you have a rotary pump.

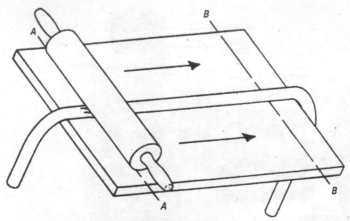

**Figure 6-39** Tubing pump basic principles same as tube being rolled with rolling pin. *(Courtesy TAT Engineering)*

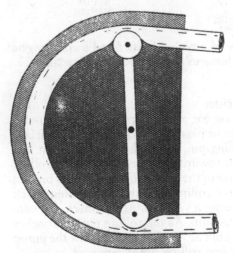

**Figure 6-40** Roller squeezing action. *(Courtesy TAT Engineering)*

Since the pump is a rolling valve that is always closed, what occurs on the suction side of the pump has no influence on what occurs on the discharge side of the pump, and vice versa. Clearly, a tubing pump cannot be any better than the tubing it employs. If the tubing has a maximum pressure limitation of 40 psi that is the pressure limit for the pump in which it is used. The ability of the tube to rebound under various conditions will determine its suction lift. If, for example, the tubing comes back to shape to draw a near-perfect vacuum, the pump will be self-priming to 33 feet at sea level. Conversely, if the tubing cannot come back to shape at a suction lift of 15 feet with water, that is the pump's limit.

## Lubricating the Tubing

Abrading of the tubing is not crucial, particularly if the tubing is lubricated with silicon grease or some other lubricant. Tubing life depends upon how well the compound withstands repeated flexing. Since the number of rollers in a pump determines the number of flexes per revolution, it is best to keep the number of rollers at a minimum. One roller would be ideal, except that the mechanical configuration makes it expensive to build and cumbersome to change the tubing, so two rollers becomes the optimum for long life. More than two rollers means, that you flex the tubing more per revolution. The only advantage of the multi-roller pump is the ability to use less precision in the manufacture because you have more than one roller squeezing the tube all the time.

## Variable Speed

The constant kneading of the rollers against the tubing causes the tubing to heat up. Generally, if the pump speeds exceed 200 rpm, the tubing becomes heated to a point at which it tends to fail rapidly. Most pumps come with a speed-reducing gear train to deliver greatly increased tubing life.

## Types Available

There are various types of tubing pumps available. Flow rates from $1/4$ cc to nearly 60 gallons per hour are available with suction lifts to 28 feet, vacuums to 30 inches of mercury. Discharge pressures are up to 40 psi and 100 psi with special tubing. Viscosities handled can be up to 10,000 centipoises.

Heads may be equipped with $1/8$-, $1/4$-, $3/8$-, or $1/2$-inch inside diameter (ID) extruded tubing and operated up to 200 rpm. Table 6-2 shows basic flow rates for various sizes of tubing.

In the tubing pump, only the tubing contacts the chemicals. Therefore, the proper tubing has to be selected for the type of chemical to be pumped. Tables 6-3 and 6-4 show the tubing materials available and the types of chemicals that can be handled by each type.

Figures 6-41 and 6-42 show the heavy-duty high-vacuum tubing pump head. At 200 rpm it can pump 438.1 gallons per hour with $3/4$-inch ID tubing.

## Table 6-2   Basic Flow Rates

| Tubing ID | CC/Revolution |
| --- | --- |
| $1/8$ in. | 1.6 |
| $1/4$ in. | 6.0 |
| $3/8$ in. | 11.0 |
| $1/2$ in. | 18.0 |

## Table 6-3   Chemical Resistance Data for TAT Industrial and Chemical Pumps

| Chemical | Extruded Tubing |
| --- | --- |
| Acetic Acid | Tb, Tr, S, H, Nr |
| Acetic Anhydride | N, H |
| Acetone | Nr |
| Aluminum Sulfate | Tb, Tr, N, G, B, H, U |
| Alums | Tb, Tr, N |
| Ammonia | Tr, S, N, U |
| Ammonium Chloride | Tb, Tr, N, B, H, V |
| Ammonium Hydroxide | Tr, S, U, N, H, V, U |
| Ammonium Sulfate | Tb, Tr, N, G, B, H, U |
| Amyl Acetate | V |
| Aniline | Nr |
| ASTM Fuels | U, V |
| Barium Carbonate | Tb, Tr, N, B, U |
| Barium Hydroxide | Tb, Tr, N, G, B, H, U |
| Beet Sugar Liquors | Tb, Tr, N, G, B, H |
| Black Sulfate Liquors | Tb, Tr, N, B, H |
| Bleaching Liquors | Tb, Tr, H |
| Boric Acid | Tb, Tr, N, G, B, H, U |
| Bromine, Anhydrous | V, U |
| Butane | N, B, H, U |
| Butyl Alcohol | Tb, Tr, N, G, B, H, U |
| Cadmium Cyanide | N, H |
| Calcium Acetate | B |
| Calcium Chloride | Tb, Tr, N, G, B, H, U |
| Calcium Hypochloride | Tb, Tr, H, Nr |
| Carbon Bisulfide | Nr, V |
| Carbon Tetrachloride | V |

*(continued)*

## Table 6-3  (continued)

| Chemical | Extruded Tubing |
|---|---|
| Chlorine Gas, Wet | Tr, V |
| Chloracetic Acid | N, H |
| Chlorobenzene | V |
| Chloroform | V |
| Chromic Acid | Tb, Tr, H, Nr |
| Citric Acid | G, H, U |
| Copper Cyanide | Nr |
| Copper Chloride | Tb, Tr, N, B, H |
| Copper Sulfate | Tb, Tr, N, G, B, H |
| Cottonseed Oil | Tb, Tr, U, N, V |
| Creosote | B, V |
| Developer Solution | Tb, Tr, N, B |
| Dibutyl Phosphate | Tb, Tr, N, B |
| Diethyl Phthalate | Ts, B |
| Diesel Fuels | B, U |
| Diethyl Amine | Nr |
| Diethylene Glycol | Nr |
| Ethyl Alcohol | Tb, Tr, S, N, G, H, V |
| Ethylene Glycol | Tb, Tr, N, B, H, V, U |
| Ferric Chloride | Tb, Tr, S, B, H, U |
| Ferric Sulfate | Tb, Tr, N, G, B, H, U |
| Fluoboric Acid | Tb, Tr, G, H |
| Fluosilicic Acid | Tb, Tr, N, G, H |
| Formaldehyde (70°) | Tb, Tr, N, B, H, V, U |
| Formic Acid | Tb, Tr, N, H |
| Freon 11, 12, 22, 113, 114 | H, V |
| Fuel Oil | Ts, U, B, V |
| Furfarat | Nr |
| Gasoline | Ts, B, V, U |
| Glucose | Tb, Tr, N, G, B, H |
| Glycerine | Tb, Tr, U, N, G, B, H |
| Hydraulic Oils | Ts, N, B, H, V, U |
| Heptane | U, N, B, H |
| Hydrochloric Acid (Dil.) | Tb, Tr, N, H, V, U |
| Hydrochloric Acid (Con.) | Tr, H, U |
| Hydrofluoric Acid | Tr, H, V |

(continued)

## Table 6-3 (continued)

| Chemical | Extruded Tubing |
|---|---|
| Hydrogen Peroxide (Dil.) | Tb, Tr, H, Nr, U |
| Hydrogen Peroxide (Con.) | Nr, V, U |
| Hydrogen Sulfide | Tb, Tr, N, H |
| Isopropyl Alcohol | N, H |
| JP-4, 5, -6 (Jet Fuel) | V |
| Kerosene | Ts, U, B, V |
| Lactic Acid | Tr, N, H, U |
| Lemon Oil | B |
| Lime Slurry | N |
| Linseed Oil | Tb, Tr, U, N, B |
| Lubricating Oils | Ts, U, B, V |
| Magnesium Chloride | Tb, Tr, N, G, B, H, U |
| Magnesium Hydroxide | Tr, N, G, B, N, U |
| Mercuric Chloride | Tr, B, H |
| Mercury | Tb, Tr, U, N, B, H |
| Methyl Alcohol | Tb, Tr, S, N, B, H, U |
| Methyl Ethylketone | Nr |
| Mineral Oil | U, N, B |
| Mixed Acids | V |
| Naptha | U, V |
| Nickel Acetate 10 percent | N |
| Nickel Chloride | Tb, Tr, N, B |
| Nickel Nitrate | Tb, Tr, N |
| Nitric Acid 30 percent | Tb, Tr, H, Nr |
| Nitric Acid 60 percent | Tr, Nr |
| Nitric Acid 70 percent | V |
| Nitrobenzene | Nr |
| Oleic Acid | Ts, V, Nr, U |
| Oleum, 25 percent | Nr, V |
| Olive Oil (80°F) | N |
| Palmitic Acid | B, Nr |
| Perchloroethylene | V |
| Phenol | S, Nr, V |
| Phosphoric Acid | Tb, Tr, S, N, H, V |

(continued)

## Table 6-3 (continued)

| Chemical | Extruded Tubing |
|---|---|
| Pickling Bath Sulfuric-Nitric | H, V |
| Pickling Bath Nitric-Hydrofluoric | H, V |
| Plating Solutions Brass, Cadmium, Copper, Lead, Gold, Tin, Nickel, Silver, Zinc | Tr, N, V |
| Plating Sol. (Chrome) | Tr, H |
| Picric Acid | N, B, H |
| Potassium Dichromate | H, U |
| Potassium Hydroxide | Tr, N, B, H, U |
| Propyl Alcohol | Tb, Tr, N, B, H |
| Pyridine | Nr |
| Skydrol 500 | Nr |
| Soap Solution | Tb, Tr, S, N, B, H |
| Sodium Carbonate | Tb, Tr, S, N, G, B, U |
| Sodium Chloride | Tb, Tr, S, N, G, B, U |
| Sodium Dichromate | Nr, U |
| Sodium Hydroxide | Tr, S, N, G, U |
| Sodium Hypochlorite | Tb, Tr, H, Nr, V |
| Sodium Thiosulfate | N, G, B |
| Soybean Oil | N, V |
| Stannic Chloride | G, Nr |
| Sulfuric Acid to 50 percent | Tb, Tr, N, H, V |
| Sulfuric Acid over 50 percent | H, V |
| Sulfurous Acid | Tb, Tr, H, Nr |
| Sulfuric Acid Foaming 20 percent Oleum | V |
| Sulfur Trioxide | Nr |
| Tannic Acid, 10 percent | Tb, Tr, N, H |
| Tartaric Acid | Tb, Tr, N, G, H |
| Trichloroethylene | V |
| Triethanolamine | N, B, H |
| Trinitrotoluene | N |
| Tung Oil | N, B |
| Ticresyl Phosphate | Nr, V |
| Turpentine | Tb, Tr, U, B, V |
| Zinc Chloride | Tb, Tr, H, U |
| Zinc Sulfate | Tb, Tr, N, G, B, H, U |

*Note:* This table indicates recommended elastomer and hose pipe materials for specific chemical solutions. Special approved compositions are available for foods, beverages and bio-chemicals.

## Table 6-4 Chemical Resistance Data for TAT Industrial and Chemical Pumps

| Chemical | High-Cap Pumps | |
| --- | --- | --- |
| | Tube | Hose Pipes |
| Acetic Acid | H, Nr | SS, Te |
| Acetic Anhydride | N, H | SS, Te |
| Acetone | Nr | SS, Te |
| Aluminum Sulfate | N, B, H, | P, SS, Te |
| Alums | N | P, SS, Te |
| Ammonia | N | SS, Te |
| Ammonium Chloride | N, B, H, V | P, Te |
| Ammonium Hydroxide | N, H, V | P, SS, Te |
| Ammonium Sulfate | N, B, H | P, Te |
| Amyl Acetate | V | Te |
| Aniline | Nr | SS, Te |
| ASTM Fuels | V | SS, Te |
| Barium Carbonate | N, B | SS, P, Te |
| Barium Hydroxide | N, B, H | P, SS, Te |
| Beet Sugar Liquors | N, B, H | P, SS, Te |
| Black Sulfate Liquors | N, B, H | P, SS, Te |
| Bleaching Liquors | H | P, Te |
| Boric Acid | N, B, H | |
| Bromine, Anhydrous | V | Te |
| Butane | N, B, H | P, SS, Te |
| Butyl Alcohol | N, B, H | SS, Te |
| Cadmium Cyanide | N, H | SS, Te |
| Calcium Acetate | B | SS, Te |
| Calcium Chloride | N, B, H | P, Te |
| Calcium Hypochloride | Nr, H | P, Te |
| Carbon Bisulfide | Nr, V | SS, Te |
| Carbon Tetrachloride | V | SS, Te |
| Chlorine Gas, Wet | V | P, Te |
| Chloracetic Acid | N, H | P, Te |
| Chlorobenzene | V | SS, Te |
| Chloroform | V | SS, Te |
| Chromic Acid | Nr, H | Te |
| Citric Acid | H | P, SS |

*(continued)*

**Table 6-4  (continued)**

| Chemical | High-Cap Pumps | |
| --- | --- | --- |
| | Tube | Hose Pipes |
| Copper Cyanide | N, B | P, Te |
| Copper Chloride | N, B, H | P, Te |
| Copper Sulfate | N, B, H | P, SS, Te |
| Cottonseed Oil | N, B, V | P, SS, Te |
| Creosote | B, V | SS, Te |
| Developer Solution | N, B | P, SS, Te |
| Dibutyl Phosphate | Nr | Te |
| Diethyl Phthalate | Nr | Te |
| Diesel Fuels | N, B | SS, Te |
| Diethyl Amine | B | SS, Te |
| Diethylene Glycol | Nr | SS, Te |
| Ethyl Alcohol | N, H, V | P, SS, Te |
| Ethylene Glycol | N, B, H, V | P, SS, Te |
| Ferric Chloride | N, B, H | P, Te |
| Ferric Sulfate | B, H | P, SS, Te |
| Fluoboric Acid | H | P, Te |
| Fluosilicic Acid | N, H | P, Te |
| Formaldehyde (70°) | N, B, H, V | P, SS, Te |
| Formic Acid | N, H | P, SS, Te |
| Freon 11, 12, 22, 113, 114 | H, V | SS, Te |
| Fuel Oil | N, B, NV | P, SS, Te |
| Furfarat | Nr | SS, Te |
| Gasoline | B, V | P, SS, Te |
| Glucose | N, B, H | P, SS, Te |
| Glycerin | N, B, H | P, SS, Te |
| Hydraulic Oils | N, B, H, V | SS, Te |
| Heptane | N, B, H | P, SS, Te |
| Hydrochloric Acid (Dil.) | N, H, V | P, Te |
| Hydrochloric Acid (Con.) | H | P, Te |
| Hydrofluoric Acid | H, V | Te |
| Hydrogen Peroxide (Dil.) | Nr, H | P, SS, Te |
| Hydrogen Peroxide (Con.) | Nr, V | Te |
| Hydrogen Sulfide | N, H | P, SS, Te |

(*continued*)

**Table 6-4 (continued)**

| Chemical | High-Cap Pumps | |
|---|---|---|
| | Tube | Hose Pipes |
| Isopropyl Alcohol | N, H | SS, Te |
| JP-4, 5, -6 (Jet Fuel) | V | SS, Te |
| Kerosene | N, B, V | P, SS, Te |
| Lactic Acid | N, H | Te |
| Lemon Oil | B | SS, Te |
| Lime Slurry | N | Te |
| Linseed Oil | N, B | P, SS, Te |
| Lubricating Oils | N, B, V | P, SS, Te |
| Magnesium Chloride | N, B, H | P, Te |
| Magnesium Hydroxide | N, B, H | P, SS, Te |
| Mercuric Chloride | N, B, H | P, SS, Te |
| Mercury | N, B, H | P, SS, Te |
| Methyl Alcohol | N, B, H | P, SS, Te |
| Methyl Ethylketone | Nr | SS, Te |
| Mineral Oil | N, B | P, SS, Te |
| Mixed Acids | V | Te |
| Naphtha | V | P, SS, Te |
| Nickel Acetate 10 percent | N (100*) | SS, Te |
| Nickel Chloride | N, B | P, Te |
| Nickel Nitrate | N | P, SS, Te |
| Nitric Acid 30 percent | Nr, H, V | SS, Te |
| Nitric Acid 60 percent | Nr, V | SS, Te |
| Nitric Acid 70 percent | V | SS |
| Nitrobenzene | Nr | Te |
| Oleic Acid | Nr | P, SS, Te |
| Oleum, 25 percent | Nr, V | Te |
| Olive Oil 80°F | N | SS, Te |
| Palmitic Acid | B, Nr | P, SS, Te |
| Perchloroethylene | V | SS, Te |
| Phenol | Nr, V | SS, Te |
| Phosphoric Acid | N, H, V | P, Te |
| Pickling Bath Sulfuric-Nitric | V | SS, Te |
| Pickling Bath Nitric-Hydrofluoric | H, V | P, Te |

*(continued)*

## Table 6-4  (continued)

| Chemical | High-Cap Pumps | |
| | Tube | Hose Pipes |
| --- | --- | --- |
| Plating Solutions Brass, Cadmium Copper, Lead Gold, Tin, Nickel, Silver, Zinc | N, V | P, Te |
| Plating Solution Chrome | H | P, Te |
| Picric Acid | N, B, H | SS, Te |
| Potassium Dichromate | H | P, Te |
| Potassium Hydroxide | N, B | P, SS, Te |
| Propyl Alcohol | N, B, H | P, SS, Te |
| Pyridine | Nr | Te |
| Skygdrol 500 | Nr | SS, Te |
| Soap Solution | N, B, H | P, SS, Te |
| Sodium Carbonate | N, B | P, SS, Te |
| Sodium Chloride | N, B, H | P, Te |
| Sodium Dichromate | Nr | P, Te |
| Sodium Hydroxide | N, V | P, SS, Te |
| Sodium Hypochlorite | Nr, H, V | P, Te |
| Sodium Thiosulfate | N, B | P, SS, Te |
| Soybean Oil | N, H, V | SS, Te |
| Stannic Chloride | Nr | P, Te |
| Sulfuric Acid to 50 percent | N, H, V | P, Te |
| Sulfuric Acid over 50 percent | H, V | P (75°)Te |
| Sulfurous Acid | Nr, H | P, Te |
| Sulfuric Acid, Foaming 20 percent Oleum | V | Te |
| Sulfur Trioxide | Nr | P |
| Tannic Acid, 10 percent | N, H | P, SS, Te |
| Tartaric Acid | N, H | P, SS, Te |
| Triethanolamine | N, B, H | P, SS, Te |
| Trinitrotoluene | N | SS, Te |
| Tung Oil | N, B | SS, Te |
| Tricresyl Phosphate | Nr, V | Te |
| Turpentine | B, V | P, SS, Te |
| Xylene | V | SS, Te |
| Zinc Chloride | H | P, Te |
| Zinc Sulfate | N, B, H | P, SS, Te |

*Note:* This table indicates recommended elastomer and hose pipe materials for specific chemical solutions.

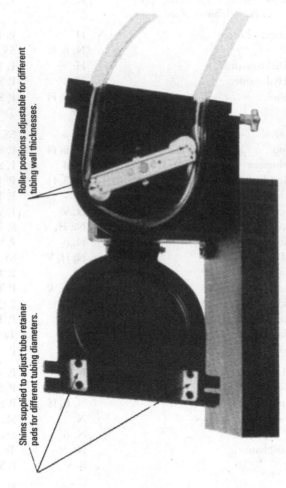

**Figure 6-41** Pump head opens for easy changing of the tubing size. *(Courtesy TAT Engineering)*

Roller positions adjustable for different tubing wall thicknesses.

Shims supplied to adjust tube retainer pads for different tubing diameters.

**Figure 6-42** Heavy duty high-vacuum pump head. *(Courtesy TAT Engineering)*

The rotary hand pump reverses rotation for returning excess fluid or to drain pump. Figure 6-43(a) shows the hand pump and Figure 6-43(b) shows the pump mounted on a plastic drum adaptor for safe, easy transfer of corrosive liquids.

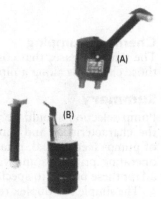

### Air Driven Pump
The pump head can also be driven by an air motor (see Figure 6-44). The air motor driven pump is shown with air gage and adjustable regulator for variable speeds and an on/off switch. The capacity of this type of pump is up to 450 gallons per hour and 5 feet suction lift with 10 psi discharge pressure.

**Figure 6-43** (a) Hand-powered rotary capacity pump. (b) Steel and plastic drum adapter for the hand-operated pump.

*(Courtesy TAT Engineering)*

### Multi-Tube Pumps
Figure 6-45 illustrates the high vacuum pump that employs four extruded tubes. Some pumps have used as many as seven tubes. Adding tubes is a versatile way to gain capacity without increasing pump speed, to pump different fluids at the same time, or to pump solutions in fixed ratios—all in one pump.

**Figure 6-44** High capacity air motor driven pump. *(Courtesy TAT Engineering)*

**Figure 6-45** Multi-tube pump. *(Courtesy TAT Engineering)*

## Chemical Pumping

The fluids or gases that contact the pump limit its ability to move those chemicals along a piping arrangement.

## Summary

Pump selection should be based on a thorough understanding of the characteristics and fundamental principles of the basic types of pumps (centrifugal, rotary, reciprocating, and so on). The basic operating principles and various design characteristics or features adapt these pumps to specialized or unusual service conditions.

The simplex or duplex reciprocating piston-type pumps are well-adapted for the service requirements found in tanneries, sugar refineries, bleacheries, and so on. They are used for high-pressure service on water supply systems in country clubs, dairies, and industrial plants.

The general-purpose rotary gear-type pumps are designed to handle either thick or thin liquids, and they are designed to operate smoothly in either direction of rotation with equal efficiency. The thicker liquids (such as roofing materials and printing inks), as well as fuel oils, gasolines, and similar thin liquids can be handled by these pumps. The rotary gear-type pump is also adapted to pressure lubrication, hydraulic service, fuel supply, or general transfer work (which includes the pumping of clean liquids).

The self-priming motor-mounted centrifugal pump has several applications, including lawn sprinklers, swimming pools, booster service, recirculation, irrigation, dewatering, sump and bilge, liquid fertilizers, and chemical solutions. Centrifugal pumps are also adapted to solids-handling applications (such as sanitary waste, sewage lift stations, treatment plants, industrial waste, general drainage, sump service, industrial process service, food processing, and chemical plants).

The chemical and processing industries employ centrifugal pumps extensively in handling a wide variety of corrosive and abrasive liquids. The service life of pumps has been extended substantially by the use of rubber-lined pump parts for handling various liquids used in the chemical and paper industries. Pyrex or glass pumps are used to pump acids, milk, fruit juices, and other acid solutions. Therefore, the liquid being pumped is not contaminated by the chemical reactions between the liquid and the material in the pump parts.

Centrifugal pumps are also used to pump either raw sewage or sludge. The handling of sludge from a sewage treatment plant is more difficult than handling raw sewage, because larger quantities of various solids are present. The impeller is designed differently for handling raw sewage and sludge (depending on which is to be handled).

Magma pumps are designed to handle crude mixtures (especially of organic matter) that are in the form of a thin paste. For example, in the sugar making process, heavy confectionery mixtures and non-liquids are involved. These pumps are designed without inlet valves. The valves are not needed, because the liquid flows by gravity to the pump, and their function is performed by the piston of the reciprocating pump.

Other special service pumps include sump pumps, irrigation pumps, diaphragm-type pumps (which may be either closed-diaphragm or open-diaphragm pumps), and shallow-well or deep-well pumps (including the jet-type and submersible pumps). Turbine pumps and rubber impeller pumps are also finding wide use in industry and in marine applications.

## Review Questions

1. What factors determine pump selection?
2. What are the typical applications for reciprocating piston-type pumps?
3. What are the typical applications for centrifugal pumps?

4. What types of pumps are suitable for the chemical and the processing industry?

5. What design feature is important in the selection of pumps for raw sewage and sludge?

6. What is the chief factor in determining whether to select a shallow-well or a deep-well pump?

7. What type of liquids can the rotary gear-type pump handle?

8. What type of pump would you select for a swimming pool?

9. What types of pumps are used widely in the paper industry?

10. What does the term *magma* mean?

11. Where does the sump pump operate?

12. What types of pumps are utilized in irrigation systems?

13. Where are rubber impeller pumps used?

14. What are some advantages of diaphragm pumps?

15. Where are turbine pumps used?

# Part III

# Hydraulics

# Chapter 7

# Hydraulic Accumulators

An *accumulator* is a storage device that stores liquid under pressure, so that it can serve as a reservoir and regulator of power. A simple accumulator is sometimes used in a household water system. It usually consists of a tee with a side branch pipe that is capped. The air that is trapped in the side branch pipe is compressed, and then acts like a compressed spring. As a faucet is opened or closed quickly, a sudden change in pressure and flow occurs. The trapped air acts as a cushion (or shock absorber) to prevent water hammering in the piping system.

In industrial hydraulics, the accumulator is an important component in many hydraulic systems. A *hydraulic accumulator* is designed to accumulate energy that is to be expended intermittently, which enables the pump, operating under uniform load, to meet an intermittent or fluctuating demand for power. However, there are a number of other uses for accumulators in industry.

The hydraulic accumulator can be compared to the more common storage battery in that energy is stored until it is needed. The hydraulic accumulator stores *hydraulic energy*, and the storage battery (simply called the *accumulator* in England) stores electrical energy.

## Basic Construction and Operation

Following are the essential parts (see Figure 7-1) of the weight-loaded accumulator:

- Plunger or ram
- Cylinder
- Weights

In the accumulator, a plunger is placed within a vertical cylinder that is closed at the lower end and provided with a stuffing box at the upper end. Weights are secured in position at the upper end of the plunger to produce the desired pressure. An outlet and an inlet for the liquid are located at the lower end of the cylinder.

As shown in Figure 7-1, water is forced into the cylinder by means of the force pump, which causes the weighted plunger to rise. The force or pressure from the weighted plunger on the water is transmitted to the machines operated by the system, as long as the plunger has not reached the lower end of its stroke.

During periods when there is no demand for power, the plunger is prevented from rising too high and leaving the cylinder by means of stops that arrest the motion of the plunger when it arrives at the

317

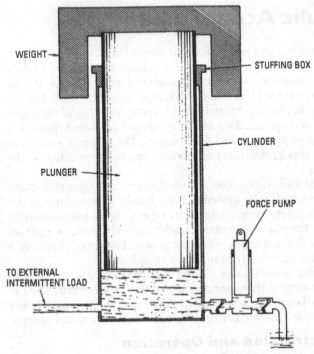

WEIGHT

STUFFING BOX

CYLINDER

PLUNGER

FORCE PUMP

TO EXTERNAL
INTERMITTENT LOAD

**Figure 7-1** The essential parts of a weight-loaded hydraulic accumulator.

upper end of its stroke. If there is a demand for power and the load is small, the force pump can supply the required quantity of water by continuing to operate. However, if the load is large and more power is demanded than the pump can supply, a portion of the power is supplied by the slow descent of the plunger inside the accumulator. When the load is removed or power is no longer demanded, the force pump continues to operate, gradually filling the cylinder and causing the plunger to rise to the upper end of its stroke. Thus, a quantity of energy equal to the quality of energy expended in the descent of the plunger is accumulated, or stored, in the cylinder.

## Types of Accumulators
Following are the three general types of hydraulic accumulators:

- Weight-loaded
- Spring-loaded
- Air- or gas-type

## Weight-Loaded

In the weight-loaded accumulator (see Figure 7-2), the dead weight on the plunger or ram may be cast iron, steel, concrete, water, or other heavy material. The weight-loaded accumulator may be either a direct or an inverted type (see Figure 7-3). In the direct type of accumulator, the plunger is movable; in the inverted type, the cylinder is movable.

**Figure 7-2**  Weight-loaded accumulators. The weighted-tank type of accumulator (left) and the cast iron weighted type of accumulator (right) store hydraulic fluid under constant pressure during periods of off-peak demand and return it to the system to meet peak demands with minimum pump capacity.

In the larger direct-type accumulators, the plunger may be provided with a yoke at its upper end. Two rods are suspended in the ends of the yoke, and the required number of weights (ring-type weights) is threaded onto the rods (see Figure 7-4). This type of accumulator may be provided with a 24-inch plunger, and it may be weighted to develop 600 pounds of pressure per square inch. The accumulator is self-contained. Therefore, it requires no frame or guideposts. A desirable feature is that the packing is readily accessible at the top of the cylinder. An undesirable feature of the inverted type of accumulator is that it is difficult to adjust or renew the packing in the stuffing box at the lower end of the cylinder.

Stops or lugs are provided to prevent the plunger from over-traveling its stroke (see Figure 7-5). As shown in Figure 7-5, four stops or lugs are provided in the lower end of the cylinder, and similar lugs are provided on the inner portion of the upper end of the cylinder to prevent plunger over-travel. Since both the plunger

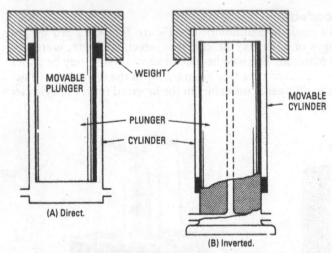

MOVABLE
PLUNGER

WEIGHT

MOVABLE
CYLINDER

PLUNGER

CYLINDER

(A) Direct.

(B) Inverted.

**Figure 7-3** The direct type (a) and the inverted type (b) of weight-loaded accumulator. The plunger is movable in the direct type, and the cylinder is movable in the inverted type.

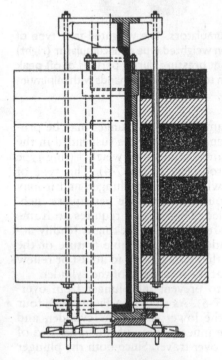

**Figure 7-4** Construction details of a weight-loaded direct-type accumulator. Ringed weights are threaded onto the two rods suspended at the ends of the yoke at the upper end of the hydraulic cylinder.

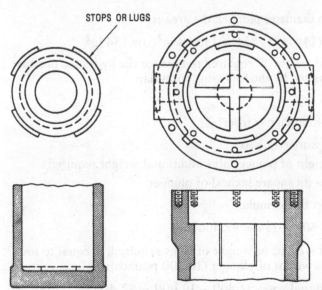

**Figure 7-5**  Construction details of the direct type of accumulator, showing lugs on the lower end of the plunger (left) and on the upper end of the cylinder (right), which prevent over-travel of the plunger.

and the cylinder are provided with lugs or stops, when the plunger is inserted into the cylinder, it is necessary to turn the plunger until the lugs are in position to pass. Then the plunger is turned (one-eighth turn) and fastened in position. Therefore, the ends of the lugs register, to prevent over-travel by the plunger. Suitable provision is made to prevent turning of the yoke.

**Weight Required**
A number of cast iron rings can be used as weights to develop the required hydraulic pressure (see Figure 7-4). Calculations for determining the weight required to develop a given hydraulic pressure are illustrated in the following problem.

**Problem**
In a direct-type accumulator, the 14-inch diameter plunger weighs 10,000 pounds. What additional weight is required to develop a hydraulic pressure of 600 psi?

**Solution**
The cross-sectional area of the plunger can be found by the following formula:

$$a = d^2 \times 0.7854$$

For a 14-inch diameter plunger, the area is:

$$a = (14)^2 \times 0.7854 = 153.94 \text{ in}^2, \text{ or } 154 \text{ in}^2$$

Then the total weight $W$ required to balance the hydraulic pressure can be determined by the following formula:

$$W = Pa$$

in which the following is true:

$P$ is pressure required

$W$ is weight of plunger plus additional weight required

$a$ is area (in square inches) of plunger

Substituting in the formula:

$$W = 600 \times 154 = 92,400 \text{ lb}$$

Additional weight, or weight of rings required, is equal to total load $W$ minus weight of plunger (10,000 pounds):

$$\text{additional wt} = 92,400 - 10,000 = 82,400 \text{ lb}$$

**Plunger Dimensions Required**
In the design of weight-loaded accumulators, it is necessary to determine the dimensions of the plunger or ram for a given set of conditions. It should be noted that less weight is required to balance the hydraulic pressure as the diameter of the plunger decreases. Also, if adequate vertical space is available, less weight is required as the length of the plunger stroke increases.

If one of the dimensions of the plunger is fixed, the other dimension can be determined for a given displacement (volume displaced by the plunger per stroke), as shown in the following problem.

**Problem**
If the diameter of the plunger is 14 inches, what length of stroke is required for the plunger in an accumulator having a displacement of 250 gallons?

**Solution**
Since 1 gallon of water displaces a volume of 231 cubic inches, the total displacement is:

$$\text{displacement} = 250 \times 231 = 57,750 \text{ in}^3$$

The plunger area is:

$$(14)^2 \times 0.7854 = 154 \text{ in}^2 (\text{approximately})$$

The formula for determining length of stroke is:

$$\frac{\text{displacement}}{\text{area of plunger} \times 12} \text{ (for length of stroke, ft)}$$

Substituting in the formula:

$$\text{length of stroke} = \frac{57,750}{154 \times 12} = 31.25 \text{ ft}$$

A *pressure booster* or *intensifier* (see Figure 7-6), or differential type of accumulator, is a hydraulic device that is used to convert low pressure to high pressure. Its function is similar to that of an electrical transformer that converts low-voltage current to high-voltage current.

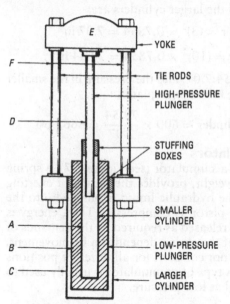

**Figure 7-6** Basic construction of a pressure booster or intensifier, which is a differential type of accumulator. The intensifier is a hydraulic device that is used to convert a liquid under low pressure to a liquid under high pressure.

Labels in figure: YOKE, TIE RODS, HIGH-PRESSURE PLUNGER, STUFFING BOXES, SMALLER CYLINDER, LOW-PRESSURE PLUNGER, LARGER CYLINDER

To operate an intensifier, water is supplied at low pressure to a piston in a large cylinder that, in turn, operates a ram in a smaller cylinder. As shown in Figure 7-6, the basic construction of the intensifier consists of a smaller cylinder and a larger cylinder. The smaller cylinder *A* is part of the plunger *B* that fits inside the larger cylinder *C*. The smaller high-pressure plunger *D* fits inside the smaller cylinder *A*. The low-pressure plunger *B* acts as a stationary ram for the smaller cylinder *A*. The latter plunger is attached to the yoke *E* which is, in turn, attached to the larger cylinder C by means of the tie rods F.

In actual operation, the force pump (not shown in Figure 7-6) forces the low-pressure plunger B upward, thereby exerting pressure on the water in the smaller cylinder A that is acted upon by the smaller plunger D. This pressure is intensified in the smaller cylinder A in proportion to the ratio of cross-sectional areas of the two plungers, as shown in the following calculations.

**Problem**

If the diameter of the smaller plunger is 3 inches and the diameter of the larger plunger is 10 inches, what is the pressure in the smaller cylinder if the pressure in the larger cylinder is 600 psi?

**Solution**

The areas of the smaller and the larger cylinders are:

$$\text{Area smaller cylinder} = (3)^2 \times 0.7854 = 7.07 \text{ in}^2$$

$$\text{Area larger cylinder} = (10)^2 \times 0.7854 = 78.54 \text{ in}^2$$

Then the plunger ratio is 78.54:7.07, and the pressure in the smaller cylinder can be found by the following equation:

$$\text{pressure in small cylinder} = 600 \times \frac{78.54}{7.07} = 6665 \text{ psi}$$

## Spring-Loaded Accumulators

In the spring-loaded type of accumulator (see Figure 7-7), a spring (or springs), rather than a weight, provides the means of exerting pressure on the liquid. As the hydraulic liquid is pumped into the accumulator, the plunger or piston is compressed. Thus, energy is stored in the spring, and it is released as required by the demands of the system. Since the force of the spring depends on its movement, the pressure on the liquid is not constant for all different positions of the plunger or piston. This type of accumulator is usually used to deliver small quantities of oil at low pressure.

## Air- or Gas-Type Accumulators

Water and the other liquids that are used in hydraulic systems are nearly incompressible. This fact means that a large increase in hydraulic pressure decreases the volume of the liquid only slightly. On the other hand, a large increase in air or gas pressure results in a large decrease in the volume of the air or gas. Relatively speaking, the hydraulic liquids are less elastic (or spring-like) than air or gas, which indicates that the liquids cannot be used effectively to store energy. Therefore, this type of accumulator uses air or gas (instead of a weight or a mechanical spring) to provide the spring-like action.

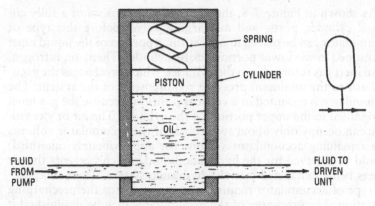

**Figure 7-7**  Basic parts of a spring-loaded accumulator.

Air or gas accumulators can be divided into two types: *nonseparator* and *separator*.

**Nonseparator Type**

In this type of accumulator (see Figure 7-8), a so-called free surface exists between the liquid and the air or gas. As an increased quantity of liquid is pumped into the accumulator, the air or gas above the liquid is compressed still further. The energy stored in the compressed air or gas is released as it is needed to meet the requirements of the system.

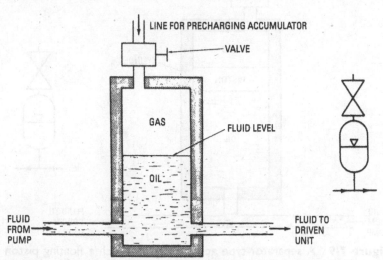

**Figure 7-8**  A nonseparator-type air or gas accumulator.

**326** Chapter 7

As shown in Figure 7-8, the accumulator consists of a fully enclosed cylinder, ports, and a charging valve. Before this type of
accumulator can be placed in operation, a portion of the liquid must
be trapped in the lower portion of the cylinder. Then, air, nitrogen,
or an inert gas is forced into the cylinder, which precharges the accumulator to the minimum pressure requirements of the system. The
accumulator is mounted in a vertical position, because the gas must
be retained in the upper portion of the cylinder. The air or gas volume can occupy only about two-thirds of the accumulator volume.
The remaining accumulator volume (or approximately one-third)
should be reserved for the hydraulic liquid, which prevents the air
or gas being exhausted into the hydraulic system. The nonseparator type of accumulator requires a compressor for the precharging
operation. The precharge of the accumulator may be diminished if
aeration (mixing) of the liquid and the air or gas occurs. If the liquid
absorbs the air or gas, the accumulator does not function properly.

### Separator Type
A freely floating piston serves as a barrier between the liquid and
the air or gas in the separator type of accumulator (see Figure 7-9).

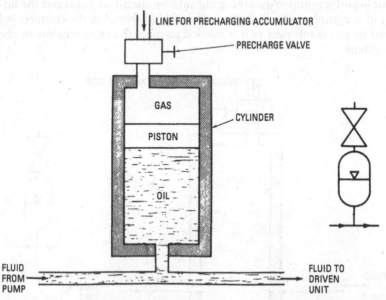

**Figure 7-9** A separator-type accumulator in which a floating piston
separates the air or gas and the working hydraulic fluid.

The piston and packing separate the liquid and the air or gas in the cylinder. The accumulator is precharged with high-pressure air or gas on one side of the piston, and the hydraulic liquid or fluid on the opposite side.

Figure 7-10 shows the cross section of a piston-type accumulator. Note the seals that are employed for high pressures. This type of accumulator may be mounted in any position that is convenient to meet the needs of the installation. If the accumulator is to be mounted on the shell, it should not be clamped in such a way that the clamps prohibit thermal expansion, or cause distortion to the shell. Always mount the accumulator so that the gas valve is readily available.

Table 7-1 shows an accumulator-sizing chart for a 1-gallon accumulator.

**Figure 7-10** Piston-type accumulator.

(Courtesy Parker-Hannifin Corp.)

# Table 7-1 Accumulator Sizing Table

| Gas Precharge Pressure — psi (gage) | Operating Pressure—psi (gage) | | | | | | | | | | | | | | |
|---|---|---|---|---|---|---|---|---|---|---|---|---|---|---|---|
| | 100 | 200 | 300 | 400 | 500 | 600 | 700 | 800 | 900 | 1000 | 1100 | 1200 | 1300 | 1400 | 1500 |
| 100 | | 95.6 | 136.0 | 159.0 | 174.0 | 185.0 | 193.0 | 200.0 | 205.0 | 209.0 | 212.0 | 216.0 | 219.0 | 221.0 | 223.0 |
| | | 124.0 | 169.0 | 192.0 | 207.0 | 216.0 | 223.0 | 228.0 | | | | | | | |
| 200 | | | 63.3 | 99.0 | 123.0 | 140.0 | 152.0 | 162.0 | 171.0 | 178.0 | 183.0 | 188.0 | 192.0 | 196.0 | 199.0 |
| | | | 84.5 | 128.0 | 155.0 | 173.0 | 186.0 | 195.0 | 203.0 | 289.0 | 215.0 | 219.0 | 222.0 | 225.0 | 228.0 |
| 300 | | | | 47.8 | 78.6 | 101.0 | 116.0 | 131.0 | 141.0 | 150.0 | 158.0 | 164.0 | 170.0 | 175.0 | 179.0 |
| | | | | 65.9 | 104.0 | 130.0 | 147.0 | 163.0 | 175.0 | 183.0 | 191.0 | 197.0 | 203.0 | 207.0 | 211.0 |
| 400 | | | | | 37.8 | 64.9 | 85.4 | 101.0 | 114.0 | 125.0 | 134.0 | 142.0 | 149.0 | 155.0 | 160.0 |
| | | | | | 51.5 | 86.6 | 112.0 | 131.0 | 145.0 | 157.0 | 167.0 | 175.0 | 182.0 | 188.0 | 194.0 |
| 500 | | | | | | 31.7 | 55.4 | 74.0 | 88.8 | 101.0 | 112.0 | 121.0 | 129.0 | 136.0 | 142.0 |
| | | | | | | 43.4 | 74.4 | 97.9 | 116.0 | 131.0 | 143.0 | 153.0 | 162.0 | 159.0 | 175.0 |
| 600 | | | | | | | 27.1 | 48.1 | 64.8 | 79.5 | 91.6 | 102.0 | 111.0 | 119.0 | 126.0 |
| | | | | | | | 37.2 | 65.1 | 87.0 | 105.0 | 119.0 | 131.0 | 142.0 | 150.0 | 158.0 |
| 700 | | | | | | | | 23.9 | 42.6 | 58.5 | 71.8 | 83.4 | 93.5 | 102.0 | 110.0 |
| | | | | | | | | 33.0 | 58.0 | 78.5 | 97.9 | 109.0 | 121.0 | 131.0 | 148.0 |
| 800 | | | | | | | | | 21.0 | 38.6 | 53.0 | 65.5 | 76.6 | 84.6 | 86.2 |
| | | | | | | | | | 29.0 | 52.5 | 71.3 | 87.5 | 101.0 | 113.0 | 123.0 |
| 900 | | | | | | | | | | 19.1 | 34.8 | 48.2 | 60.3 | 70.5 | 80.1 |
| | | | | | | | | | | 26.3 | 47.6 | 65.4 | 80.9 | 94.1 | 105.0 |
| 1000 | | | | | | | | | | | 17.3 | 30.6 | 44.7 | 56.0 | 65.6 |
| | | | | | | | | | | | 23.6 | 43.6 | 60.6 | 75.4 | 87.8 |
| 1100 | | | | | | | | | | | | 15.7 | 29.5 | 41.2 | 52.2 |
| | | | | | | | | | | | | 23.5 | 48.4 | 58.4 | 70.5 |
| 1200 | | | | | | | | | | | | | 14.6 | 27.4 | 38.6 |
| | | | | | | | | | | | | | 20.5 | 37.8 | 52.6 |
| 1300 | | | | | | | | | | | | | | 13.6 | 25.5 |
| | | | | | | | | | | | | | | 18.9 | 35.1 |
| 1400 | | | | | | | | | | | | | | | 12.8 |
| | | | | | | | | | | | | | | | 17.6 |
| 1500 | | | | | | | | | | | | | | | |
| 1600 | | | | | | | | | | | | | | | |
| 1700 | | | | | | | | | | | | | | | |
| 1800 | | | | | | | | | | | | | | | |
| 1900 | | | | | | | | | | | | | | | |
| 2000 | | | | | | | | | | | | | | | |
| 2100 | | | | | | | | | | | | | | | |
| 2200 | | | | | | | | | | | | | | | |

(Courtesy Parker-Hannifin Corporation

1-gallon size (231-cubic-inch capacity) performance table (adiabatic and isothermal)
Maximum Gas Capacity (piston against hydraulic cap)—266 in$^3$
Maximum Oil Capacity (piston against gas cap)—231 in$^3$
Adiabatic—Light Type
Isothermal—Bold Type

| 1600 | 1700 | 1800 | 1900 | 2000 | 2100 | 2200 | 2300 | 2400 | 2500 | 2600 | 2700 | 2800 | 2900 | 3000 | Gas Precharge Pressure—psi (gage) |
|---|---|---|---|---|---|---|---|---|---|---|---|---|---|---|---|
| 225.0 | 226.0 | 228.0 | 230.0 | | | | | | | | | | | | 100 |
| 203.0 | 206.0 | 208.0 | 210.0 | 212.0 | 214.0 | 215.0 | 217.0 | 218.0 | 219.0 | 221.0 | 222.0 | 223.0 | 224.0 | 225.0 | 200 |
| 231.0 | | | | | | | | | | | | | | | |
| 183.0 | 186.0 | 190.0 | 192.0 | 195.0 | 197.0 | 199.0 | 201.0 | 203.0 | 205.0 | 207.0 | 208.0 | 210.0 | 211.0 | 213.0 | 300 |
| 214.0 | 217.0 | 220.0 | 222.0 | 224.0 | 226.0 | 228.0 | 230.0 | | | | | | | | |
| 164.0 | 169.0 | 173.0 | 177.0 | 179.0 | 182.0 | 185.0 | 187.0 | 190.0 | 192.0 | 194.0 | 196.0 | 198.0 | 200.0 | 201.0 | |
| 198.0 | 202.0 | 205.0 | 208.0 | 211.0 | 213.0 | 216.0 | 218.0 | 220.0 | 222.0 | 224.0 | 225.0 | 227.0 | 228.0 | 229.0 | |
| 147.0 | 152.0 | 156.0 | 161.0 | 164.0 | 168.0 | 172.0 | 175.0 | 177.0 | 180.0 | 182.0 | 184.0 | 186.0 | 188.0 | 190.0 | 500 |
| 181.0 | 186.0 | 190.0 | 194.0 | 198.0 | 201.0 | 204.0 | 207.0 | 209.0 | 212.0 | 213.0 | 215.0 | 217.0 | 219.0 | 220.0 | |
| 132.0 | 139.0 | 142.0 | 147.0 | 151.0 | 155.0 | 159.0 | 162.0 | 164.0 | 168.0 | 171.0 | 173.0 | 176.0 | 178.0 | 180.0 | 600 |
| 165.0 | 171.0 | 175.0 | 180.0 | 185.0 | 188.0 | 192.0 | 195.0 | 198.0 | 201.0 | 203.0 | 205.0 | 208.0 | 210.0 | 212.0 | |
| 117.0 | 123.0 | 129.0 | 134.0 | 138.0 | 142.0 | 146.0 | 150.0 | 154.0 | 157.0 | 160.0 | 163.0 | 165.0 | 168.0 | 170.0 | 700 |
| 148.0 | 155.0 | 161.0 | 167.0 | 171.0 | 176.0 | 180.0 | 184.0 | 187.0 | 190.0 | 193.0 | 196.0 | 198.0 | 201.0 | 203.0 | |
| 102.0 | 109.0 | 115.0 | 121.0 | 126.0 | 131.0 | 135.0 | 138.0 | 142.0 | 146.0 | 149.0 | 152.0 | 155.0 | 158.0 | 160.0 | 800 |
| 131.0 | 139.0 | 147.0 | 153.0 | 158.0 | 163.0 | 168.0 | 172.0 | 176.0 | 180.0 | 183.0 | 186.0 | 189.0 | 192.0 | 194.0 | |
| 88.3 | 95.7 | 102.0 | 109.0 | 114.0 | 119.0 | 124.0 | 128.0 | 132.0 | 136.0 | 140.0 | 143.0 | 146.0 | 149.0 | 153.0 | 900 |
| 115.0 | 124.0 | 132.0 | 139.0 | 145.0 | 151.0 | 156.0 | 161.0 | 165.0 | 169.0 | 173.0 | 176.0 | 179.0 | 182.0 | 185.0 | |
| 74.7 | 82.6 | 89.9 | 96.5 | 103.0 | 108.0 | 113.0 | 118.0 | 122.0 | 126.0 | 130.0 | 134.0 | 136.0 | 140.0 | 143.0 | 1000 |
| 96.9 | 109.0 | 117.0 | 125.0 | 132.0 | 138.0 | 144.0 | 149.0 | 154.0 | 158.0 | 163.0 | 167.0 | 170.0 | 173.0 | 176.0 | |
| 61.7 | 70.0 | 77.6 | 84.8 | 91.2 | 97.4 | 103.0 | 108.0 | 112.0 | 116.0 | 120.0 | 124.0 | 128.0 | 131.0 | 135.0 | 1100 |
| 82.4 | 93.1 | 103.0 | 111.0 | 119.0 | 126.0 | 132.0 | 138.0 | 143.0 | 146.0 | 152.0 | 157.0 | 161.0 | 164.0 | 167.0 | |
| 48.9 | 57.5 | 65.9 | 73.4 | 80.3 | 81.6 | 92.2 | 97.8 | 103.0 | 107.0 | 112.0 | 116.0 | 120.0 | 123.0 | 126.0 | 1200 |
| 66.3 | 79.5 | 88.0 | 97.4 | 106.0 | 113.0 | 120.0 | 127.0 | 132.0 | 138.0 | 143.0 | 147.0 | 152.0 | 156.0 | 159.0 | |
| 35.9 | 45.2 | 54.3 | 62.2 | 69.6 | 76.2 | 82.1 | 88.0 | 93.0 | 98.0 | 103.0 | 107.0 | 111.0 | 114.0 | 118.0 | 1300 |
| 49.2 | 61.7 | 73.1 | 83.2 | 92.5 | 101.0 | 108.0 | 115.0 | 121.0 | 127.0 | 132.0 | 137.0 | 142.0 | 146.0 | 150.0 | |
| 23.9 | 34.0 | 43.1 | 51.1 | 59.0 | 65.9 | 72.5 | 78.5 | 83.6 | 89.0 | 93.8 | 98.0 | 103.0 | 107.0 | 110.0 | 1400 |
| 33.0 | 46.6 | 58.5 | 69.2 | 79.3 | 88.3 | 96.2 | 103.0 | 110.0 | 116.0 | 122.0 | 127.0 | 132.0 | 136.0 | 141.0 | |
| 11.7 | 22.4 | 31.9 | 40.7 | 48.9 | 56.0 | 62.8 | 69.2 | 75.0 | 80.3 | 85.6 | 89.8 | 94.6 | 98.5 | 103.0 | 1500 |
| 16.5 | 30.9 | 43.9 | 55.3 | 66.0 | 75.6 | 81.4 | 92.0 | 98.9 | 105.0 | 112.0 | 117.0 | 123.0 | 128.0 | 132.0 | |
| | 11.2 | 21.3 | 30.3 | 38.6 | 46.2 | 53.2 | 59.8 | 66.0 | 71.5 | 76.6 | 82.1 | 86.7 | 91.1 | 95.2 | 1600 |
| | 15.7 | 29.5 | 41.7 | 52.6 | 62.7 | 71.8 | 80.3 | 88.0 | 94.9 | 101.0 | 108.0 | 113.0 | 118.0 | 124.0 | |
| | | 10.4 | 20.0 | 28.7 | 36.4 | 44.2 | 50.8 | 57.5 | 63.4 | 68.6 | 73.9 | 79.0 | 83.3 | 87.7 | 1700 |
| | | 14.6 | 27.7 | 39.6 | 50.0 | 80.1 | 68.6 | 77.1 | 84.5 | 92.2 | 97.8 | 104.0 | 109.0 | 114.0 | |
| | | | 9.9 | 19.1 | 27.4 | 34.8 | 42.3 | 48.9 | 54.8 | 60.6 | 66.0 | 71.0 | 75.6 | 80.3 | 1800 |
| | | | 13.8 | 26.6 | 37.7 | 47.8 | 57.7 | 65.9 | 73.6 | 81.2 | 88.2 | 94.4 | 100.0 | 106.0 | |
| | | | | 9.3 | 18.3 | 26.3 | 33.5 | 40.4 | 46.6 | 52.7 | 58.0 | 63.3 | 68.4 | 72.9 | 1900 |
| | | | | 13.0 | 25.3 | 36.2 | 48.0 | 55.0 | 63.4 | 71.3 | 78.0 | 84.9 | 91.2 | 96.8 | |
| | | | | | 9.1 | 17.5 | 25.0 | 32.5 | 38.8 | 45.2 | 50.8 | 56.0 | 61.2 | 66.0 | 2000 |
| | | | | | 12.5 | 24.2 | 34.6 | 44.4 | 52.9 | 61.2 | 68.6 | 75.2 | 81.9 | 88.0 | |
| | | | | | | 8.5 | 16.5 | 23.9 | 30.6 | 37.2 | 43.1 | 48.9 | 54.0 | 59.1 | 2100 |
| | | | | | | 12.0 | 22.9 | 33.0 | 42.0 | 50.6 | 58.7 | 66.0 | 72.8 | 79.0 | |
| | | | | | | | 8.3 | 16.0 | 23.2 | 29.5 | 35.9 | 41.8 | 47.1 | 52.4 | 2200 |
| | | | | | | | 11.7 | 22.1 | 31.9 | 40.7 | 49.0 | 56.6 | 63.8 | 70.7 | |
| | | | | | | | | 8.0 | 15.4 | 22.1 | 28.5 | 34.1 | 40.2 | 45.2 | 2300 |
| | | | | | | | | 11.2 | 21.3 | 30.6 | 39.2 | 47.1 | 54.7 | 61.7 | |
| | | | | | | | | | 7.7 | 14.6 | 21.3 | 27.4 | 33.3 | 38.8 | 2400 |
| | | | | | | | | | 10.6 | 20.5 | 29.2 | 37.8 | 45.8 | 52.9 | |
| | | | | | | | | | | 7.2 | 14.1 | 20.5 | 26.6 | 32.5 | 2500 |
| | | | | | | | | | | 10.2 | 19.7 | 28.5 | 36.7 | 44.4 | |
| | | | | | | | | | | | 6.9 | 13.6 | 19.7 | 25.5 | 2600 |
| | | | | | | | | | | | 9.8 | 18.9 | 27.2 | 35.4 | |
| | | | | | | | | | | | | 6.7 | 13.3 | 18.9 | 2700 |
| | | | | | | | | | | | | 9.6 | 18.3 | 26.3 | |

## Note

Pressure change of the gas is inversely proportional to its change in volume. When 100 cu. in. of gas, originally at 1000 psia (pounds per square inch absolute), is compressed to 50 cu. in. volume, pressure will be 2000 psia if gas temperature is kept constant. This is *isothermal* performance.

Compression and expansion of the gas cause heating and cooling which increase and decrease pressure in addition to the effect of volume change. If gas were perfectly insulated to prevent giving up any of this extra heat to or through the metal in which it is contained for picking up heat when cooled, performance would be *adiabatic*. Here, 100 cu. in. of gas originally at 1000 psia is compressed to 61.2 cu. in. to build up to 2000 psia. Therefore, less oil can enter the accumulator. The single-wall, all-metal construction of the Parker accumulator transmits heat rapidly to diminish this effect.

Actual performance will lie between isothermal and adiabatic. Rapid operation would approach adiabatic figures; slow operation would approach isothermal. The isothermal figures are usually employed; allowance as usually made for reserve capacity will be adequate to include effect of temperature changes resulting from compression and expansion.

In the double-shell construction of the cylindrical accumulator shown in Figure 7-11, the pressure-balanced inner shell contains a floating position piston and serves as a separator between the

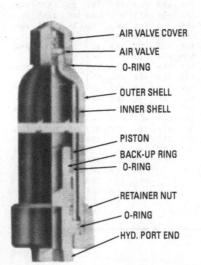

AIR VALVE COVER
AIR VALVE
O-RING

OUTER SHELL
INNER SHELL

PISTON
BACK-UP RING
O-RING

RETAINER NUT

O-RING

HYD. PORT END

**Figure 7-11** Cutaway view of a piston-type hydraulic accumulator.
*(Courtesy Superior Hydraulics, Division of Superior Pipe Specialties)*

**Figure 7-12** Bladder-type hydraulic accumulator.

pre-charged air or gas and the hydraulic fluid. The outer shell serves as a gas container. A coolant for the working area of the inner shell is provided by rapid decompression of the pre-charged air or gas, resulting from rapid discharge of the working hydraulic fluid.

The *bladder-* or *bag-type* of accumulator (see Figure 7-12) should be installed with the end that contains the air or gas at the top, so that the hydraulic fluid is not trapped when discharging. A gas valve that opens into the shell is located at one end of the shell. A plug assembly containing an oil port and a poppet valve is mounted on the opposite end of the shell.

As shown in Figure 7-13, the bladder or bag is pear shaped and is made of synthetic rubber. The bladder, including the molded air stem, is fastened by means of a locknut to the upper end of a seamless steel shell that is cylindrical in shape and spherical at both ends.

In the *diaphragm-type* accumulator (see Figure 7-14), two steel hemispheres are locked together, and a flexible rubber diaphragm is clamped around the periphery. The gas and hydraulic oil pressures are equal, because the separating member is flexible. The air or gas acts as a spring that is compressed as the pressure increases.

## Shock Absorbers or Alleviators

A hydraulic shock absorber or alleviator (see Figure 7-15) is used in conjunction with hydraulic systems and pipelines to absorb shocks caused by the sudden halting of the flow of a liquid. If sudden shocks occur in the hydraulic line, they are absorbed safely and smoothly by means of a packed spring-loaded plunger that rises to relieve sudden pressure surges.

Figure 7-16 shows the basic construction of a shock absorber or alleviator. A spring-loaded plunger moves inside a cylinder that is long enough to provide sufficient travel for it to function properly.

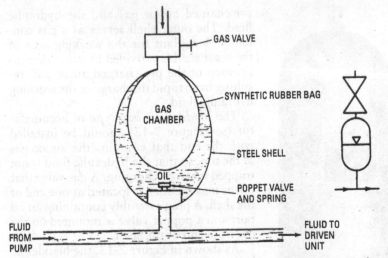

**Figure 7-13** Basic construction of a bladder- or bag-type accumulator.

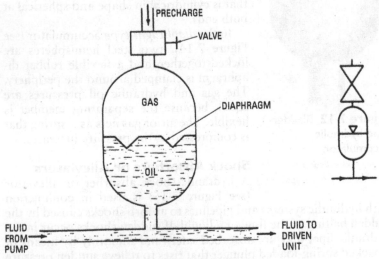

**Figure 7-14** A diaphragm-type accumulator.

**Figure 7-15**   A shock-absorbing alleviator. Hydraulic system shocks are absorbed safely and smoothly by means of a packed spring-loaded plunger that rises to relieve sudden pressure surges.

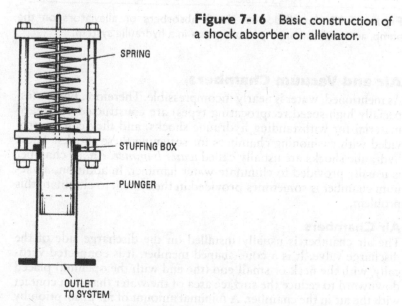

**Figure 7-16**   Basic construction of a shock absorber or alleviator.

SPRING

STUFFING BOX

PLUNGER

OUTLET
TO SYSTEM

If a sudden shock occurs in the hydraulic line, the piston moves upward in the cylinder to another position *A* (dotted lines in the diagram). The movement of the piston does not permit the liquid to escape.

Shock absorbers or alleviators are installed in the hydraulic system at the pump, accumulator, and hydraulic press (see Figure 7-17). They absorb the shock caused by the sudden closing of a valve or the sudden halting of the weight in a weight-loaded accumulator. A violent hammering action may damage the fittings and piping. The entire hydraulic system can be fully protected against shock to increase its effectiveness.

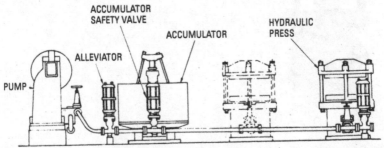

**Figure 7-17** Installation of shock absorbers or alleviators on the pump, accumulator, and hydraulic press in a hydraulic system.

## Air and Vacuum Chambers

As mentioned, water is nearly incompressible. Therefore, pumps (especially high-speed reciprocating types) are constructed of rugged material for withstanding hydraulic shocks, and they can be provided with cushioning chambers for softening these shocks. These hydraulic shocks are usually called *water hammer*. An air chamber is usually provided to eliminate water hammer. In addition, a vacuum chamber is sometimes provided in the system to eliminate this problem.

### Air Chambers

The air chamber is usually installed on the discharge side of the discharge valve. It is a cone-shaped member. It is connected vertically, with the neck or small end (the end with the opening) placed downward to reduce the surface area of the water that is in contact with the air in the chamber. A minimal amount of air absorption by the water occurs if the contact surface area is reduced. If absorption occurs, the air chamber fills with water gradually, until it is rendered ineffective for cushioning of the shocks.

Figure 7-18 shows the location of the air chamber in relation to the pump. Since the water is under pressure in the discharge chamber, the air in the air chamber is compressed during each discharge stroke of the pump piston. When the piston stops its movement, momentarily, at the end of the discharge stroke and during the return stroke, the air in the chamber expands slightly to produce a gradual halting of the flow of water. Thus, the valves are permitted to seat easily and without shock (see Figure 7-19).

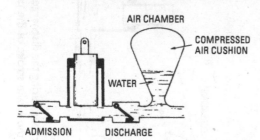

**Figure 7-18** Location of the air chamber in relation to the pump in a water system.

Air chambers are useless unless they are provided with a device for keeping them supplied with air. At pressures above 300 psi, water absorbs air so rapidly that an air-charging device is necessary. A small petcock can be installed near the pump in the admission line. By opening the petcock slightly, a small volume of air and the water are admitted to the pump; thus, the air chamber is supplied with enough additional air to redeem the loss of air absorbed by the water. During the admission stroke, the additional air is drawn into the line through the petcock. During the discharge stroke the air is discharged from the cylinder and enters the air chamber (see Figure 7-20).

The size of air chambers may vary considerably, depending on the type of service. For boiler-feed and many types of service pumps, the volume of the air chamber should be two or three times the piston displacement of a simplex pump and one to two times the displacement of a duplex pump. If the piston speed is unusually high (as in fire pumps, for example), the volume of the air chamber should be approximately six times the piston displacement.

## Vacuum Chambers

A vacuum chamber (see Figure 7-21) is sometimes installed in the admission line to a pump, especially if the line is lengthy and the

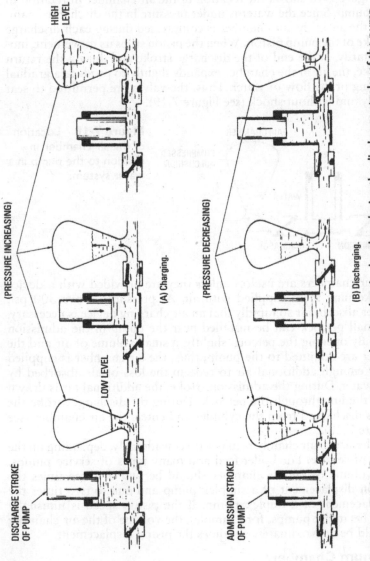

**Figure 7-19** Progressive steps in charging the air chamber during the discharge stroke of the pump (top) and discharging the air in the air chamber during the intake stroke of the pump (bottom).

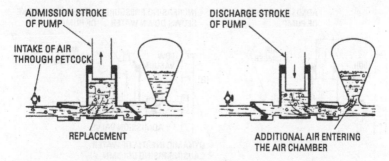

AIR INTAKE CYCLE

**Figure 7-20** Method of supplying additional air to replace air that is absorbed by the water.

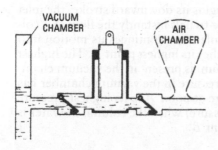

**Figure 7-21** Location of the vacuum chamber in the inlet line to the pump.

resistance to water flow is high. Once the column of water in the admission line is in motion, it is quite important (especially at high speeds and with long intake) to keep the water at full flow and to stop its flow gradually when its flow is halted. This is accomplished by means of a vacuum chamber placed in the inlet line to the pump.

The operating principle of the vacuum chamber is nearly opposite that of the air chamber—it facilitates the changing of a continuous flow to intermittent flow. The flowing column of water compresses the air in the vacuum chamber at the end of the pump stroke. When the piston starts to move, the air expands (thereby creating a partial vacuum), which aids the piston in returning the column of water to full flow.

Figure 7-22 shows the basic operation of the vacuum chamber. As the water is drawn into the cylinder on the upward stroke of the cylinder, the least volume of water (of highest vacuum) is present in the vacuum chamber; the receding column of water reaches the

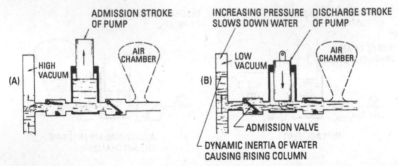

**Figure 7-22** Basic operation of the vacuum chamber at the end of the admission stroke of the pump (top) and at the end of the discharge stroke (bottom).

low point *A*. When the piston begins its downward stroke, the inlet valve closes. Since it is impossible to halt instantly the flow of a column of water, the water in the inlet pipe continues its motion into the vacuum chamber until it reaches its highest point *B*. The highest volume of water (or lowest vacuum) is present in the vacuum chamber at this point. The absolute pressure in the vacuum chamber has increased from its lowest pressure at point *A* to its highest pressure (perhaps above atmospheric pressure) when the column of water is brought to rest at its highest point *B*.

## Accumulator Circuits

Many hydraulic circuits can be designed to make effective use of an accumulator.

In Figure 7-23, an accumulator, *D* is employed to act as a shock absorber.

In a high-pressure, large-volume system similar to that shown, the sudden shifting of a four-way, three-position control valve *B* to the center position may result in high oil shock waves or a hammering effect. The violent hammering action may damage the fittings and piping, sometimes to the point of leakage. Leakage at the joints may occur, which causes a messy situation. High shock is detrimental to other components of the system like pump *E* and relief valve *C*. The accumulator is capable of absorbing the shock, thus protecting the system.

Figure 7-24 shows an accumulator *D* that acts as a power-saver. Cylinder *A* provides the loading force, as on an aluminum rolling mill operation, in which the force moves only a short distance. When

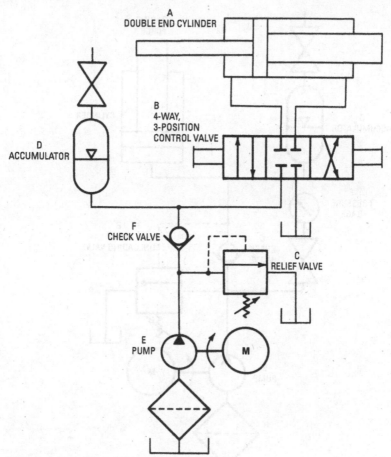

**Figure 7-23** Circuit of a system in which accumulator acts as a shock absorber. *(Courtesy Sperry Vickers, Div. Sperry Rand Corp.)*

the oil pressure is built up on the face of the cylinder piston, the accumulator can supply the loading force for a period. During this period, it is unnecessary to keep the pump *E* delivering at high pressure. An unloading relief valve *B* is provided to return the pump delivery to the reservoir at low pressure. During the unloading process, the pump discharge pressure is at a low level, while the accumulator pressure is at a high level to provide the loading force. Thus, the accumulator acts as a power-saving device.

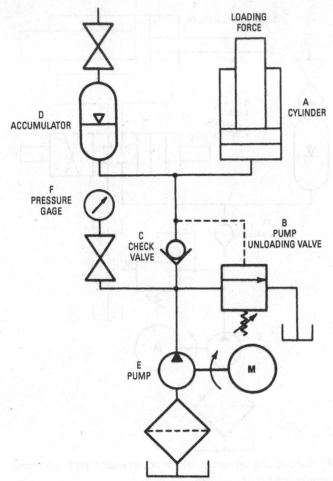

**Figure 7-24** Circuit diagram of a system that includes a power-saving accumulator.

Figure 7-25 depicts a circuit diagram in which accumulators (*E*) are employed to provide a reserve power supply in the event of an electric power failure. In some applications, a movement must take place, even in the event of a complete power failure. Among such applications are the closing or opening of large gates on a power dam, actuating large gate-type valves, or operation of doors and feed mechanism on gas-fire heat-treating furnaces.

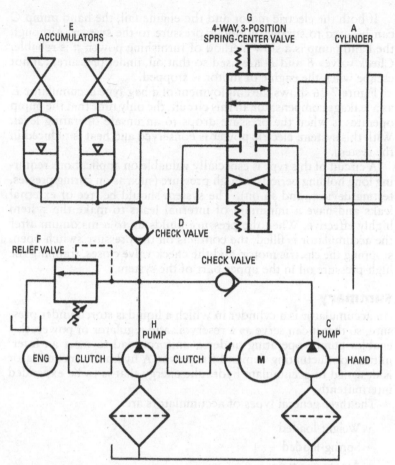

**Figure 7-25** Circuit designed to provide a reserve power supply in event of an electric power failure.

In the circuit shown, in normal operation, the clutch connected to the electric motor is engaged and the fixed delivery pump $H$ is actuated by the electric motor to deliver oil to the system. If an electric power failure should occur, the clutch on the side of the electric motor is disengaged, and the clutch on the side of the gasoline engine is engaged after the engine is started. Then the pump is actuated and oil is pumped to the system. Accumulators $E$ are placed in the system as a reserve source of oil under pressure.

If both the electric motor and the engine fail, the hand pump $C$ can be used to supply oil under pressure to the system. Although the hand pump is a slow method of furnishing power, it is reliable. Check valves $B$ and $D$ are used so that oil, under pressure, cannot escape when the engine or motor is stopped.

Figure 7-26 shows the employment of a bag-type accumulator $E$ as a leakage compensator. In this circuit, the only time that the pump operates is when the pressure drops to an unsafe operating level. With this system, electric power is conserved and heat is reduced in the system.

A circuit of this type is especially valuable on applications requiring long holding periods or high pressure (such as on curing presses, testing devices, and so on). The system should be free of external leaks and have a minimum of internal leaks to make the system highly effective. When the pressure builds up to a maximum after the accumulator is filled, the contacts on the pressure switch open, stopping the electric motor, and the check valve closes, trapping the high-pressure oil in the upper part of the system.

## Summary

An accumulator is a cylinder in which a liquid is stored under pressure, so that it can serve as a reservoir and regulator of power; this enables pumps operating under a uniform load to meet an intermittent or fluctuating demand for power. A hydraulic accumulator is designed to accumulate hydraulic energy that is to be expended intermittently.

The three general types of accumulators are:

- Weight-loaded
- Spring-loaded
- Air- or gas-type

A pressure booster or intensifier is a type of accumulator that is used to convert low pressure to high pressure. A shock absorber or alleviator is also used in conjunction with hydraulic systems and pipelines to absorb sudden shocks caused by the sudden halting of the flow of a liquid.

The air chamber is provided in a hydraulic system to cushion or soften hydraulic shocks or water hammer. When an air chamber is used in the system, it is usually installed on the discharge side of the discharge valve.

In addition to the air chamber, a vacuum chamber is sometimes installed in the admission line to a pump, especially if the line is

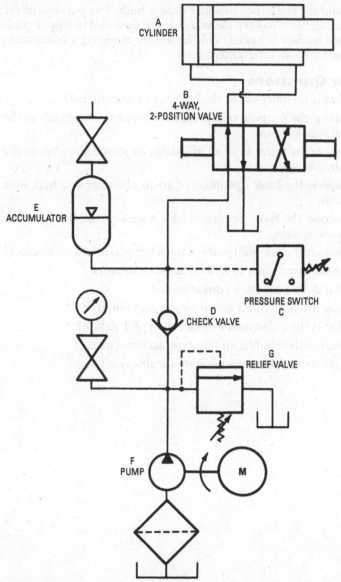

**Figure 7-26** The use of a bag-type accumulator as a leakage compensator.

lengthy and the resistance to water flow is high. The purpose of the vacuum chamber is to keep the water at full flow and to stop it gradually when its flow is halted. This facilitates changing a continuous flow of water to an intermittent flow.

## Review Questions

1. What is the purpose of the hydraulic accumulator?

2. What is the purpose of a pressure booster or intensifier in the hydraulic system?

3. What is the purpose of an alleviator or shock absorber in the hydraulic system?

4. Describe the basic operation of an air chamber in a hydraulic system.

5. Describe the basic operation of a vacuum chamber in a hydraulic system.

6. What are the essential parts of the weight-loaded accumulator?

7. List the three types of hydraulic accumulators.

8. What does a pressure converter do?

9. Draw the symbol for an air or gas accumulator.

10. What is the cubic inch displacement of 1 gallon?

11. Describe the bladder or bag type accumulator.

12. What is another name for a shock absorber?

# Chapter 8

# Power Transmission

Early internal-combustion engines were lacking in power and torque flexibility, in comparison with the steam engines of the period. With a mechanically connected driving means, it was impossible for the internal-combustion engine to attain the smoothness and flexibility of the steam engine, because the internal-combustion engine depends on a series of quick successive explosions in the cylinder to produce power and torque, and on the momentum stored in a heavy flywheel to keep it operating between explosions. Also, the steam engine was capable of delivering rapid, smooth acceleration of the load to maximum speed within its power range, without pauses or slowdowns for shifting gears, as was required with the internal-combustion engine. All these factors contributed to the development of hydraulic drives for internal-combustion engines.

Figure 8-1 shows the various types of power transmissions and their overall efficiency.

| POWER TRANSMISSIONS | | COMPONENTS | WEIGHT/ HORSEPOWER | COST/ HORSEPOWER | OVERALL EFFICIENCY |
|---|---|---|---|---|---|
| MECHANICAL | DIRECT | SHAFTS<br>CLUTCHES<br>SPUR GEARS<br>BRAKES | 3.0-9.4 | $6.00-8.00 | –2% PER MESH<br>–2% PER U-JOINT |
| MECHANICAL | PLANETARY | SHAFTS<br>CLUTCHES<br>PLANETARY GEARS<br>BRAKES | 6.9-14.0 | $6.00-12.00 | –2% PER MESH<br>–2% PER U-JOINT |
| ELECTRICAL | AC | ALTERNATOR<br>MOTOR<br>CONTROL GEAR | 47.4-148.0 | $28.00-138.00 | 65%-85% |
| ELECTRICAL | DC | GENERATOR<br>MOTOR<br>CONTROL GEAR | 44.0-420.0 | $39.00-226.00 | 70%-90% |
| PNEUMATIC | HOT GAS | JET EXHAUST<br>GAS MOTORS<br>VALVES | | | |
| PNEUMATIC | COMPRESSED GAS | COMPRESSOR<br>GAS MOTOR<br>VALVES | 20.0-240.0 | $25.00-160.00 | 30%-70% |
| HYDRAULIC | HYDRODYNAMIC | IMPELLER<br>REACTOR | | | |
| HYDRAULIC | HYDROSTATIC | PUMP-SPLIT, VARIABLE, FIXED<br>MOTOR (FLUID) VARIABLE, FIXED<br>VALVES-CONTROL | 3.0-11.2 | $10.90-16.80 | PUMPS-90%–96%<br>MOTORS-87%–92% |

NOTE: COMBINATIONS OF THE ABOVE TYPES ARE OFTEN USED. I.E. HYDROMECHANICAL

**Figure 8-1** Typical representative values.

## Hydraulic Drives

A *hydraulic drive* (also called a *fluid drive* or *liquid drive*) is a flexible hydraulic coupling. It is a means of delivering power from a prime mover to a driven member through a liquid medium—with no mechanical connection. A pioneer type of hydraulic drive (see Figure 8-2) was invented and built by H. E. Raabe in 1900. In construction, a separate hydraulic motor (similar to a common gear-type pump) was attached to each rear wheel, thereby permitting differential rotation on curves.

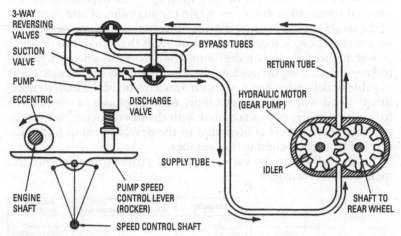

**Figure 8-2** A pioneer type of hydraulic drive invented and built by H. E. Raabe in 1900.

In the Figure 8-1, one of the motors, the connecting piping, the reversing valves, the pump, and the pump drive are shown. In the actual operation of the system, an eccentric on the engine shaft provides an oscillating motion to the rocker that operates the pump, forcing oil through the system as indicated by the arrows in the diagram. If the two reversing valves are turned 90 degrees in unison, the flow of oil and the motion of the car are reversed. Variable speed is obtained by shifting the position of the fulcrum of the rocker, which varies the stroke of the pump. A multi-cylinder pump is used to avoid pulsations and a jerky movement to the car.

## Basic Operating Principles
The hydraulic drive mechanism consists of the following three essential parts:

- Driver
- Follower (sometimes called runner)
- Casing or housing

The power from the engine is delivered to the driver. It is then transmitted (flexibly) to the follower through the hydraulic medium.

In actual operation of the hydraulic drive, the power is transmitted from the driver to the follower because of effect of circulation of the hydraulic medium or oil. The circulation of the hydraulic oil is caused by the difference in the centrifugal force set up in the driver and the centrifugal force set up in the follower.

It should be remembered that centrifugal force (see Figure 8-3) is the force that acts on a body moving in a circular path, tending to force it farther from its axis of rotation. Hydraulic oil *does not circulate* if both the driver and the follower rotate at the *same speed*, because the centrifugal force in the driver is equal to the centrifugal force set up in the follower. Hence, the oil does not tend to

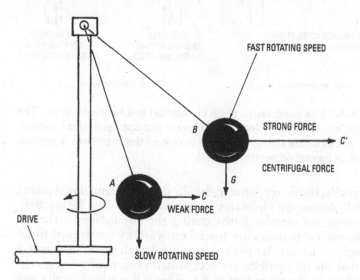

**Figure 8-3** Centrifugal force as the basic operating principle of the common fly-ball type of engine governor. At slow speed, the ball is forced outward to position A in its circular path. If the engine load decreases, the engine speed increases, and the increased centrifugal force moves the wall outward to position B in its circular path. The increased centrifugal force is indicated by the lengths of the vectors C and C'. In each instance, the downward pull of gravity G is constant and is overcome by centrifugal force. Similarly, in the hydraulic drive, centrifugal force is greater in the high-speed driver than in the follower rotating at a slower speed. This difference in the centrifugal force causes the hydraulic oil to circulate.

circulate, because neither force is excessive. This also indicates that the hydraulic oil *does circulate* only when the centrifugal force set up in the driver is *greater than* the centrifugal force set up in the follower (see Figure 8-4). To obtain this condition, it is necessary for the driver to rotate at a higher speed than the follower. That is because the intensity of the centrifugal force depends on the speed of rotation.

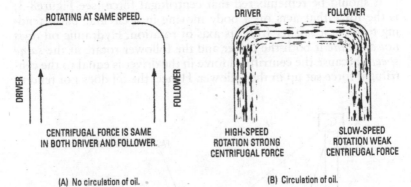

ROTATING AT SAME SPEED.

DRIVER    FOLLOWER

DRIVER    FOLLOWER

CENTRIFUGAL FORCE IS SAME
IN BOTH DRIVER AND FOLLOWER.

HIGH-SPEED
ROTATION STRONG
CENTRIFUGAL FORCE

SLOW-SPEED
ROTATION WEAK
CENTRIFUGAL FORCE

(A) No circulation of oil.

(B) Circulation of oil.

**Figure 8-4** The basic cause of oil circulation in a hydraulic drive. The centrifugal forces in the driver and follower are opposed, and hydraulic oil does not circulate unless the force in one of the members is greater, the result of increased rotation speed.

Essentially, the driver in the hydraulic drive is a centrifugal pump, and the hydraulic oil circulates in circular paths (see Figure 8-5). In traversing the circular paths during the circulation of the oil, the hydraulic oil particles are forced outward by centrifugal force. The length of the circular path increases for each revolution, which means that the oil particles are accelerated tangentially (tangential acceleration) as shown in Figure 8-6. This is illustrated by the successive positions $A, B, C,$ and $D$ of an oil particle as it is forced outward from the hub to the rims of the adjacent vanes of the driver. As shown in the diagram, the distance traveled by an oil particle in one revolution increases as the oil particle moves outward to positions $C$ and $D$. Similarly, the *tangential velocity* increases. This means that there is *tangential acceleration*, which is indicated by the increasing distances $Aa, Bb, Cc,$ and $Dd$ from the axis of rotation.

## Tangential Acceleration

Both tangential acceleration and tangential deceleration require an expenditure of energy (supplied by the engine). Most of the

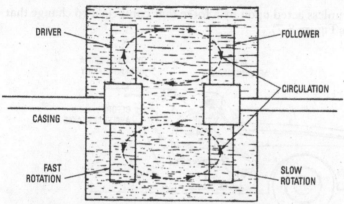

**Figure 8-5** Circular path of hydraulic oil particles in circulating from the driver to the follower. Centrifugal force forces the oil particles outward from the hub to the rim, and they return at the hub.

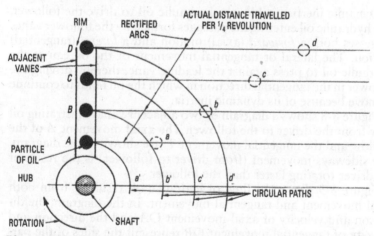

**Figure 8-6** Tangential acceleration as the hydraulic oil particles move outward from the hub to the rims of the adjacent vanes of the driver.

tangentially accelerated oil in the driver is converted at the follower into torque during deceleration. An expenditure of energy is required to tangentially accelerate the oil, because the oil particles press against the vanes of the driver as they move outward. Strictly speaking, the vane presses against the oil particles to overcome the dynamic inertia resulting from acceleration of its tangential velocity as the oil moves outward from the hub to the rim of the vane. Dynamic inertia causes a moving body to remain in a state of uniform motion, unless acted upon by a force that compels it to change that state (see Figure 8-7).

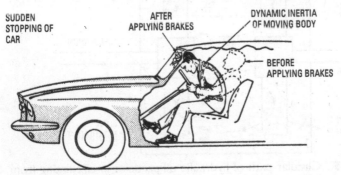

**Figure 8-7** Dynamic inertia.

Dynamic inertia enables the hydraulic oil to drive the follower. The hydraulic oil entering the passages formed by the follower vanes possesses both a *forward* (axial) motion and a *lateral* (tangential) motion. The lateral or tangential movement of the oil causes the hydraulic oil to press against the leading vane, thereby driving the follower in the tangential direction in which the oil tends to continue to move because of its dynamic inertia.

Figure 8-8 shows a diagram of two adjacent vanes illustrating oil flow from the driver to the follower. The axial movement $A$ of the oil flow and the tangential movement $T$ are shown in the diagram. The sideways movement (from driver to follower) is the result of the driver rotating faster than the follower.

Figure 8-9 shows the flow of hydraulic oil resulting from both axial movement and tangential movement. In the diagram, the direction and velocity of axial movement $OA$ and the direction and velocity of tangential movement $OB$ represent the sides of the parallelogram $OABR$. Then the diagonal $OR$ represents the direction and velocity of the actual oil flow resulting from the two component movements. Since the diagonal $OR$ of the parallelogram represents the actual direction of oil flow, the oil strikes the leading vane of the follower at an angle $AOR$. The normal thrust $AR$ on the leading vane of the follower produces torque that tends to rotate the vane. Since the tangential direction and velocity of the oil $OB$ is equal to the thrust $AR$ on the vane of the follower, tangential acceleration produces torque that tends to rotate the follower.

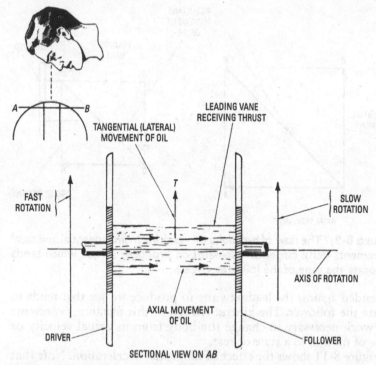

**Figure 8-8**  Sectional diagram (bottom) with a portion of the rims of adjacent vanes cut away (on line AB in upper left), showing the axial movement A and the tangential movement T of hydraulic oil.

## Tangential Deceleration

As mentioned, an expenditure of energy is required either to increase the velocity of a moving body or to slow it down (see Figure 8-10). For example, work is required by a locomotive to increase the speed of a train to its top speed. The kinetic energy (stored capacity resulting from the momentum of the moving body) acquired thereby is expended by applying the brakes to slow the speed of the train. In this instance, the energy is converted to heating of the brake shoes.

Tangential deceleration also produces torque that tends to rotate the follower. On entering the follower, the hydraulic oil is forced (by excessive forward centrifugal force) to flow from the rim to the hub, which decelerates its tangential velocity. In the meantime, the tangential component (see *AR* in Figure 8-9) of the kinetic energy originally possessed by the hydraulic oil on entering the follower is

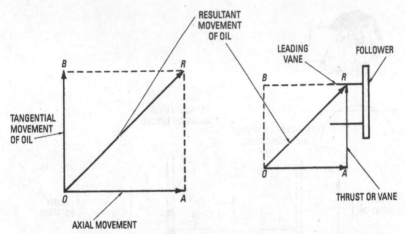

**Figure 8-9** The flow of hydraulic oil resulting from tangential and axial movements (left); tangential acceleration produces torque which tends to rotate the vane of the follower (right).

expended against the leading vane to produce torque that tends to rotate the follower. The kinetic energy, in this instance, represents the work necessary to change the body from its actual velocity or state of motion to a state of rest.

Figure 8-11 shows the effect of tangential deceleration. Note that in positions $A, B, C,$ and $D$ in the follower, the oil particle is moving toward the hub, and that these positions are the same distance from the axis in tangential deceleration as they were in tangential acceleration. This indicates tangential deceleration of the hydraulic oil produces a driving force or thrust on the leading vane of the follower.

## Types of Hydraulic Drives

The basic operating principle of the hydraulic drive is used in many different types of transmissions. In all these drives, the two rotating members are the driver and the follower (or runner).

### Fluid Drive

Figure 8-12 shows a section diagram showing details of construction of a typical fluid drive. Stamped, pressed, and forged parts are used in construction of the fluid drive. The hubs for the impeller housing and the follower (runner) are forged. The outer housing and the follower disk are made of pressed cold-rolled steel, and the vanes (22 in the impeller and 24 in the follower) are made of stamped cold-rolled steel (see Figure 8-13). The vanes are assembled

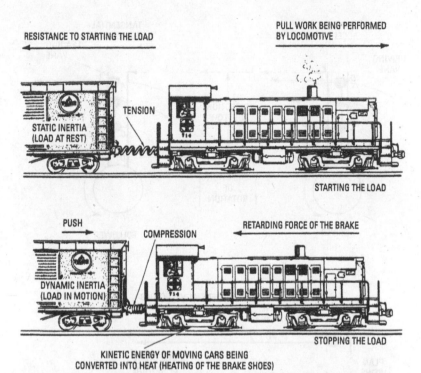

RESISTANCE TO STARTING THE LOAD

PULL WORK BEING PERFORMED BY LOCOMOTIVE

TENSION

STATIC INERTIA (LOAD AT REST)

STARTING THE LOAD

PUSH

COMPRESSION

RETARDING FORCE OF THE BRAKE

DYNAMIC INERTIA (LOAD IN MOTION)

STOPPING THE LOAD

KINETIC ENERGY OF MOVING CARS BEING CONVERTED INTO HEAT (HEATING OF THE BRAKE SHOES)

**Figure 8-10** Kinetic energy. Work is required by the locomotive to start the load (top), and work is required for the moving cars to stop the load, which tends to remain in motion (bottom). Note that the work required to stop the load is converted to heating of the brake shoes.

permanently onto the impeller and follower disks by spot welds on each vane. The follower disk is welded permanently to the follower hub. The follower is mounted in the impeller on a ball bearing, and is located in the forward portion of the assembly. The follower is supported in the assembly by the transmission drive pinion shaft. A low-viscosity mineral oil is used in the fluid coupling. It provides the lubrication required by the bearing enclosed within the coupling, and the oil pours at the lowest temperature anticipated.

As shown in Figure 8-12, the two rotating elements are the driver and the follower (runner). The difference or *slip* in rotating speeds of the two members is approximately 1 percent in average driving conditions at normal speeds over level roads. On a long, difficult pull, the percentage of slip is greater, and it is 100 percent when the car is stopped and in gear with the engine idling.

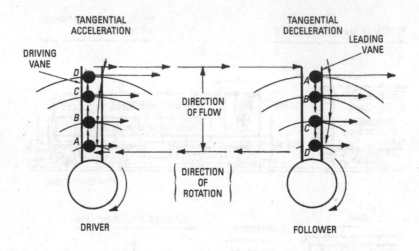

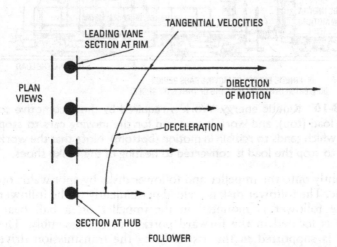

**Figure 8-11** Application of force to the driver to obtain tangential acceleration of the hydraulic oil (upper left); and why tangential deceleration produces torque or a tendency to rotate the follower (upper right). The effect of tangential deceleration is illustrated further in the lower diagram.

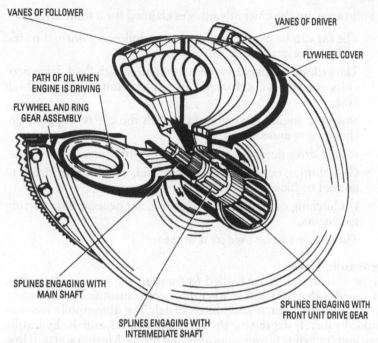

VANES OF FOLLOWER

VANES OF DRIVER

FLYWHEEL COVER

PATH OF OIL WHEN
ENGINE IS DRIVING

FLY WHEEL AND RING
GEAR ASSEMBLY

SPLINES ENGAGING WITH
MAIN SHAFT

SPLINES ENGAGING WITH
FRONT UNIT DRIVE GEAR

SPLINES ENGAGING WITH
INTERMEDIATE SHAFT

**Figure 8-12** Construction details of a typical fluid drive.

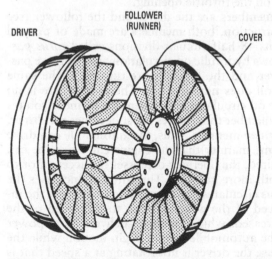

DRIVER

FOLLOWER
(RUNNER)

COVER

**Figure 8-13** Appearance of members of a fluid drive.

Following are the chief advantages claimed for a fluid drive:

• The car can be placed in high gear in following normal traffic without declutching or shifting gears.

• The kick-down overdrive in conjunction with fluid drive provides virtually an automatic two-speed transmission in high gear.

• Since the engine does not stall when the car is stopped, declutching is unnecessary.

• A fluid drive dampens engine torsion vibrations.

• Gear shifting or operation of the clutch pedal is eliminated in normal traffic conditions.

• Declutching or shifting out of gear is not necessary in starting the engine.

• The engine can be used as a brake.

## Hydraulic Drive

A type of hydraulic drive (called *hydramatic* by the manufacturer), combined with a fully automatic four-speed transmission, eliminates the conventional clutch and clutch pedal. The automobile is set in motion by merely depressing the accelerator pedal, and the hydraulic drive unit transmits power smoothly and firmly. Motion starts in low gear, shifting successively to second, third, and fourth gears in order. Shifting is dependent on the throttle opening.

The two rotating members are the driver and the follower (see Figure 8-14). In construction, both members are made of pressed steel and each contains 48 half-circular divisions called *torus passages*. Figure 8-15 shows hydraulic oil circulation around these passages. When the driver and the follower are turning at the same speed, the hydraulic oil does not circulate. Therefore, there is no torque produced. The oil circulates only when one member rotates at a faster speed. The member that rotates faster becomes the *driver*, and the other is the *driven* member. Thus, the automobile can drive the engine—in the same manner that the engine can drive the car. If more power is required, the operator depresses the accelerator to provide the engine with more gasoline. This, in turn, increases the speed of the driver; this circulates more hydraulic oil to meet the increased load. The speed of the driver relative to the speed of the driven member is increased, which, in effect, delivers more power to the rear wheels. The automobile cannot begin motion while the engine is idling, because the driver is not rotating at a speed that is fast enough to overcome the static inertia of the car.

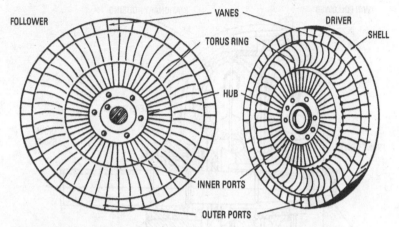

**Figure 8-14** The driver and follower in a hydramatic drive.

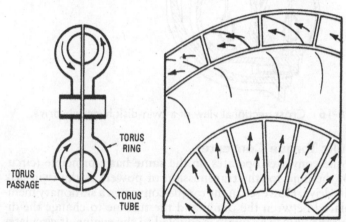

**Figure 8-15** Cross-sectional diagram showing circulation of hydraulic oil in the driver and follower (left); circulation of oil in the rotor is illustrated further (right).

## Twin-Disk Hydraulic Drive

In the twin-disk hydraulic drive (see Figure 8-16), two parallel units deliver power to the same drive shaft. The basic parts are: (1) the front driver; (2) twin followers (two followers); and (3) the rear driver. This drive is sometimes called a hydraulic clutch and hydraulic power takeoff.

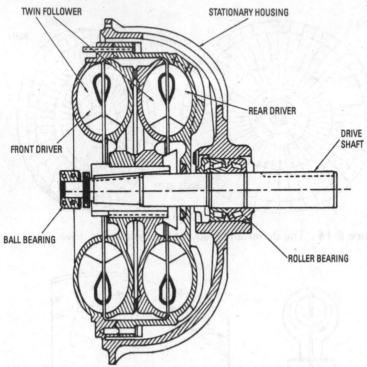

**Figure 8-16** Cross-sectional view of a twin-disk hydraulic drive.

## Hydraulic Torque Converter

The torque converter operates on the same basic principle (circulation of the fluid for the transmission of power) as the twin-disk hydraulic drive. However, in the torque converter, a stationary member is placed between the pump and the turbine to change the direction of the fluid. The pump is coupled to the engine. It circulates the fluid. The centrifugal pump imparts velocity to the fluid, which becomes the transmission medium for the power delivered by the engine.

As the fluid is forced against the turbine blades and the blades mounted on the stationary housing, its energy is manifested in the form of torque and speed. As a result of the action of the fluid on the turbine and stationary blades, the engine or input torque is multiplied increasingly as the speed of the output shaft is reduced. The engine cannot be stalled, because there is no mechanical connection between the engine and the turbine.

In construction of the hydraulic torque converter (see Figure 8-17), two sets of stationary blades are located between the three stages of the turbine. The stationary blades that redirect the flow of fluid are mounted in the stationary housing. The housing is fastened either to the engine or to a solid base. The blades cannot move when the fluid is forced against them. Therefore, the fluid flow is redirected and the resistance of the fluid to this change. The result is torque multiplication or an increase in torque at each turbine stage. This also comes with a corresponding decrease in speed of the output shaft.

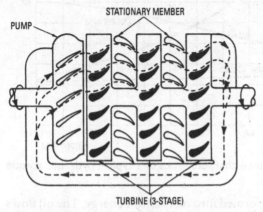

**Figure 8-17** Basic operating principle of a hydraulic torque converter.

As shown in Figure 8-18, the curve indicates the output torque obtained with an engine delivering 500 lb-ft of torque at 1800 rpm, which is the broad, nearly flat curve, peaking at approximately 85 percent. As the output shaft-speed approaches nearly two-thirds, the speed of the engine (depending on losses in the system), the engine torque, and the converter torque become equal.

## Hydrostatic Transmission Systems

A *hydrostatic transmission* (or *hydraulic displacement-type transmission*) is a combination of two interconnected positive-displacement units: a pump and a motor. This is in contrast to the hydrodynamic type of transmission.

The pump may be located at a considerable distance from the motor, or the two units (pump and motor) may be combined in a single housing. The rotary hydraulic pump transforms mechanical energy into fluid pressure energy. The high-pressure oil is then delivered to a rotary hydraulic motor. In the motor, the fluid

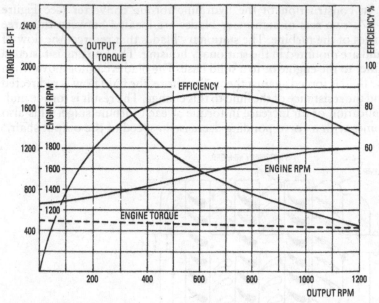

**Figure 8-18** Performance characteristics of a typical hydraulic torque converter.

pressure energy is transformed into mechanical energy. The oil flows from the pump to the motor. It then flows from the motor to the pump.

This type of transmission has a number of advantages:

- Output shaft speed can be maintained at a constant speed for a variable-speed input.
- Speed and direction of the output shaft can be controlled accurately and remotely.
- Constant power output can be maintained over a wide range of speeds.
- Automatic torque control at the output shaft can be maintained.
- Output shaft speed can be varied in minute steps.
- Output shaft can be reversed quickly and without shock.
- Power consumption can be kept at a low level.
- Automatic overload protection can be maintained.

Figure 8-19 illustrates the basic circuits for standard hydraulic transmissions. In Figure 8-19(a), a variable-displacement pump can

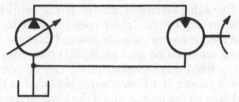

(A) Diagram of a variable-displacement pump connected to a fixed-displacement motor.

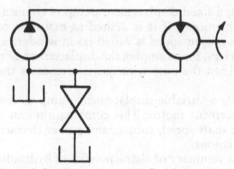

(B) Diagram of a fixed-displacement pump and a fixed-displacement motor.

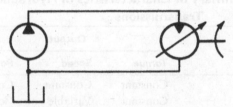

(C) Diagram of a fixed-displacement pump and a variable-displacement motor.

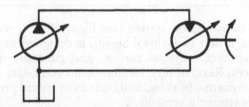

(D) Diagram of a variable-displacement pump and a variable-displacement motor.

**Figure 8-19** Basic circuits for standard hydraulic transmission.

be connected to a fixed-displacement motor. If the workload is constant, a variation in the output shaft speed results in an almost constant output torque and in variable power. Constant-torque transmissions are used on machine tool feeds and conveyors.

In Figure 8-19(b), a fixed-displacement pump can be connected to a fixed-displacement motor. If a flow control is placed in the line to vary the rate of oil flow to the motor, and if the load on the output shaft is constant, the torque is constant and the power can be varied with the speed.

In Figure 8-19(c), a fixed-displacement pump is connected to a variable-displacement motor. If it is desired to maintain constant power output as the output speed is varied (as in winders and machine tool spindle drives, for example), the displacement of the motor can be varied. Then the output torque decreases as the speed increases.

In Figure 8-19(d), a variable-displacement pump is connected to a variable-displacement motor. This combination can provide variations in output shaft speed, torque, and power characteristics of hydraulic transmissions.

Table 8-1 shows a summary of characteristics of hydraulic transmissions.

**Table 8-1    Summary of Characteristics of Hydraulic Transmissions**

| Displacement | | Output | | |
|---|---|---|---|---|
| Pump | Motor | Torque | Speed | Power |
| Fixed | Fixed | Constant | Constant | Constant |
| Variable | Fixed | Constant | Variable | Constant |
| Fixed | Variable | Variable | Variable | Variable |
| Variable | Variable | Variable | Variable | Variable |

The Dynapower transmission system (see Figure 8-20), manufactured by Hydreco, a unit of General Signal, is designed to produce constant horsepower, constant torque, and constant speed within close tolerances. Reversibility, instantaneous response, stepless speed variation, dynamic braking, and built-in overload protection make the drive extremely versatile.

The Dynapower hydrostatic transmission utilizes a closed hydraulic system. The inlet and discharge ports of a variable-displacement axial piston pump are connected to the discharge and inlet ports of an axial piston motor (fixed or variable) by hose or tubing.

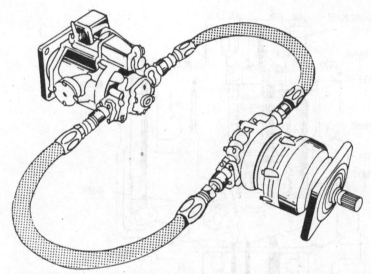

**Figure 8-20** Dynapower hydrostatic transmission system. Dynapower is the registered trademark of the New York Air Brake Company for its hydrostatic transmission and related components. The transmission system consists of a closed hydraulic system. *(Courtesy Hydreco, Div. of General Signal)*

The filter, reservoir, charge pump, and valves are also part of the transmission. The charge-pump and check valves are located in the cover of the pump, and the other valves are built into the cover of the motor. This feature minimizes the plumbing and hardware required in the system.

In operation, the hydraulic oil flows from the pump (see Figure 8-21) to the motor through the main system lines and causes the motor to rotate when the variable cam in the pump is moved from neutral position. The discharge oil from the motor returns directly to the pump inlet. The motor reverses when the cam in the pump is moved through neutral to the opposite direction.

Oil is supplied to the charge pump by a line direct from the reservoir, and it discharges through one of the two check valves to the low-pressure side of the system. This provides makeup oil and inlet pressurization to the closed system. Pressurization is maintained by the low-pressure relief valve, which also relieves excess oil from the charge pump back to the reservoir.

The system is protected from over-surges or extremely high starting pressures by the pilot-operated high-pressure relief valve. The valve discharges to the low-pressure side of the system, to prevent depletion of oil in the closed system.

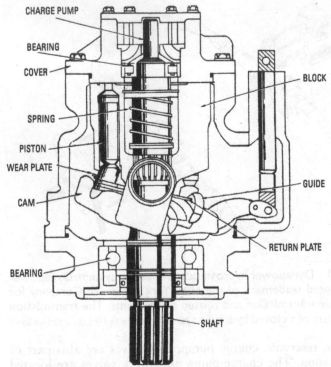

**Figure 8-21** Cutaway view showing the basic parts of the pump in a Dynapower hydrostatic transmission system.

*(Courtesy Hydreco, Div. of General Signal)*

A pressure-actuated shuttle valve directs both high pressure and low pressure to the respective relief valves. The shuttle valve moves toward the low-pressure side when one line is high pressure. This, in turn, ports the high pressure to the high-pressure relief valve and the low-pressure side to its relief valve. The shuttle valve is also used to direct high-pressure relief valve discharge to the low-pressure side of the system.

The output of standard production machines can be made suitable to specific needs by selecting the correct control. A wide variety of interchangeable types of controls are available. Following are controls used on the Dynapower hydrostatic transmission system:

- *Manual*—The cam angle is determined by the position of a simple control rod linked directly to the pump or motor cam.

A hand wheel on a threaded rod is attached for fine positioning of the pump or motor cam.

- *Low-pressure hydraulic control*—Charge pump pressure (approximately 100 psi) is utilized by the control to position the cam and to provide over-center travel. The actuating arm of the control requires 2 to 4 pounds of effort for its operation.

- *Input torque limiting control*—Stalling of the engine is prevented under all conditions.

- *Constant-speed control*—Input speeds can be varied while maintaining constant output speed. The output varies with demand.

- *Pressure-compensated motor control*—Motor speed is increased automatically by decreasing the cam angle in the motor, as pressure decreases. Conversely, by increasing the cam angle in the motor as the pressure increases, lower speed and higher output torque are obtained.

- *Pressure-compensator pump control*—This control has been made necessary by the hydrostatic transmission, although it is not normally a component of the system itself. Full flow (maximum cam angle) is maintained until a predetermined system pressure is attained. The flow is then halted while the pressure is maintained. Flow resumes to compensate for system leakage (or at system demand). A manual cutoff of flow and pressure can be used to reduce wear and horsepower demand on the system.

The transmission as shown in Figure 8-22 is designed to operate at speeds up to 4200 rpm and pressures up to 4500 psi. This transmission provides a smooth, infinite control of speed, direction, and dynamic braking with the touch of a single lever control. Use of this type of transmission is found on garden tractors, snowmobiles and other pleasure vehicles, materials-handling equipment, and on other applications requiring variable speed.

Hydrostatic hydraulic drives have been developed for many mobile and industrial applications (see Figure 8-23). These drives are used on machine tools, centrifugal pumps, production machines, winches, paper and textile rolls, surface and center windows, conveyors, reels, calendaring equipment, printing presses, fans and blowers, agitators, mixers, disintegrators, materials-handling machines, plastic extruders, and dynamometers.

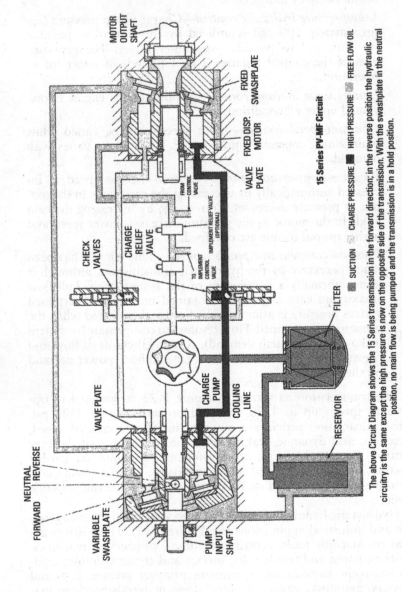

**Figure 8-22** Sundstrand Hydro-Transmission. *(Courtesy Sundstrand Hydro-Transmission, Div. of Sundstrand Corp.)*

The above Circuit Diagram shows the 15 Series transmission in the forward direction; in the reverse position the forward circuitry is the same except the high pressure is now on the opposite side of the transmission. With the swashplate in the neutral position, no main flow is being pumped and the transmission is in a hold position.

**15 Series PV-MF Circuit**

■ HIGH PRESSURE    □ FREE FLOW OIL
▨ CHARGE PRESSURE
░ SUCTION

MOTOR OUTPUT SHAFT

FIXED SWASHPLATE

FIXED DISP. MOTOR

VALVE PLATE

CHECK VALVES

CHARGE RELIEF VALVE

IMPLEMENT RELIEF VALVE (OPTIONAL)

FROM CONTROL VALVE

TO IMPLEMENT CONTROL VALVE

FILTER

CHARGE PUMP

COOLING LINE

RESERVOIR

VALVE PLATE

NEUTRAL

FORWARD    REVERSE

VARIABLE SWASHPLATE

PUMP INPUT SHAFT

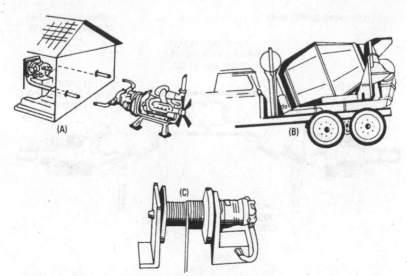

**Figure 8-23** Application of the hydrostatic transmission: (a) remote drive; (b) mixing-drum drive; and (c) winch drive.
*(Courtesy Hydreco, Div. of General Signal)*

## Hydraulic Adjustable-Speed Drive

The Vickers hydraulic adjustable-speed drives (see Figure 8-24) are also based on the hydrostatic principle. Power is transmitted by means of an electrically driven variable-displacement pump, providing pressure and a flow that is converted to usable power by a hydraulic motor (see Figure 8-25). The hydraulic fluid is returned to the pump, completing the cycle without entering a tank or sump. In the diagram, both the pump and the motor are fitted with stroke controls. The pump control can be adjusted to obtain an

**Figure 8-24** Vickers hydraulic adjustable-speed drives are based on the hydrostatic principle. *(Courtesy Sperry Vickers, Div. of Sperry)*

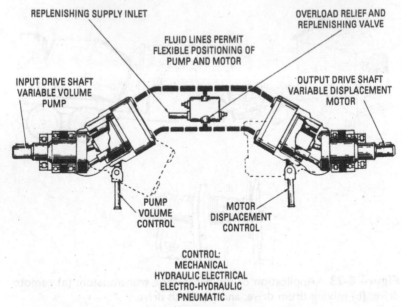

REPLENISHING SUPPLY INLET

OVERLOAD RELIEF AND
REPLENISHING VALVE

FLUID LINES PERMIT
FLEXIBLE POSITIONING OF
PUMP AND MOTOR

INPUT DRIVE SHAFT
VARIABLE VOLUME
PUMP

OUTPUT DRIVE SHAFT
VARIABLE DISPLACEMENT
MOTOR

PUMP
VOLUME
CONTROL

MOTOR
DISPLACEMENT
CONTROL

CONTROL:
MECHANICAL
HYDRAULIC ELECTRICAL
ELECTRO-HYDRAULIC
PNEUMATIC

**Figure 8-25** Diagram of a hydraulic adjustable-speed drive. Both the pump and the motor are fitted with stroke controls.

*(Courtesy Sperry Vickers, Div. of Sperry Rand Corp.)*

output-speed range from minimum to intermediate speed, with constant torque throughout the range. The fluid motor can be adjusted to obtain speeds above this range (with a decrease in torque).

Hydraulic adjustable-speed drives provide either a direct or an indirect drive for many industrial applications, including hoists and winches, food machinery, and processing equipment. The adjustable-speed transmission shown in Figure 8-26 is used on many fractional-horsepower machinery drives; with constant output torque characteristics, it can be started under full-load torque from zero rpm in either direction of rotation. Overload protection permits stalling of the unit without damage, and the output speed is smooth and step-less over its entire range.

## Farm Tractor Applications

The hydraulic transmission is used extensively on farm tractors. A large-capacity variable-displacement pump is used to provide remote control of equipment (see Figure 8-27). The pump is also used to provide power steering and power braking, which have become standard equipment on many farm tractors. In addition, remote

**Figure 8-26** An adjustable-speed transmission.

*(Courtesy Sperry Vickers, Div. of Sperry Rand Corp.)*

cylinders, a three-point hitch, and a power-differential lock are all parts of the hydraulic system.

On tillage operations, instant shifting (without stopping) permits ground travel speeds to be maintained. A constant load on the engine is retained, and the engine rpm stays within the operating range (for greater economy and efficiency). Power takeoff equipment can be kept working at full capacity when harvesting heavy crops by merely moving the shift lever to maintain the proper ground speed (including neutral position).

Action that requires direction reversal can be provided without clutching—and without stopping. This is ideal for operations that require constant changing in direction (such as stacking and loading).

Figure 8-28 shows a large farm tractor that makes use of a considerable amount of hydraulics. It has hydrostatic power steering, hydrostatic power wet disk brakes, a hydraulic power-actuated clutch, and even a hydraulically adjusted seat.

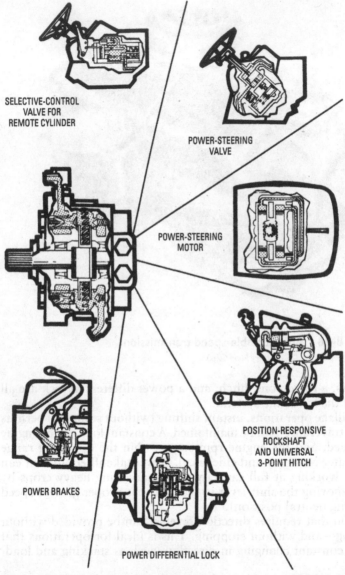

SELECTIVE-CONTROL
VALVE FOR
REMOTE CYLINDER

POWER-STEERING
VALVE

POWER-STEERING
MOTOR

POSITION-RESPONSIVE
ROCKSHAFT
AND UNIVERSAL
3-POINT HITCH

POWER BRAKES

POWER DIFFERENTIAL LOCK

**Figure 8-27** A single-pump hydraulic system can be used on a farm tractor to provide power for steering, braking, position control for a three-point hitch, and a power-differential lock.

**Figure 8-28**    International 1066 Turbo Tractor.

Figure 8-29 shows a loader that finds many applications in industry. This machine has a hydrostatic four-wheel drive with infinitely variable zero to 8 mph forward-reverse ground speeds. The unit can be slowed to a very low speed, reversed, or accelerated to full speed in either direction without changing engine rpm, and without having to change gears.

On each side of the loader is a hydrostatic drive pump that eliminates chattering or hopping at extra low speeds.

Figure 8-30 shows an earthmover that makes use of a very large amount of hydraulic equipment.

## Pumps for Robots
In recent years, more and more manufacturing concerns have been using industrial robots, and reliable pumps are crucial to the success of such machinery. When a robot calls for, say, 10 gallons of oil, a pump must be able to deliver it instantaneously, to avoid operational efficiency or damage.

### Robots
Robots need some type of power to cause them to function. To make the arm or any other part of the system move, it is necessary

**Figure 8-29** Skid steer pay-loader. *(Courtesy Caterpillar Tractor Co.)*

**Figure 8-30** Hydraulic equipment is important on large earth-moving machinery.

372

to develop the power in a usable form and in some type of readily available unit.

Following are three are ways used to power a robot and its manipulators:

* Hydraulic systems
* Pneumatic systems
* Electrical systems

Of course, electricity is used to make the pneumatic and hydraulic systems operate. However, there is also an all-electric type of drive available in the form of electric motors being directly attached to the manipulator or moving portion of the robot.

### Hydraulics

Of the three types of drive systems for robots, the hydraulic system (see Figure 8-31) is capable of picking up or moving the heaviest loads. That is why it is used. The hydraulically powered robot is ideally suited for spray painting where electric systems can be hazardous.

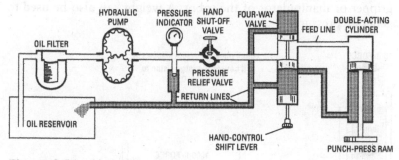

**Figure 8-31** Hydraulic system.

The automobile uses a hydraulic system for braking. As the brake pedal is depressed, it puts pressure on a reservoir of liquid that is moved under pressure to the brake cylinders. The brake cylinders then apply pressure to the pads that make direct contact with the rotors that are attached to the four wheels. By applying the pressure to the pads in varying amounts, forward motion of the car is either stopped or slowed, as desired. The hydraulic system used for the robot is similar to the baking system of an automobile.

*Hydra* is the Greek word for water. However, oil, not water, is used in robot drive systems. (Some water-based hydraulic fluids are used in foundry and forging operations.) This oil presents the

**374** Chapter 8

pressure to the correct place at the right time in order to move the manipulator or grippers. Hydraulic pressure in a hydraulic system applies force to a confined liquid. The more force applied to the oil, the greater the pressure on the liquid in the container.

**Pressure**
The force applied to a given area is called *pressure*. Pressure is measured in pounds per square inch. Pressure can be abbreviated as lb/in$^2$ or psi.

The SI metric unit for pressure is the *pascal*. The pascal is one newton per square meter (*newton* is the unit of force). Because the pascal is such a small unit, kilopascals (kPa) and megapascals (MPa) are used. For example, a typical automobile tire pressure is usually not more than 35 psi.

There are several ways to develop the pressure in a robot. The pressure developed in a hydraulic circuit in terms of how much push or force is applied per unit area (see Figure 8-32). Pressure results because of a load on the output of a hydraulic circuit or because of some type of resistance to flow. A pump used to create fluid flow through the piping to the point where it causes movement of the gripper or manipulator of the robot. A weight can also be used to generate pressure.

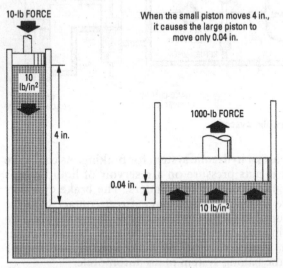

**Figure 8-32** Hydraulic pressure changes. The 20-pound force can cause 1000 pounds to be lifted, but notice the distance traveled of both the force and the object being moved.

One of the disadvantages of using the hydraulic system is its inherent problem with leaks. While fluid is under pressure, it has a tendency to seek the weakest point in the system and then run out. This means that each joint along the piping must be able to withstand high pressure without leaking. It also means that the moving part of the robot (the manipulator) must be able to contain the fluid while it is under pressure. In most cases, the O-rings are not able to contain the fluid and leaks occur. O-rings are very important in sealing the moving end of the manipulator or terminating end of the fluid line. This is one of the maintenance problems associated with robots used for picking up heavy objects. The hydraulics system is used to pick up the weight. It is then typically dependent on industrial air pressure lines equal to at least 700 kPa. One lb/in$^2$ is equal to 6.896 kPa, and 1 kPa is equal to 1.45 lb/in$^2$.

The pressure applied to a hydraulic system is in terms of how much push or force is applied to a container of oil, as shown in Figure 8-32. There are several ways to develop the pressure needed in a robot.

The pressure developed in a hydraulic circuit is in terms of how much push or force is applied per unit area. Pressure results because of a load on the output of a hydraulic circuit or because of some type of resistance to flow.

A pump is used to create fluid flow thorough the piping to the point where it causes movement of the gripper or manipulator of the robot. A weight can also be used to generate pressure.

Pressure build-up in a hydraulic system can be a problem. It is sometimes necessary to prevent its continuing to increase beyond the capacity of the pipes and containers. That is where a relief valve becomes important. A relief valve is used as an outlet when the pressure in the system rises beyond the point where the system can handle it safely. The valve is closed as long as the pressure is below its design value. Once the pressure reaches the point of design for the relief valve, it opens. Then the excess fluid must be returned to the reservoir or tank. This means a relief valve and the return of the fluid to tank must be part of the system.

Hydraulic systems need filters to keep the fluid clean. Even small contaminants can cause wear quickly. The filter must take out contaminants that measure only microns in size. The filters must be changed regularly to keep the fluid and system in operating condition.

Pressures of several thousand psi are not uncommon in the operation of a robot. This means that a leak as small as a pinhole can be dangerous. If you put your hand near a leak, it is possible that the fluid being emitted though the hole would cut off your hand before you feel it happen. However, this will not occur at pressures below 2000 psi. High pressures are very dangerous.

Additional volume for the operation of the robot may be obtained by pressuring the surge tank with a gas. An accumulator can only be charged to the main system pressure. In some cases, the robot will require a large volume of fluid for rapid motion than the pump alone can provide. In other cases, a surge tank on the high-pressure side of the line is charged with a gas in a flexible container. This flexible container is inside the tank and expands. High pressure of the fluid compresses the gas when there is excess pressure available. Then, when the system needs the fluid for the rapid motion, the pressure drops slightly and the gas in the tank expands to force the extra fluid into the system.

Hydraulic motors mounted on a manipulator are operated by signals from a transducer. This is either to open the valve for a given length of time, or to keep it closed. The control system is necessary to do it. Hydraulic motors and cylinders are compact and generate high levels of force and power. They make it possible to obtain exact movements very quickly (see Figure 8-33).

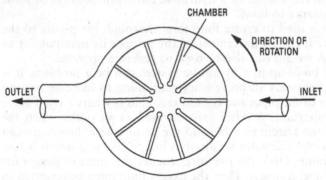

**Figure 8-33** Hydraulic motor. Pressure on the hydraulic fluid causes the motor to turn in a counter-clockwise direction. Reversing the direction of fluid flow reverses the direction of the motor rotation.

**Pumps**

The hydraulic system must have a pump to operate. The pump converts mechanical energy, usually supplied by an electric motor.

Hydrostatic pumps are further classified into gear pumps and vane pumps. These pumps are called on to deliver a constant flow of fluid to the manipulator. The gear pump produces its pumping action by creating a partial vacuum when the gear teeth near the inlet unmeshes. It causes fluid to flow into the pump to fill the vacuum. As the gears move, they cause the fluid or oil to be moved around the outside of the gears to the outlet hole at the top. The meshing of the gears at the outlet end creates a pushing action that forces the oil out the hole and into the system.

The vane pump gets its name from its design, as does the gear pump. The vanes cause the fluid to move from a large volume area to a small volume area. By pushing the fluid to a smaller area, the pressure on the fluid increases since the fluid itself it not compressible (see Figure 8-34).

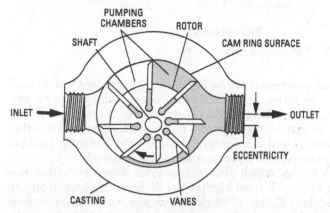

**Figure 8-34** Vane-type pump/motor. Note also that this hydraulic pump can be used as a motor when driven by the hydraulic fluid under pressure.

A vane pump can be used to supply low-to-medium pressure ranges, capacity, and speed ranges. They are used to supply the manipulator with the energy to lift large loads. Pressures can reach as high as 2000 psi with about 25 gallons of fluid per minute (gpm) moving in the system.

Piston-type rotary devices are also used to increase the pressure in a hydraulic system. The pistons retract to take in a large volume of fluid and then extend to push the fluid into the high-pressure output port. There are usually seven or nine cylinders. These vary with the different configurations of machines. The inline piston pumps are good for robot applications. They have a very high capacity, and their speed ranges are from medium to very high. Pressures can get up over 5000 psi (see Figure 8-35).

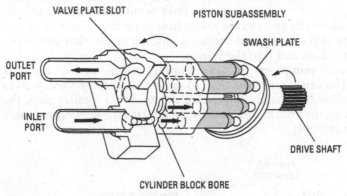

**Figure 8-35** Piston-type rotary pump.

*Variable-displacement piston pumps* have been used well in such operations. The units, also called *atmospheric inlet piston pumps*, come in a variety of designs and capacities, but they all work basically the same way. A sensor in the pump senses when oil or other liquid is required and is engineered to deliver it instantly. In other words, the pump senses a pressure drop and responds.

One such pump, which also incorporates sound reduction features, is the Quietpak from Hydreco, a division of General Signal. It uses a standard C-face 1780 electric motor; variable pump flow is available to 30 gpm with a variety of controls, including pressure compensation, load sensing, and horsepower control. In fuel displacement systems, flows to 50 gpm are available with pressures to 3,000 psi. Flow control valves are available to suit application requirements. Figure 8-36 shows the circuit.

## CCS Systems

Hydreco also makes a so-called CCS system, which is a hybrid of closed-center and flow-demand systems. It has the advantages of both. It consists of an elemental circuit (see Figure 8-37) that includes a parallel control valve and an atmospheric inlet variable-displacement pump with a control mounted on it. Fluid pressure on the pump control will reduce its fluid delivery.

The CCS system works simply. For example, a tractor/digger/loader hydraulic system will have a loader and a backhoe

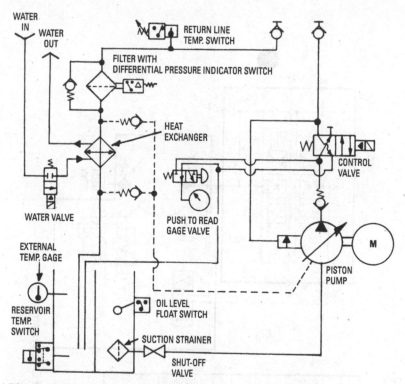

**Figure 8-36**  Circuit of the Quietpak from Hydreco.

valve. The two control valves (parallel or tandem) can be plumbed in parallel while maintaining the signal circuit in series. Regardless of the number of valves, there will be only one signal line for each pump (auxiliary valves may also be added). A steering valve, for example, could be supplied from a priority valve, which could include the ability to limit pump stroke if only the steering is in use. This is done by allowing partial signal flow to continue through the system while steering. The balance of the signal flow could be restricted by an implement valve downstream, urging the pump to increase flow.

The system, according to Hydreco, works very well in machines where contact flow is not a requirement. Indeed, the longer the period between full flow, the more appropriate the CCS feature. Equipment such as excavators, backhoes, and cranes are good candidates for this system.

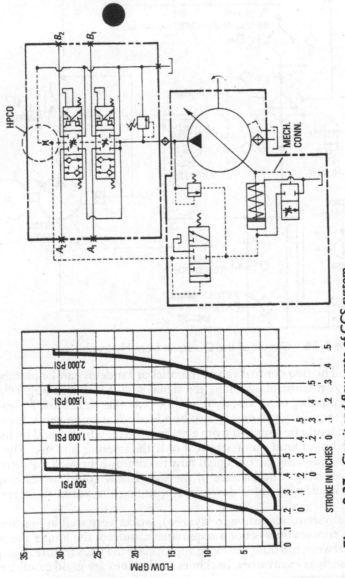

**Figure 8-37** Circuit and flow rate of CCS system.

## Summary

A hydraulic drive (also called fluid drive or liquid drive) is a flexible hydraulic coupling. It is a means of delivering power from a prime mover to a driven member through a liquid medium—with no mechanical connection.

The hydraulic drive mechanism consists of three essential parts: the driver,) follower, and casing (or housing). The power from the engine is delivered to the driver. It is then transmitted (flexibly) to the follower through the hydraulic medium.

Centrifugal force acts on a body moving in a circular path, tending to force it farther from its axis of rotation. For hydraulic oil to circulate, it is necessary for the driver to rotate at a higher speed than the follower, because the intensity of the centrifugal force depends on the speed of rotation.

Both tangential acceleration and tangential deceleration require an expenditure of energy (supplied by the engine). Most of the tangentially accelerated oil in the driver is converted at the follower into torque during deceleration. Any expenditure of energy is required to tangentially accelerate the oil, because the oil particles press against the vanes of the driver as they move outward.

The Dynapower hydrostatic transmission system is designed to produce constant horsepower, constant torque, and constant speed within close tolerances. Reversibility, instantaneous response, stepless speed variation, dynamic braking, and built-in overload protection make the drive extremely versatile.

Hydrostatic hydraulic drives have been developed for many mobile and industrial applications. These drives are used on machine tools, centrifugal pumps, production machines, winches, paper and textile rolls, surface and center winders, conveyors, reels, calendaring equipment, printing presses, fans and blowers, agitators, mixers, disintegrators, materials-handling machines, plastic extruders, and dynamometers.

Hydraulic adjustable-speed drives are also based on the hydrostatic principle. Either a direct or an indirect drive is provided for many applications (including hoists and winches, food machinery, and processing equipment).

The hydraulic transmission is also used extensively on farm tractors. The hydraulic system is used to provide remote control of equipment, power steering, and power braking. Action that requires direction reversal can be provided without clutching—and without stopping. This is ideal for operations that require constant changing in direction (such as stacking and loading).

## Review Questions

1. What is the basic operating principle of a hydraulic drive?
2. What are the essential parts of a hydraulic drive mechanism?
3. What conditions are necessary for hydraulic oil to circulate in a hydraulic drive?
4. Why is a torque a result of both tangential acceleration and tangential deceleration of the hydraulic oil?
5. List the advantages of a fluid drive.
6. How does construction of the torque converter differ in construction from the other hydraulic drives?
7. List the advantages of the hydrostatic transmission.
8. List six typical applications of the hydrostatic transmission.
9. List the advantages of the hydraulic transmission for farm tractor applications.
10. What is another name for a hydraulic drive?
11. What is *dynamic inertia*?
12. What is *kinetic energy*?
13. What are the two rotating elements in a fluid drive?
14. What does *declutching* mean?
15. What is a *torque converter*?
16. What are the two interconnected positive displacement units in a hydrostatic transmission?
17. Draw the diagram for a fixed displacement pump and a variable displacement motor.
18. What type of transmission uses a closed hydraulic system?
19. What type of transmission does the cement mixer-drum drive utilize?
20. Where are hydraulic adjustable-speed drives used?

# Chapter 9

# Hydraulic Power Tools

Hydraulic power transmission methods have contributed to the replacement of pulleys and belts for electric motor drive units. In many instances, hydraulically controlled circuits and systems will operate and control machine tools and machinery in industry. The hydraulic circuit includes a hydraulic pumping unit, necessary control valves, and the hydraulic actuator that consists of cylinder, motor, or rotary actuator.

Figure. 9-1 shows a mobile aerial tower that uses a large amount of hydraulic equipment to accomplish its operation. An engine power take-off drives a hydraulic pump that stores energy in an accumulator. The system operating pressure is 1200 psi. The lifts are powered hydraulically by closed-center, parallel-circuit systems. The turntable [see Figure 9-1(a)] is actuated by a hydraulic motor drive.

Figure 9-2 shows a hydraulically operated load-unload mechanism as used in conjunction with a modern machine tool in a high-production plant.

Figure 9-3 illustrates a 28-ton hydraulic trim press. The operator's hands must remain on the cycle-start buttons to start and maintain the cycle. If the operator removes one hand, the press ceases to operate. If both hands are removed, not only will the cycle stop, but also the shuttle will return to its start position. This is a desired safety feature on presses. Note how the cycle-start buttons have recessed heads to make them foolproof.

## Hydraulic Circuits

The hydraulic motor in a hydraulic circuit is operated by hydraulic oil under pressure. The function of the hydraulic motor is similar to that of a prime mover (such as an electric motor or a steam engine). The hydraulic pump is used to supply hydraulic oil under pressure to the hydraulic motor. This is in the same manner that steam under pressure is piped to the steam engine. The hydraulic pump can be positioned near the motor, combined with the motor in a single unit, or installed at a location remote from the motor.

Figure 9-4 shows a hydraulic power unit to which is mounted the hydraulic valve panel. This eliminates a lot of piping and presents a neat, compact unit. Note that the electric motor and hydraulic pump are mounted in a vertical position.

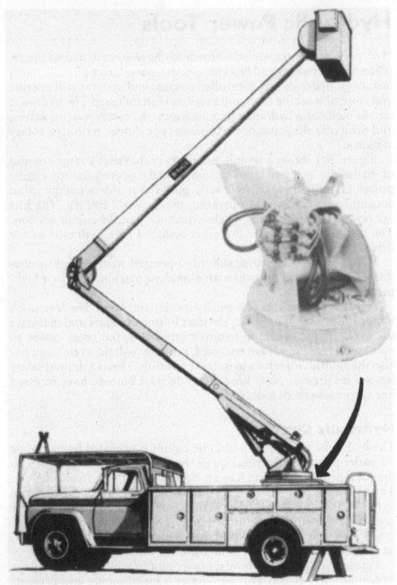

**Figure 9-1** Aerial towers with hydraulic actuation are used by utility companies and fire departments. *(Courtesy Mobile Aerial Towers, Inc.)*

**Figure 9-2**    View of a hydraulically operated load-unload mechanism.

## Hydraulic Motors

In the hydraulic pump, mechanical energy is converted to liquid pressure energy. In the hydraulic motor, liquid pressure energy is converted to mechanical energy.

Hydraulic motors may be classified as either *constant-displacement* or *variable-displacement* motors. In the constant-displacement motor, speed changes are accomplished by varying the volume of oil that flows through the motor. Speed changes in the variable-displacement motor are accomplished by varying displacement of the motor, in addition to controlling the supply of hydraulic

**Figure 9-3** A 28-ton hydraulic trim press.

oil from the pump. A wider range of speed can be obtained from the variable-displacement type of motor.

### Constant-Displacement Motors

Figure 9-5 shows a sectional diagram with the basic construction of a typical constant-displacement motor. Hydraulic oil under pressure enters the cap-end section. It is forced into the cylinders through suitable openings in a port plate. A single circular valve mounted

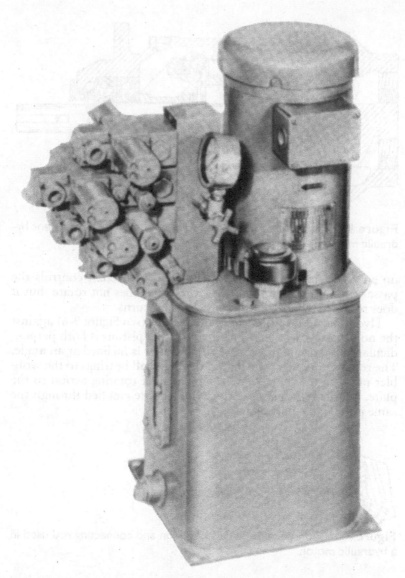

**Figure 9-4**  Hydraulic pump unit with control valves.

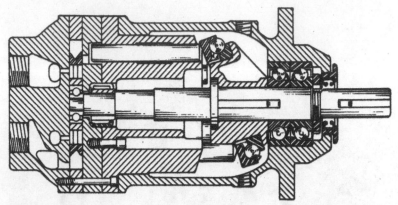

**Figure 9-5** Basic construction of a typical constant-displacement hydraulic motor.

on an eccentric stud formed on the end of the shaft controls the passage of oil through the ports. The valve does not rotate, but it does receive a gyrating motion as the shaft turns.

Hydraulic oil pressure forces the pistons (see Figure 9-6) against the non-rotating wobbler. The thrust of the pistons is both perpendicular and tangential, because the wobbler is inclined at an angle. The resultant force is transmitted through ball bearings to the wobbler plate on the shaft, thereby imparting a rotating action to the plate. On the return stroke, the cylinders are emptied through the same ports in the port plate.

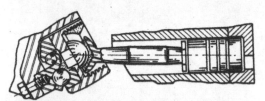

**Figure 9-6** Construction details of piston and connecting rod used in a hydraulic motor.

### Variable-Displacement Motors
The piston stroke can be varied in the variable-displacement type of hydraulic motors by changing the angle at which the wobbler is inclined. The axial piston-type motor can be either a constant-displacement or a variable-displacement motor, depending on the

angle between the cylinder axis and the output shaft. If this angle can be varied (either manually or automatically), the motor is a variable-displacement motor.

Figure 9-7 shows the basic operation of an axial piston-type motor. Hydraulic oil under high pressure enters the stationary valve plate. This oil forces the pistons outward, thus rotating the output shaft.

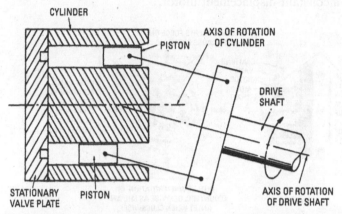

**Figure 9-7**  Basic operating principle of an axial piston-type motor.

## Types of Hydraulic Motors

The rotary hydraulic motor is used in many applications. The hydraulic motor may be a gear-type, vane-type, or piston-type motor. These motors are similar in construction to the gear-type, vane-type, and piston-type pumps.

### Gear-Type

The basic principle of operation of the gear-type motor (see Figure 9-8) is nearly the same as the reverse action of a gear-type pump. The hydraulic oil under high pressure enters at the inlet, pushes each of the gears, and then flows outward. The load is usually connected to only one of the gears. In most instances, the gears are of the spur type. However, they may be the helical or herringbone type in some motors. A gear-type motor is a constant-displacement motor.

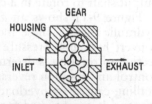

**Figure 9-8**  Basic operating principle of a gear-type hydraulic motor.

## Vane-Type

Figure 9-9 shows the basic operating action of a rotary vane-type motor. Oil under high pressure enters at the inlet, exerts pressure on the vanes to turn the rotor, and passes through to the outlet. Either fluid pressure or springs are required as a means of holding the vanes against the contour of the housing at the start of rotation. Centrifugal force holds the vanes against the housing. The vane-type motor is a constant-displacement motor.

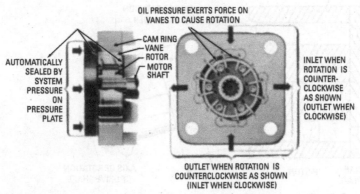

**Figure 9-9**   Basic operating principle of a rotary vane-type hydraulic motor. *(Courtesy Sperry Vickers, Inc., Div. of Sperry Rand Corp.)*

## Piston-Type

Piston-type motors are available in both radial and axial designs. Figure 9-10 shows a *radial piston-type constant-displacement* hydraulic motor. The cylinder rotates around a fixed pintle. Hydraulic oil under high pressure enters the upper ports of the pintle; this forces the pistons to move outward, causing the cylinder and the output shaft to rotate in a clockwise direction.

Figure 9-11 shows an *axial piston-type constant-displacement* hydraulic motor. These motors are used in hydraulic circuits to convert hydraulic pressure to rotary mechanical motion. The direction of rotation is determined by the path of the oil flow. Speed control and rotation reversals are accomplished easily and simply. Stalling caused by overloading does not damage this type of motor, and it can be used for dynamic braking. If proper overload-relief valve settings are used in the system, the operation can be continuous, intermittent, continuously reversing, or stalled without damage to the motor. The mounting position is not restricted, except that the drain line must be connected to the reservoir, so that

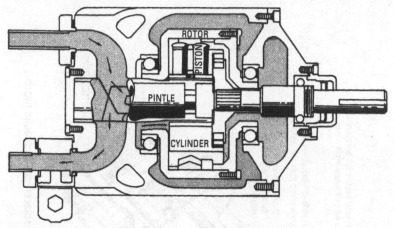

**Figure 9-10** Construction of a radial piston-type constant-displacement hydraulic motor. *(Courtesy the Oilgear Company)*

the motor case is filled with hydraulic fluid during all operations. The internal parts of the motor depend on the hydraulic oil for lubrication.

In the Vickers *fixed-displacement* hydraulic motors (see Figure 9-12), a continuous flow of hydraulic oil under pressure is converted to rotary mechanical motion. The cylinder block is offset relative to the drive shaft, which causes the pistons to traverse their respective cylinder bores. As each piston moves away from the valve plate under pressure, the opposite bottomed piston moves toward the valve plate, exhausting the spent hydraulic oil. Since the nine pistons perform the same operation in succession, the acceptance of hydraulic oil under pressure is continuous, and the conversion of the rotary motion is very smooth.

In the Vickers *variable-displacement* axial piston-type motors (see Figure 9-13), hydraulic oil under high pressure enters at the inlet port, passes through the pintle, yoke, valve block, and inlet of the valve plate, and then passes into the cylinder. The oil under high pressure pushes the pistons away from the valve plate, causing the cylinder and the output shaft to rotate. The hydraulic oil leaves through the outlet port of the valve plate, passing through the yoke, outlet pintle, and outward through the discharge flange. The angle between the cylinder axis and the output shaft can be varied from $7\frac{1}{2}$ to 30 degrees. Therefore, the displacement can be varied from a minimum quantity to four times the minimum quantity.

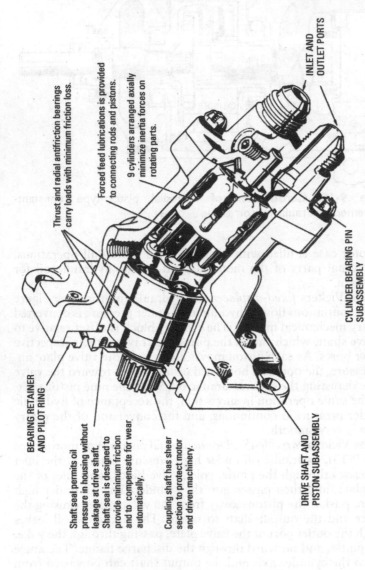

Thrust and radial antifriction bearings carry loads with minimum friction loss.

Forced feed lubrications is provided to connecting rods and pistons.

9 cylinders arranged axially minimize inertia forces on rotating parts.

INLET AND OUTLET PORTS

CYLINDER BEARING PIN SUBASSEMBLY

BEARING RETAINER AND PILOT RING

DRIVE SHAFT AND PISTON SUBASSEMBLY

Shaft seal permits oil pressure in housing without leakage at drive shaft. Shaft seal is designed to provide minimum friction and to compensate for wear automatically.

Coupling shaft has shear section to protect motor and driven machinery.

**Figure 9-11** Cutaway view of an axial piston-type constant-displacement hydraulic motor.

*(Courtesy Sperry Vickers, Div. of Sperry Rand Corp.)*

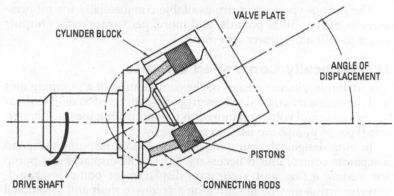

**Figure 9-12** Basic principle of operation of fixed-displacement piston-type hydraulic motor. *(Courtesy Sperry Vickers, Div. of Sperry Rand Corp.)*

**Figure 9-13** Cutaway view of a variable-displacement axial piston-type hydraulic motor. *(Courtesy Sperry Vickers, Div. of Sperry Rand Corp.)*

The piston-type motors are available commercially for oil pressures as high as 5000 pounds, and more, per square inch. Outputs range to 150 horsepower and more.

## Hydraulically Controlled Circuits

The hydraulic circuit consists of the combination of a pumping unit and the necessary control valves properly connected to deliver oil at the pressure and volume required to the motivating means. Numerous types of pumps are used to supply power.

In some designs, the pumping unit consists of the pump only, and a separate control unit is necessary. A variable-displacement pump for feeding action and a constant-displacement pump for rapid-traverse action may be mounted on a common shaft and assembled in a compact housing.

## Combination Pump and Control Valve Unit

Figure 9-14 is a sectional diagram showing a hydraulic pump with the control valves in the same casing. A single shaft is mounted in two large antifriction-type bearings. The constant-displacement pump is mounted on the end of the shaft. A hardened and ground roller that is keyed to the shaft rotates in positive contact with a rotor or ring. The ring turns in its housing or roller bearings.

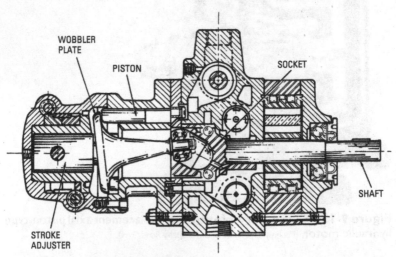

**Figure 9-14** Basic construction of a hydraulic motor having main control valves in the same housing.

The shaft extends through the main control valve section. A socket bearing (with its axis placed at an angle to the axis of the shaft) is provided at the end of the shaft. The ball-bearing end of the shank of a wobbler plate is borne in the socket. The pistons contact the wobbler plate near its outer edge.

As the shaft rotates, a wobbling action (without rotation) is imparted to the wobbler plate. The stroke of the piston is varied by the forward movement of the stroke adjuster, which more nearly aligns the axis of the wobbler plate shank and the shaft, to reduce the wobble. The stroke of the piston can be changed either manually or automatically.

## Remote Directional Control Valves

A typical hydraulically controlled circuit may consist of a pumping unit, directional control valves, functional control valves, and a cylinder, to produce a complete operating cycle. The pumping unit and the main control valves can be remote pilot-operated or remote electrically operated. Figure 9-15 shows a circuit using ANS Graphic Symbols that make use of both pilot-operated and solenoid-operated directional control valves. This type of circuit could be used on a machine tool.

This is how the circuit functions. To operate the circuit, the operator positions the work piece on the machine table, which is equipped with a magnetic chuck. The operator depresses pushbutton *PB-1*, which energizes solenoid *X* of four-way valve *H*. This directs pilot pressure to pilot *A* of four-way valve *G*. The spool shifts, allowing oil to flow to the blind end of the feed cylinder *D*. The feed cylinder piston rod moves the machine table forward at a rapid rate, until the cam on the feed table contacts the cam roller of cam-operated flow control valve *E*. As the cam roller is depressed, the exhaust oil flow is shut off, and the exhaust oil meters through the speed control portion at a rate determined by the needle setting. When the roller rides off the cam, the cylinder operates at full speed. When the piston rod of cylinder *D* reaches the end of its stroke, the limit switch *B* is contacted, energizing the solenoid *Y*, shifting the spool of valve *H* to its original position, and directing oil to the pilot *B*. The spool of valve *G* shifts to direct oil to the rod end of cylinder *D*, whereupon the piston and rod retract at a rapid rate. The operator unloads the work piece, reloads, and is ready for the next cycle.

The pilot control valve *H* controls the action of valve *G*. On equipment that provides a central station for the operator, pushbuttons may be used to save on installation costs. Skip feed valves are used on milling machines and other large machine tools, where

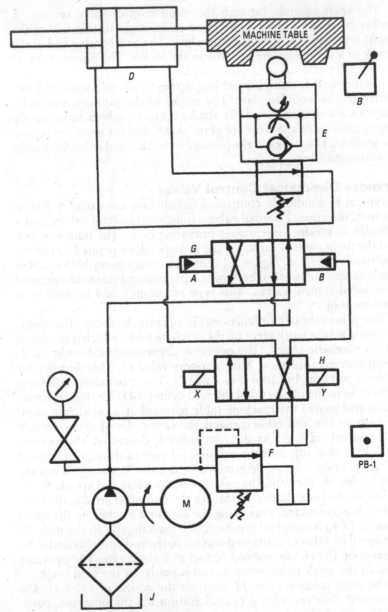

**Figure 9-15** Hydraulic circuit with skip feed arrangement.

there may be a considerable space between the machining area and the free traverse area. On some applications, there may be several machining areas.

By installing additional controls, dwell periods can be set up at various places during the stroke of the cylinder (either on the outstroke or return stroke).

An accumulator-type power unit is used in conjunction with a hand-operated, four-way, three-position directional control valve to actuate the chucking cylinder on a machine tool (see Figure 9-16).

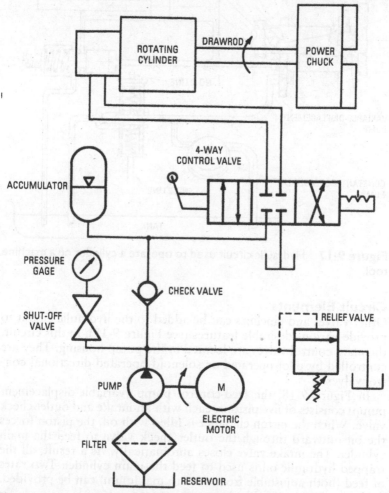

**Figure 9-16**  Hydraulic system for power chucking.

## Operation of a Cylinder on a Machine Tool

Figure 9-17 shows a typical hydraulic circuit for operating a cylinder on a machine tool. This circuit is controlled by varying the oil pressure from the constant-displacement pump.

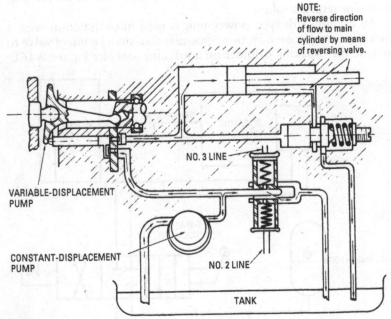

**Figure 9-17** Hydraulic circuit used to operate a cylinder on a machine tool.

## Circuit Elements

Other valves and pipelines can be added to the hydraulic circuit to provide various desirable features (see Figure 9-18). In this circuit, the main control valves are located in the pump housing. They are controlled by pilot-operated or solenoid-operated directional control valves.

In Figure 9-18, the feed-control pump (variable-displacement pump) consists of five pistons, each with an intake and outlet check valve. When the piston chamber is filled with oil, the piston forces the oil outward through the outlet check valve to feed the main cylinder. The intake valve closes automatically. As a result, all the trapped hydraulic oil is used to feed the main cylinder. Two rates of feed (both adjustable from zero to maximum) can be provided. Some pumps can provide three rates of feed.

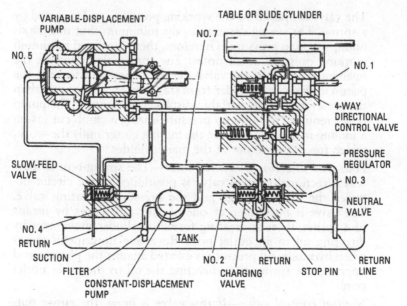

**Figure 9-18** Hydraulic circuit with additional controls for operation of a cylinder on a machine tool.

A feed adjustment for the two rates of feed is provided on the outside of the pump housing by two knobs (one for fast feed and the other for slow feed).

The constant-displacement pump in Figure 9-18 produces the rapid-traverse action. It is a self-priming rotary pump consisting of an external rotor, an internal rotor, and a crescent. This pump also provides oil for charging the variable-displacement pump.

The main control valves located in the pump housing are controlled by pilot-operated directional control valves, or electrical solenoid-operated directional control valves. These auxiliary valves are connected to the main control valves by pipelines, and they are actuated by dogs located on a moving member of the machine. The dogs trip the pilot-operated directional control valves, or open and close limit switches, to operate the electrical solenoid-operated directional control valves. The main control valves are as follows:

- *Pressure-regulating valve*—This valve is used on the return side of the main cylinder. A slight backpressure is produced, enough to provide a steady feed under no load. This valve also blocks the return of the hydraulic oil to the tank when in a climb cut or when a drill breaks through the work.

The valve is opened by the working pressure, and is closed by a spring. The spring determines the minimum (but not maximum) working pressure. Therefore, there is a fixed minimum working pressure at all times. The spring also controls the opening required in the valve during a climb cut to keep the piston in the main cylinder from traveling at a faster rate than the preset rate of the variable-displacement pump. The pump is not required to build up pressure during a climb cut (as on a milling machine), because the milling cutter pulls the work, which forces the oil out of the main cylinder.

- *Four-way directional control valve*—The spring-return four-way directional control valve is positioned in the circuit between the main cylinder and the pressure-regulating valve. The valve is positioned at one of the valve bores by means of a spring, thereby creating forward cylinder movement by directing oil to an outlet port. Reversing movement occurs when hydraulic oil pressure is exerted against the plunger end opposite the spring end, directing the oil to the other outlet port.

- *Neutral control valve*—If this valve is open, the entire output of the constant-displacement pump returns to the tank. When the circuit is in "feed" and "rapid-traverse" positions the neutral control valve is closed and it remains closed. It stays closed during the entire cycle. A small hole in the valve pump permits oil to return to the tank through a pilot line to the pilot-operated valve. When the pilot-operated valve prevents the escape of oil through this hole, the neutral control valve closes.

- *Charging-valve*—When it is open, this valve creates the pressure required to charge the variable-displacement piston-type pump with oil from the constant-displacement pump, and permits the surplus oil to return to the tank at charging pressure. When the valve is closed, the entire output of the constant-displacement pump is forced through the check valves on the variable-displacement pump to the actuated unit, producing the rapid-traverse action. A small hole in the valve plunger permits oil to return to the tank through a pilot line to the pilot valve. The pilot-operated directional control valve prevents the escape of oil through the small hole to close the charging valve.

- *Slow-feed valve*—With this valve, two preset rates of feed can be obtained. When the valve is open, a fast feed results. A slower rate of feed is obtained when the valve is closed. An intermediate feed can be obtained while the valve is open for a third rate of feed.

* *Pressure-relief valve*—When an actuated machine member feeds against a positive stop or during overload, this relief valve serves as a safety valve. A preset pressure setting is used for the valve.

The pipelines (see Figure 9-18) serve various purposes. The pipeline from the pilot-operated directional control valve to the tank is not numbered in the illustration. It returns the excess oil to the tank.

A pipeline (see No. 1 in Figure 9-18) is connected to each end of the spring-return four-way directional control valve to control the position of the valve spool. Actuation of the pilot- or solenoid-operated directional control valve admits pressure to one of the lines, and opens, simultaneously, the other line to the tank. This moves the four-way valve spool to the desired position.

In some four-way directional control valves, the spool is held in position by a spring for forward travel. Then only pipeline No. 1 is necessary. Oil pressure in pipeline No. 1 overcomes the resistance of the spring and shifts the valve spool to "rapid-return" position. When line No. 1 is opened to the tank, the spring returns the valve spool to its former position.

The spring end of the charging valve chamber is connected with the pilot-operated directional control valve by pipeline No. 2. When the pilot-operated directional control valve closes the line, the valve plunger in the charging valve is seated. This is accomplished by preventing passage of the oil through line No. 2 and returning to the tank. When pipeline No. 2 is open, the charging valve is open and feeding action occurs. Rapid-traverse action occurs when pipeline No. 2 is closed and the charging valve is closed.

The spring end of the neutral control valve chamber is connected to the pilot-operated directional control valve by pipeline No. 3. When the control valve closes the pipeline, the valve plunger in the neutral valve is seated. This is accomplished by preventing the escape of oil to the control valve and back to the tank. When pipeline No. 3 is closed, the neutral valve closes and either a feeding action or a rapid-traverse action occurs. When pipeline No. 3 is open, the neutral valve is open and the circuit is neutral.

Pipeline No. 4 connects the slow-feed valve to the charging valve, and pipeline No. 5 connects the chamber of the slow-feed control valve with the feed-adjusting housing.

Pipeline No. 6 is used only in conjunction with solenoid-operated directional control valves for remote pushbutton starting and for emergency-return action. Charging pressure from the constant-displacement pump is conducted through pipeline No. 7 to the remote directional control valve that, in turn, distributes the pressure to accomplish various functions.

## Operation of the Hydraulic Circuit

In the hydraulic circuit (see Figure 9-18), opening and closing of the neutral valve and the charging valve controls both the constant-displacement pump and the variable-displacement pump. The constant-displacement pump (which is self-priming) pumps hydraulic oil to a chamber that is open to the valve side of both the neutral valve and the charging valve.

In *rapid-traverse action*, all the oil from the constant-displacement pump is required, forcing it through the piston chambers in the variable-displacement pump and onward through the main cylinder. This is accomplished by actuating the pilot-operated control valve that closes pipelines No. 2 and No. 3, thereby closing both the neutral valve and the charging valve. The direction of rapid-traverse is determined by the position of the four-way directional control valve spool. On some models, the four-way directional control valve is constructed in such a manner that pipeline No. 2 is blocked automatically, to eliminate a reverse feed, when the valve spool is in the reverse rapid-traverse position.

The *feeding action* (either fast or slow) is obtained by opening the charging valve. This establishes sufficient pressure to keep the pistons in the variable-displacement pump against the wobbler plate, thereby charging the piston-type pump, which is not self-priming. The excess oil is returned to the tank through the charging valve. In "fast-feed" position, pipelines No. 2 and No. 4 are open to the tank, and pipeline No. 3 is closed. In "slow-feed" position, pipeline No. 2 to the tank is open, pipeline No. 3 is closed, and pipeline No. 4 is open to the charging pressure. The charging pressure shifts the "slow-feed" valve plunger to direct the pressure in pipeline No. 5 to the wobbler support plunger in the variable-displacement pump, which forces the plunger forward to provide the slow rate of feed by decreasing the piston travel.

The third or intermediate rate of feed can be obtained by means of a special feed-adjustment housing which contains an auxiliary wobbler plunger, feed-adjusting screw, and feed-adjusting cam used in conjunction with a solenoid-operated directional control valve. The additional directional control valve is connected to a high-pressure port in the pump housing and to a port in the feed-adjustment housing. When the solenoid-operated directional control valve is energized, pressure from the high-pressure port in the pump is admitted to the feed-adjusting housing and acts on the auxiliary wobbler plunger. De-energizing the solenoid-operated directional control valve blocks the high-pressure port and opens the port in the feed-adjustment housing to the tank. Then the feed-adjusting mechanism is free to be shifted to another rate of feed.

## Summary

The hydraulic circuit (including a hydraulic pumping unit and the necessary controls) can be used to operate and control machine tools in industry. Electric motors, along with the pulleys and belts, have been replaced, in many instances, by hydraulic power transmission units.

In the hydraulic pump, mechanical energy is converted to liquid pressure energy. In the hydraulic motor, liquid pressure energy is converted to mechanical energy.

Hydraulic motors are classified as either constant-displacement or variable displacement motors. To change the speed of the constant-displacement motor, the volume of oil flowing through the motor must be changed. Change of speed in the variable-displacement motor is accomplished by varying the displacement of the motor, in addition to controlling the supply of hydraulic oil from the pump. A wider range of speed can be obtained from a variable-displacement motor. The rotary hydraulic motors are classified as gear-type, vane-type, and piston-type. These motors are similar in construction to their respective types of hydraulic pumps.

In a typical hydraulic circuit that is used to control a machine tool, the pumping unit and a directional control valve are necessary to produce a complete operating cycle. The cycle can be controlled by placing dogs on the machine slide to actuate the arm of the directional control valve. The main control valves can be actuated by remote pilot-operated or solenoid-operated directional control valves. Additional rates of feed can be accomplished by the addition of the necessary directional control valves for performing the various machine functions.

## Review Questions

1. Explain the basic operating principle of the hydraulic motor in the hydraulic circuit.

2. What are the basic operating units in a typical hydraulic circuit for operating a cylinder on a machine tool?

3. Explain the basic difference between a constant-displacement motor and a variable-displacement motor.

4. List the various types of hydraulic motors.

5. Name five applications for hydraulic motors.

6. Name some advantages of a hydraulic motor over an electric motor.

7. Sketch a circuit showing the use of a constant-displacement motor used in conjunction with a hydraulic cylinder.

8. What will cause a hydraulic motor to lose power?

9. What purpose does a wobbler serve?

10. Draw the symbol for a four-way control valve.

11. Draw the symbol for a pressure gage.

12. Draw the symbol for an accumulator.

# Chapter 10

## Hydraulic Cylinders

The hydraulic cylinder is the component of the hydraulic system that receives the fluid (under pressure) from a supply line. The hydraulic oil in the cylinder acts on a piston to do work in a linear direction. The work that is performed is the product of the fluid pressure and the area of the cylinder bore (see Figure 10-1). The quantity of fluid delivered into the cylinder determines the speed or rate of doing work. The chief types of hydraulic cylinders are:

- Nonrotating
- Rotating

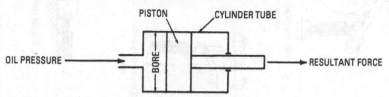

**Figure 10-1**   Schematic of a hydraulic cylinder.

### Nonrotating Cylinders

The applications for nonrotating cylinders are more common (see Figure 10-2) than for rotating cylinders. There are three types of nonrotating cylinders (see Figure 10-3):

- Double-acting
- Single-acting
- Plunger or ram-type

In the double-acting nonrotating cylinders, fluid pressure can be applied to either side of the piston. Therefore, work can be performed in either direction (see Figure 10-3a). In the single-acting nonrotating type of cylinder, the fluid pressure is applied to only one side of the piston. The piston is returned to its starting position by action of the spring in the spring-return type of single-acting cylinder (see Figure 10-3b) after the fluid pressure has been released from the piston. The plunger or ram-type of nonrotating cylinder is another type of single-acting cylinder, but it does not contain a piston (see Figure 10-3c). Either gravity or a mechanical means can

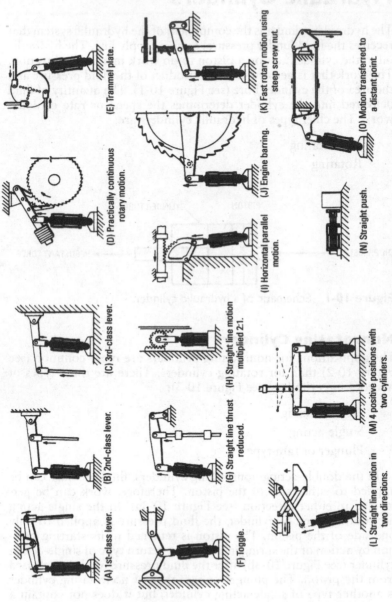

**Figure 10-2** Applications for the nonrotating type of hydraulic cylinder. *(Courtesy Rexnard, Inc, Hydraulic Component Div)*

(A) 1st-class lever.

(B) 2nd-class lever.

(C) 3rd-class lever.

(D) Practically continuous rotary motion.

(E) Trammel plate.

(F) Toggle.

(G) Straight line thrust reduced.

(H) Straight line motion multiplied 2:1.

(I) Horizontal parallel motion.

(J) Engine barring.

(K) Fast rotary motion using steep screw nut.

(L) Straight line motion in two directions.

(M) 4 positive positions with two cylinders.

(N) Straight push.

(O) Motion transferred to a distant point.

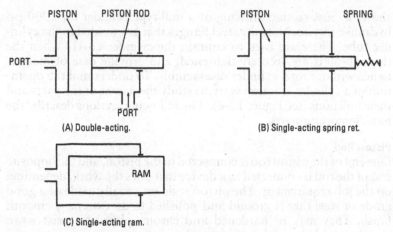

(A) Double-acting.

(B) Single-acting spring ret.

RAM

(C) Single-acting ram.

**Figure 10-3** Schematics of nonrotating cylinders: (A) double-acting; (B) single-acting spring-return; and (C) single-acting ram-type.

be used to return the piston or ram, if a spring-return is not used in the single-acting cylinder.

## Names of Parts
Figure 10-4 shows a cutaway view of a heavy-duty double-acting hydraulic cylinder. Note the heavy construction throughout, and the cushioning arrangement on each side of the piston. Figure 10-5

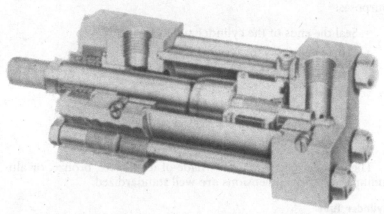

**Figure 10-4** Double-acting nonrotating type hydraulic cylinder with cushioning device at each end.

shows a cross-section drawing of a mill-type cylinder for 2000 psi hydraulic service. Note the steel flanges that are welded to the cylinder tube. These are used to contain the cylinder covers when the through-bolts are securely tightened, and provide ease of maintenance without total cylinder disassembly. To understand the operation of a cylinder, it is necessary to study the names of the parts and their functions (see Figure 10-6). The following sections describe the most important parts.

### Piston Rod

One end of the piston rod is connected to the piston, and the opposite end of the rod is connected to a device that does the work, depending on the job requirement. The piston rods are usually made of a good grade of steel that is ground and polished to an extremely smooth finish. They may be hardened and chrome-plated to resist wear. Stainless steel material is often used to resist corrosion.

### Rod Wiper

The rod wiper is used to remove foreign material from the rod as it is drawn backward into the packing. It is usually made of a durable synthetic material. A metallic scraper (see Figure 10-7) is sometimes needed to remove severe residues.

### Cylinder Covers

Each cylinder is provided with two covers: the front (*rod-end*) cover and the blank (*blind-end*) cover. The blind-end cover is sometimes a part of the tube or cylinder body. The covers are used for several purposes:

- Seal the ends of the cylinder tubes
- Provide a means for mounting
- Provide a housing for seals, rod bearings, and rod packing (front cover)
- Provide for ports of entry for the fluid
- Absorb the impact of the piston
- Provide space for a cushioning arrangement.

The cylinder covers may be made of iron, steel, bronze, or aluminum, and their dimensions are well standardized.

### Cylinder Tube

The cylinder tube may be constructed of cold-drawn seamless steel, brass, or aluminum tubing that is held to a close tolerance and

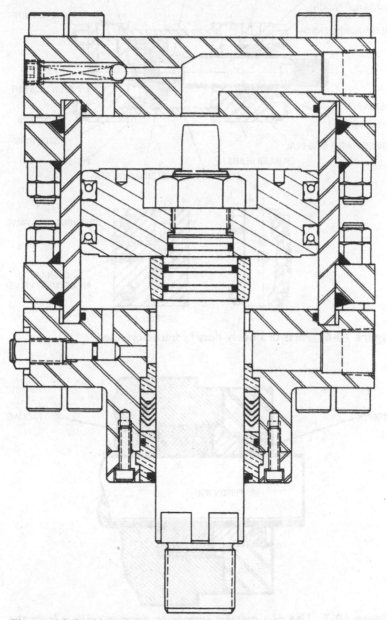

**Figure 10-5** Heavy-duty mill-type hydraulic cylinder. *(Courtesy Lynair, Inc.)*

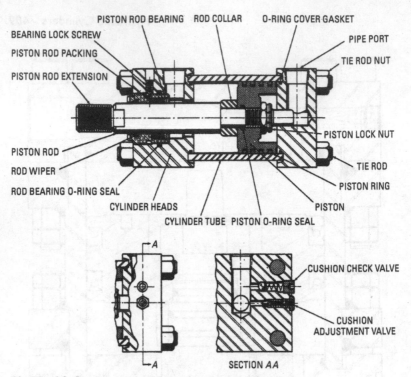

**Figure 10-6**   Parts of a heavy-duty hydraulic cylinder.

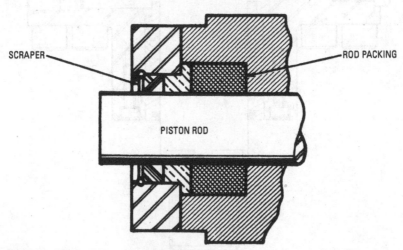

**Figure 10-7**   Use of a metallic scraper to remove residue from the piston rod.

honed to an extremely smooth finish. The sealing action is dependent largely on the finish provided in the cylinder tube. The covers are usually fastened to the cylinder tube with cover screws (see Figure 10-8).

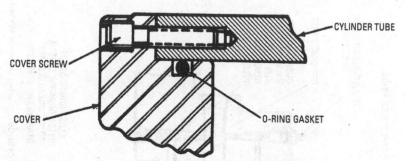

**Figure 10-8** Cover screws are used to fasten the cover to the cylinder tube.

### Piston Assembly

The function of the piston assembly in a hydraulic cylinder is similar to the function of the piston in an automobile. The piston must fit closely against the cylinder wall to provide a suitable bearing and to eliminate any possibility of extrusion of synthetic seals. Since the piston functions as a bearing, it must be constructed of a material that does not score the wall of the cylinder tube. Cast iron with high tensile strength performs well. Figure 10-9 shows one of the most common types of pistons for hydraulic service. The piston uses automotive-type rings that may be made of cast iron or bronze. The piston is designed with a relief at each end, so that small particles of dirt that may enter the system do not spring the end gland, causing the piston rings to be crushed or frozen. Chevron-type packing, cup packing, and O-rings are other types of sealing means that may be used.

### Piston Locknut

The locknut keeps the piston tightly secured to the piston rod. The need for setscrews, pins, and the other locking means, which often work loose and drop into the cylinder to cause considerable damage, is eliminated.

### Tie Rods

The cylinder is held together by tie rods. They must be strong enough to absorb the shock loads that occur when the piston contacts the cover of the cylinder.

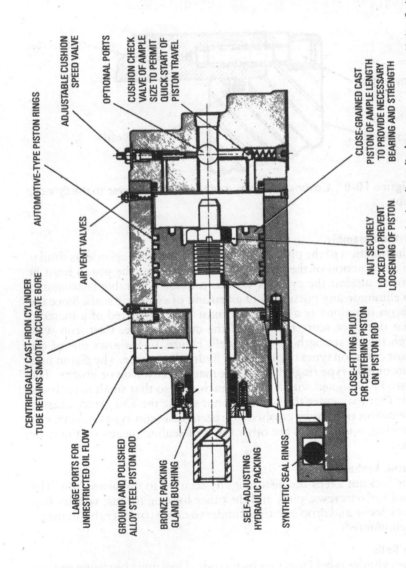

**Figure 10-9** A piston with automotive-type rings in a hydraulic cylinder. *(Courtesy Logansport Machine Co., Inc.)*

CENTRIFUGALLY CAST-IRON CYLINDER
TUBE RETAINS SMOOTH ACCURATE BORE

AUTOMOTIVE-TYPE PISTON RINGS

ADJUSTABLE CUSHION
SPEED VALVE

OPTIONAL PORTS

CUSHION CHECK
VALVE OF AMPLE
SIZE TO PERMIT
QUICK START OF
PISTON TRAVEL

CLOSE-GRAINED CAST
PISTON OF AMPLE LENGTH
TO PROVIDE NECESSARY
BEARING AND STRENGTH

AIR VENT VALVES

NUT SECURELY
LOCKED TO PREVENT
LOOSENING OF PISTON

CLOSE-FITTING PILOT
FOR CENTERING PISTON
ON PISTON ROD

LARGE PORTS FOR
UNRESTRICTED OIL FLOW

GROUND AND POLISHED
ALLOY STEEL PISTON ROD

BRONZE PACKING
GLAND BUSHING

SELF-ADJUSTING
HYDRAULIC PACKING

SYNTHETIC SEAL RINGS

**Cushion Collar and Nose**

The function of the cushion collar and nose is to alleviate shock as the piston approaches the cylinder covers. It not only alleviates the shock to the piston and the covers, but also eliminates the sudden stopping of an object that is connected to the end of the piston rod. For example, if a basket containing dishes is pushed and then stopped suddenly, the dishes will spill and break. The cushioning device eliminates this. The cushion device on the rod end side of the cylinder is known as the *cushion collar*, and the device on the blind side is called the *cushion nose*. On many applications, standard-length cushions (that have an effective length of about 1 inch) are not always long enough. In these instances, extra-long cushions should be installed. Adding length to the cushion on the blind end of the cylinder usually does not present any problem, since the blind-end cover of the cylinder can usually be drilled through and tapped. Then a pipe nipple with a cap can be screwed into the cover to receive the longer cushion nose. The rod end presents a more difficult problem, since some type of spacer must be used inside the cylinder tube to take care of the added cushion length. This shortens the stroke of the cylinder, or if the same stroke is required, the overall length of the cylinder becomes greater.

Longer cushion lengths are often required on cylinders moving a large mass on wheels, bearings, or other free-moving means at high speed. Figure 10-10 shows some of the cushioning designs that have proved very useful. Cushions 6 to 8 inches long are often necessary. Figure 10-10a shows a gradual taper on the cushion nose that, as it enters the cushion bushing, gradually closes the recess and traps the fluid. This fluid must then flow through a preset orifice. In Figure 10-10b, the cushion nose (which is only a few ten-thousandths of an inch smaller in diameter than the cushion bushing) enters the bushing, but the V groove in the nose allows the fluid to escape, until the groove is gradually closed off. Then the fluid must escape through the preset orifice. In Figure 10-10c, the holes are closed off as the cushion nose penetrates farther into the bushing until all the holes are closed off. Then, the fluid must escape through the preset orifice. The advantage of these types of cushions is that the speed is gradually reduced, instead of being reduced abruptly.

**Cushion-Adjustment Valve**

This valve works with the cushion nose and collar, and is a part of the entire cushioning arrangement. When the trapped hydraulic oil (which occurs after the cushion closes the bushing) is metered out, it must pass by the needle of the cushion-adjustment valve, which

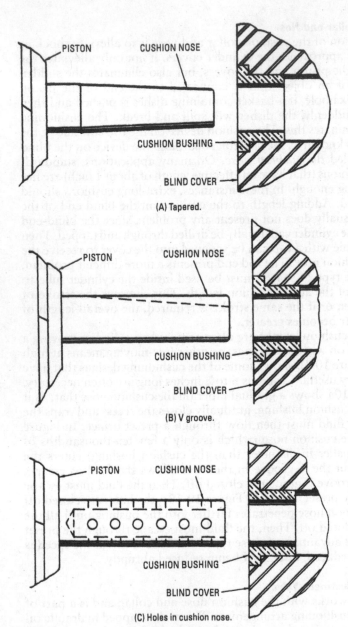

(A) Tapered.

(B) V groove.

(C) Holes in cushion nose.

**Figure 10-10**  Various types of cushion noses.

determines the size of the exhausting orifice. The trapped oil cannot flow past the ball check, because it is blocked in that direction. Some designs are made with a fixed orifice. The chief advantage of the adjustable orifice is that it can be changed for different loadings and operating pressures.

### Piston-Rod Bearing
The piston-rod bearing not only houses the rod packing, but also acts as a bearing and as a guide for the piston rod. The rod bearings are made of cast iron or a good-quality bronze.

### Rod Packing
The rod packing seals the piston rod to prevent escape of the oil from around the rod. Various designs of the piston rod packing are used, such as chevron, blocked-V, O-ring, quad-ring, and Sea-ring. Various materials (such as synthetic rubber, leather, Teflon, and nylon) are used in the packing, depending on the application.

### Cover Gaskets
These gaskets serve as a seal between the cylinder cover and the cylinder tube. When the O-ring is placed as shown in Figure 10-6, a perfect seal that is nearly indestructible is formed. The seal becomes tighter as the pressure is increased.

## Force Developed in Nonrotating Cylinders
Tremendous forces can be developed in non-rotating cylinders (see Table 10-1). Their requirements should be realized before discussing the installation of these cylinders. Table 10-1 is based on the following formulas:

$$F = PA$$

and

$$F = P(A - A_1)$$

where the following is true:

$F$ is force (theoretical force, not including friction) developed

$P$ is supply pressure (in psi)

$A$ is area (in square inches) of cylinder bore

$A_1$ is area (in square inches) of cross-section of piston rod

To determine the force developed when fluid pressure is applied to the blind end (the end opposite the piston rod) of the cylinder, the formula $F = PA$ is used, and the formula $F = P(A - A_1)$ is used

## Table 10-1 Hydraulic Cylinder "Push and Pull" Force Chart

| Cyl. Bore Dia. | Piston Rod Dia. | Cyl. Work Action | Work Area (sq. In.) | Operating Pressure (psi) | | | | | | | | | |
|---|---|---|---|---|---|---|---|---|---|---|---|---|---|
| | | | | 250 | 500 | 750 | 1000 | 1250 | 1500 | 1750 | 2000 | 2500 | 3333 |
| 1½ | | PUSH | 1.767 | 442 | 884 | 1326 | 1767 | 2209 | 2650 | 3092 | 3534 | 4418 | 5301 |
| | 5/8 | PULL | 1.460 | 365 | 730 | 1095 | 1460 | 1825 | 2190 | 2550 | 2920 | 3650 | 4380 |
| | 1 | PULL | 0.982 | 246 | 491 | 737 | 982 | 1218 | 1473 | 1719 | 1964 | 2455 | 2946 |
| 2 | | PUSH | 3.141 | 785 | 1571 | 2356 | 3141 | 3926 | 4712 | 5497 | 6282 | 7853 | 9423 |
| | 1 | PULL | 2.356 | 589 | 1178 | 1767 | 2356 | 2945 | 3534 | 4123 | 4712 | 5890 | 7068 |
| | 1⅜ | PULL | 1.656 | 414 | 828 | 1242 | 1656 | 2070 | 2484 | 2898 | 3312 | 4140 | 4968 |
| 2½ | | PUSH | 4.908 | 1227 | 2454 | 3681 | 4908 | 6135 | 7362 | 8579 | 9816 | 12270 | 14724 |
| | 1 | PULL | 4.123 | 1031 | 2062 | 3093 | 4123 | 5154 | 6185 | 7215 | 8246 | 10308 | 12369 |
| | 1¾ | PULL | 2.503 | 626 | 1252 | 1877 | 2503 | 3129 | 3755 | 4380 | 5006 | 6258 | 7509 |
| 3¼ | | PUSH | 8.295 | 2074 | 4148 | 6221 | 8295 | 10369 | 12443 | 14516 | 16590 | 20738 | 24885 |
| | 1⅜ | PULL | 6.810 | 1703 | 3405 | 5108 | 6810 | 8513 | 10215 | 11918 | 13620 | 17025 | 20430 |
| | 2 | PULL | 5.154 | 1289 | 2577 | 3866 | 5154 | 6443 | 7731 | 9020 | 10308 | 12885 | 15462 |
| 4 | | PUSH | 12.566 | 3142 | 6283 | 9425 | 12566 | 15708 | 18849 | 21991 | 25132 | 31415 | 37698 |
| | 1¾ | PULL | 10.161 | 2540 | 5081 | 7621 | 10161 | 12701 | 15242 | 17782 | 20322 | 25403 | 30483 |
| | 2½ | PULL | 7.658 | 1915 | 3829 | 5744 | 7658 | 9573 | 11487 | 13402 | 15316 | 19145 | 22974 |
| | | PUSH | 19.356 | 4909 | 9818 | 14726 | 19635 | 24544 | 29453 | 34361 | 39270 | 49088 | 58905 |

| | | | | | | | | | | | | | |
|---|---|---|---|---|---|---|---|---|---|---|---|---|---|
| 5 | 2 | PULL | 16.494 | 4124 | 8247 | 12371 | 16494 | 20618 | 24741 | 28865 | 32988 | 41235 | 49482 |
| | 3½ | PULL | 10.014 | 2504 | 5007 | 7511 | 10014 | 12518 | 15021 | 17525 | 20028 | 25035 | 30042 |
| 6 | | PUSH | 28.274 | 7069 | 14137 | 21206 | 28274 | 35343 | 42411 | 49480 | 56548 | 70685 | 84822 |
| | 2½ | PULL | 23.366 | 5842 | 11683 | 17525 | 23366 | 29208 | 35049 | 40891 | 46732 | 58415 | 70098 |
| | 4¼ | PULL | 14.088 | 3522 | 7044 | 10566 | 14088 | 17610 | 21132 | 24654 | 28176 | 35220 | 42264 |
| 7 | | PUSH | 38.485 | 9621 | 19243 | 28864 | 38485 | 48106 | 57728 | 67349 | 76970 | 96213 | 115455 |
| | 3 | PULL | 31.416 | 7854 | 15708 | 23562 | 31416 | 39270 | 47124 | 54978 | 62832 | 78540 | 94248 |
| | 5 | PULL | 18.850 | 4713 | 9425 | 14138 | 18850 | 23563 | 28275 | 33688 | 37700 | 47125 | 56550 |
| 8 | | PUSH | 50.265 | 12566 | 25133 | 37689 | 50265 | 52831 | 75398 | 87964 | 100530 | 125663 | 150795 |
| | 3½ | PULL | 40.644 | 10161 | 20322 | 30483 | 40644 | 50805 | 60966 | 71127 | 81288 | 101610 | 121932 |
| | 5¾ | PULL | 24.298 | 6074 | 12149 | 18224 | 24298 | 30373 | 36447 | 42522 | 48596 | 60745 | 72894 |
| 10 | | PUSH | 78.540 | 19635 | 39270 | 58905 | 78540 | 98175 | 117810 | 137445 | 157080 | 196350 | 235620 |
| | 4½ | PULL | 62.636 | 15659 | 31318 | 46977 | 62636 | 78295 | 93954 | 109613 | 125272 | 156590 | 187908 |
| | 5¾ | PULL | 52.573 | 13143 | 26287 | 39430 | 52573 | 65716 | 78860 | 92003 | 105146 | 131433 | 157719 |
| 12 | | PUSH | 113.10 | 28275 | 56550 | 84825 | 113100 | 141375 | 169650 | 197925 | 226200 | 282750 | 339300 |
| | 5½ | PULL | 89.34 | 22335 | 44670 | 67005 | 89340 | 111675 | 134010 | 156345 | 178680 | 223350 | 268020 |
| | 7 | PULL | 74.61 | 19653 | 37305 | 55958 | 74610 | 93263 | 111915 | 130568 | 149220 | 196525 | 223830 |
| 14 | | PUSH | 153.94 | 38485 | 76970 | 115455 | 153940 | 192425 | 230910 | 269395 | 307880 | 384850 | 461820 |
| | 7 | PULL | 115.45 | 28863 | 57725 | 86593 | 115450 | 144313 | 173175 | 202038 | 230900 | 288625 | 346350 |
| | 10 | PULL | 75.40 | 18850 | 37700 | 56550 | 75400 | 94250 | 113100 | 131950 | 150800 | 188500 | 226200 |

This chart lists theoretical values produced by various working pressures and does not allow for loss due to friction. Courtesy LynAir, Inc.

to determine the force created when fluid pressure is applied to the rod end of the cylinder. When the same force is applied either to the fluid end or to the rod end of a cylinder, the higher force is always exerted by the blind end, because it has a greater area.

Nonrotating cylinders can be specially designed to operate at pressures as high as 10,000 pounds per square inch. However, these applications are rare. Standard nonrotating hydraulic cylinders are designed for the different pressure ranges, and they should be used for applications within those ranges. These operating ranges are designated (in pounds per square inch of operating pressure for hydraulic oil) as 0–150, 0–750, 0–1500 (see Figure 10-11), 0–2000, 0–3000, and 0–5000.

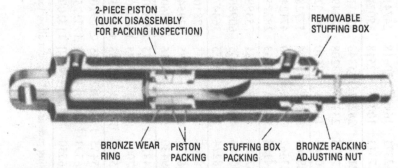

**Figure 10-11** Cutaway view of a hydraulic cylinder (1500 psi).
*(Courtesy Commercial Shearing & Stamping Company of Youngstown, Ohio)*

## Installation
Since installation is an important factor in the performance of a cylinder, the different mounting styles should be studied to obtain the best results. Of course, all mountings should be fastened securely.

### Threaded-Tube Construction
The tube is threaded on the inside diameter (ID), and the pilot of the cover is threaded (see Figure 10-12). The two parts together form the cylinder assembly. Various types of seals are used between the cover and the cylinder tube. An O-ring is shown in the illustration.

Cushion designs vary considerably. The chief purpose of the cushion is to eliminate or to reduce shock at the end of the piston travel. The cushions can be either adjustable or nonadjustable, but the adjustable cushions are more prevalent. The cushions are designed with a close fit between the metal cushion nose and the metal cushion

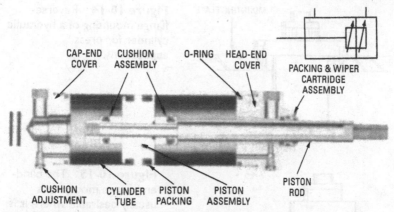

**Figure 10-12**  Cutaway view of a hydraulic cylinder with threaded-tube construction. Note the O-ring seals between the covers and the cylinder tube. *(Courtesy Pathon Manufacturing Company, Div. of Parker-Hannifin Corp.)*

Labels: CAP-END COVER, CUSHION ASSEMBLY, O-RING, HEAD-END COVER, PACKING & WIPER CARTRIDGE ASSEMBLY, CUSHION ADJUSTMENT, CYLINDER TUBE, PISTON PACKING, PISTON ASSEMBLY, PISTON ROD

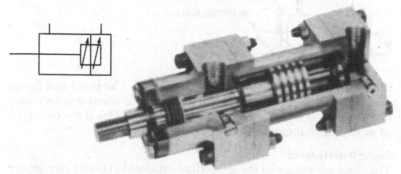

**Figure 10-13**  Cutaway view of cylinder with metal cushion sleeve and nose. The cushion sleeve floats on the piston rod. *(Courtesy Parker-Hannifin Corp.)*

orifice (see Figure 10-13). Note that the end of the cushion nose is tapered slightly. In some cushion designs, the cushion nose enters a synthetic seal to provide a positive shutoff in the cushion orifice.

### Flange-Mounted

Front–flange-mounted hydraulic cylinders are adapted to pulling applications. The front cover bears against the mounting plate that relieves the pressure on the mounting screws. Reverse-flange mounting (see Figure 10-14) is preferred for press applications, because

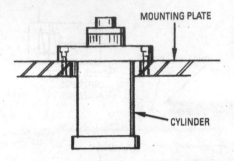

**Figure 10-14** Reverse-flange mounting of a hydraulic cylinder for press applications.

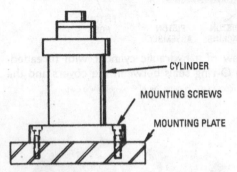

**Figure 10-15** The blind-end flange mounting is usually desirable where it is necessary to mount the cylinder in the base of a machine.

the strain on the mounting screws is relieved. The blind-end flange mounting (see Figure 10-15) is usually desirable where it is necessary to mount the cylinder in the base of a machine. This is the only type of mounting that can be used on some machines.

**Centerline-Mounted**
The chief advantage of the centerline-mounted cylinder (see Figure 10-16) is that the mounting feet or lugs are in a direct line with the

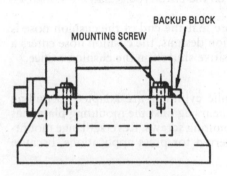

**Figure 10-16** Hydraulic cylinder with mounting lugs at the centerline.

centerline of the thrust. If the mounting bolts are keyed, full thrust on the mounting bolts is eliminated.

### Foot-Mounted

The foot-mounted method (see Figure 10-17) is commonly used for mounting non-rotating cylinders. Some torque is created, because the mounting lugs are in a different plane from that of the centerline of the thrust. Therefore, the mounting lugs should be keyed to reduce the thrust on the mounting bolts. Ample support for the piston rod should be provided on this type of cylinder, especially on cylinders with long strokes. The same type of guide should be provided for the end of the rod to prevent sagging, which causes wear on the rod bearing or damage to the piston and cylinder walls. A center support (see Figure 10-18) should be provided for extra-long cylinders to reduce sagging of the cylinder tube. This support should be aligned perfectly with the mountings at the front and rear of the machine.

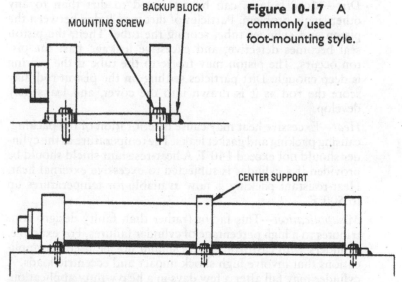

BACKUP BLOCK

MOUNTING SCREW

**Figure 10-17** A commonly used foot-mounting style.

CENTER SUPPORT

**Figure 10-18** A center support should be provided for long-stroke foot-mounted cylinders.

### Eccentric Loads

A loading problem always exists with nonrotating cylinders. If possible, these cylinders should be loaded concentrically. If this is impossible, provisions for eccentric loading should be made. The use of heavy guide rods and bearings (see Figure 10-19) is one method of compensating for eccentric loading.

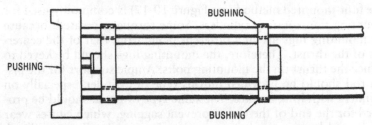

**Figure 10-19**  Method of compensating for eccentric loads.

## Causes of Failure

Several causes contribute to failure of nonrotating cylinders. Following are some major causes:

- *Dirt*—More failures can be attributed to dirt than to any other single problem. Particles of dirt may lodge between the piston and cylinder tube, scoring the tube. Then, the piston seal becomes defective, and excessive leakage past the piston occurs. The piston may freeze to the tube if the scoring is deep enough. Dirt particles settling on the piston rod may score the rod as it is drawn into the cover, and leaks may develop.

- *Heat*—Excessive heat may cause deterioration of the packing, causing packing and gasket leaks. The temperature at the cylinder should not exceed 140°F. A heat-resistant shield should be provided if a cylinder is subjected to excessive external heat. Heat-resistant packing is now available for temperatures up to 500°F.

- *Misapplication*—This factor (rather than faulty design) contributes to a high percentage of cylinder failures. For example, cylinders with cast-iron covers should not be used for applications that involve high shock impact and eccentric loads. A cylinder may fail after a few days in a heavy-duty application, but the same cylinder may provide satisfactory service on a medium-duty application for several years.

- *Misalignment, side thrust, and improperly supported eccentric loading*—Excessive wear on one side of the piston rod, leaks in the rod packing, and wear on one side of the rod bearing are early indications of the factors that lead to cylinder failure.

A bent piston rod, a broken packing gland, a scored cylinder tube, a broken cylinder cover, or a broken piston may result in serious damage.

- *Faulty mountings*—If the mounting is not secure (or is not strong enough to withstand the load produced by the cylinder), the cylinder may break loose and damage the mounting or the application with that it is connected.

## Repair and Maintenance
Suggestions for dismantling, repairing, and assembling nonrotating cylinders are as follows:

- Drain all oil from the cylinder, and dismantle in a clean location. Do not attempt to dismantle a cylinder when pressure is applied to the cylinder.
- Each part should be cleaned. Metal parts that are to be reused should be coated with a good preservative and placed in protective storage if the cylinder is to remain dismantled.
- The piston rod should be checked for straightness. If the rod is bent, it can be straightened by placing it in V blocks in a press. The piston rod should be examined for scores, indentations, and scratches. If the blemishes are not too deep, they can be removed with a fine emery cloth. However, if it is necessary to grind the rod, it can be chrome-plated to restore the diameter to its original size.
- The cover and cushion bushings should be examined for wear and finish, and they should be replaced if they are not in first-class condition.
- The cylinder tube should be repaired or replaced if it is damaged. Deep scores are difficult to repair, and it may be necessary to chrome-plate the tube.
- All the seals and gaskets should be replaced during reassembly of a cylinder. If metal piston rings are used, the manufacturer's specifications should be checked for gap clearance. If synthetic or leather seals are used on the piston, care should be exercised in placing the piston in the tube, so that the sealing surfaces are not damaged. When installing the packing, light grease makes assembly much easier.
- If a metal piston with rings is to be replaced, grind the piston concentrically with the piston rod after assembly to the rod. Grind the piston to fit the tube closely.

- Cylinders with foot-mounted covers should be assembled on a surface plate. The mounting pads of both covers should make full contact with the surface plate. Otherwise, a binding action may occur after mounting, or a mounting foot may be broken.
- The cover bolts should be tightened evenly. If O-ring or quadring gaskets are used to seal the tube and cover, tension on the cover screw is reduced to a minimum.

After assembly of a cylinder has been completed, it should be tested at low operating pressure for freedom of movement of the piston and rod, and to ensure that they are not being scored or bound. Then the pressure should be increased to the full operating range, and the cylinder checked for both internal and external leakages.

To check for internal leakage, place fluid pressure in the blindend port of the cylinder, forcing the piston to the rod end; then check the amount of fluid coming from the rod and cover port. Try placing fluid pressure in the rod-end port, moving the piston to the blind-end cover, and check for leakage at this position. If leakage is to be checked at other positions, use an external means to block the ports at these positions and check as described earlier. When excessive leakage occurs, the cylinder must be disassembled to make corrections. If metal piston rings are used in a hydraulic cylinder, the leakage varies with the operating pressure, the cylinder bore, and the oil temperature and viscosity. Leakage should be nearly nonexistent if the piston is sealed with synthetic or leather seals.

A cylinder should be mounted securely when it is reattached to a machine or fixture. It should always be remembered that these cylinders are capable of delivering considerable force. Cylinder bores may range from 1 inch in diameter to more than 30 inches in diameter. The strokes may range from less than 1 inch to more than 30 feet.

## Rotating Cylinders

A study of the names of the parts and their functions is necessary to understand the operation of a rotating cylinder. Figure 10-20 shows a rotating hydraulic cylinder.

### Names of Parts

The following sections describe the important parts of a rotating cylinder.

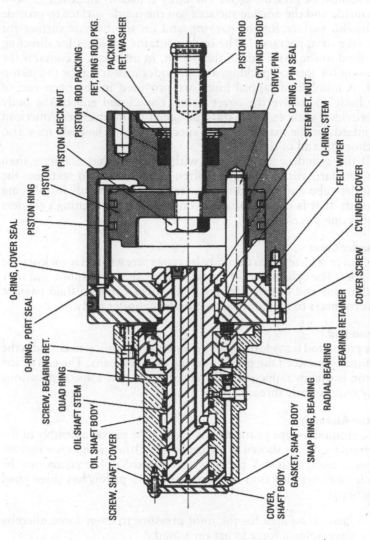

**Figure 10-20** Parts of a rotating hydraulic cylinder. *(Courtesy Logansport Machine Co., Inc.)*

## Body

The cylinder body is usually made of either cast iron or aluminum, and it must be pressure-tight. The body is usually matched on both the inside and the outside surfaces (on the inside surface to provide a smooth surface for the packing and on the outside surface for the sake of appearance). The body contains passages for directing the fluid to the front side of the piston. In addition, it contains the housing for the rod packing, and provides a bearing for the piston rod. A number of tapped holes are provided in the open end of the body to receive the cover bolts. The closed end of the body is provided with an adaptation designed to meet either American Standards or the manufacturer's specifications. The body may also anchor one end of the drive pin.

If the cylinders are designed with a stroke that is longer than the standard stroke, the body often consists of two sections: the cylinder tube and the rod-end cover. This design aids in reducing porosity that is often a problem on long-stroke rotating cylinders with a one-piece body.

## Cylinder Cover

The cover is fastened to the body by cover screws, and it encloses the cylinder. The cylinder cover carries the oil shaft assembly, and may anchor one end of the drive pin. The cover carries a fluid passage that connects to the fluid passage in the cylinder body.

## Piston Rod

The piston rod is made of a ground and polished alloy steel, and is the connector between the piston and the driven means. The end of the piston is usually tapped with female threads, but some applications may require male threads.

## Piston Assembly

An automotive-type piston ring is used in the piston assembly of the hydraulic cylinder shown in Figure 10-20. Other designs use various types of seals, such as V packing or blocked-Vs. The piston may be made of either cast iron or aluminum. The piston has three chief functions:

- Provide an area for the fluid pressure to exert force, thereby developing a force to act on a load.
- Provide a seal that prevents leaking or escaping of fluid to the exhaust side.
- Act as a guide or bearing.

### Drive Pins

These pins prevent rotation of the piston in relation to the cylinder body. Drive pins are needed because brakes are provided on many machines with rotating spindles. Without the drive pins, a sudden stopping of the spindle may cause the piston to rotate within the cylinder body, causing the piston rod to become loosened or disconnected from its connecting part. This often causes considerable trouble, and considerable time is consumed in correcting it.

### Rod Packing

Since most of the work is usually done on the in-stroke, and fluid pressure is exerted on the rod packing most of the time, the rod packing is extremely important in a rotating cylinder. The efficiency on the in-stroke is reduced greatly by any leakage past the packing. On rotating cylinders, it is a quite difficult job to make a packing change. Therefore, the packing must be durable. Synthetic rubber, impregnated leather, Teflon, and other materials are used for packing. Several designs (such as chevron, Sea rings, and hat) are used for packing.

### Cover Gasket

This gasket seals the cylinder body and the cylinder cover, and it contains holes for the cover screws and for the fluid passages between the body and the cover. Although the cover gasket can be made of a thin sheet of gasket material, it must be strong enough to prevent escape of the fluid.

### Oil Shaft Stem

This is an important part of the action of the rotating cylinder. The shaft stem must be sturdy enough to provide proper support to the remainder of the oil shaft-assembly and to withstand external force that results from the weight of the flexible hoses and pipe connectors. The oil shaft stem is made of hardened alloy steel and ground to a high finish, so that the packing is provided with a smooth bearing surface. An oil shaft assembly is often referred to as an *oil distributor*.

### Distributor Body

The distributor body functions as a housing for the shaft packing. It also functions as a housing for the means of lubrication, to provide ports of entry, and to retain the bearing. The body may be made of such materials as cast iron, cast aluminum, or cast bronze. It should be arranged to dissipate as much heat as possible from the packing and the shaft, and the body must be rugged enough to withstand the strain of the pipelines and connectors. Water-cooled distributor bodies containing large-cored passages are often necessary to effect a better cooling action where hollow oil shafts are used.

### Shaft Packing

As little friction as possible must be created by the shaft packing. However, it must provide an effective seal at shaft revolutions up to 5000 rpm and higher. Shredded lead, lead and graphite, asbestos, and other materials are used to combat the heat condition. Various packing shapes (such as formed-wedge, V, and hat types) are used.

### Bearing

A bearing is required to provide support between the shaft body and the shaft stem. Depending on the shaft design, various types of bearings, such as ball bearings, sleeve, and thrust bearings, are used. These bearings must be able to withstand considerable heat, and lubrication to the bearings is extremely important.

### Other Parts

A number of minor parts are important in the function of a rotating cylinder. Some of these parts are: retainer rings, spacers, O-ring gaskets, and screws. Incorrect assembly of these minor parts or failure to replace worn or damaged seals, gaskets, and so on, may cause the cylinder to malfunction or to fail to operate properly, resulting in increased maintenance and repair costs.

## Installation

The pressure chart shown in Table 10-1 should be studied before actual installation of a rotating cylinder is begun, if you are to visualize the great force that these cylinders can develop. Although pneumatic rotating cylinders are seldom operated at more than 90 pounds of air pressure, hydraulic rotating cylinders may be operated at pressures as high as 1000 psi.

Since the rotating cylinder may be subjected to high revolving speeds, installation is an extremely important factor. As mentioned previously, rotating cylinders are constructed with the mounting as a part of the cylinder body. On the small-diameter cylinders, the mounting is usually threaded. A number of tapped holes for mounting purposes are used on the larger cylinders.

Rotating cylinders are not mounted directly onto the machine spindle. They are mounted on an adapter that has previously been mounted onto the machine spindle. Adapters are used because the manufacturers of lathes and other machines with spindles use different end designs on the cylinder end of the machine spindle. If a manufacturer produces a dozen (or more) different spindle sizes (depending on the range of sizes of the machine), each spindle end may have different dimensions. The adapter is usually made of the same material as the body of the rotating cylinder. Therefore, an aluminum adapter is used for an aluminum cylinder.

Before installing the cylinder, it is important that the adapter is mounted properly. The adapter should bottom against the end of the spindle and then it should be locked securely. The locking action is accomplished by tightening the locking screws on a split-type adapter. On a threaded-type adapter, bronze plugs are forced inward against the threads on the spindle, and locked in place with setscrews. Two plugs spaced at a 90-degree angle from each other are often used (see Figure 10-21). After the adapter is locked securely in place, the pilot on the cylinder end should be checked for a run out with an indicator. If the run out is more than 0.002 inch, the adapter should be corrected. Since all rotating cylinders need a draw bar to make a connection to the mechanism that they operate, the draw bar is screwed into the end of the piston rod of the cylinder (see Figure 10-22). Then the cylinder is mounted onto the adapter, using the mounting screws. The adapting surfaces on the cylinder should be clean and free of burrs, since they may cause trouble. In addition, the mounting screws should not be long enough to bottom in the mounting holes in the cylinder. Otherwise, the cylinder will not fit tightly on the adapter. After the cylinder has been mounted, the indicator should be used to check the outside diameter of the cylinder for run out. If the run out is more than 0.003 inch, the cylinder mounting should be corrected.

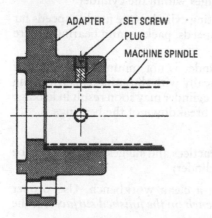

ADAPTER    SET SCREW
           PLUG
           MACHINE SPINDLE

**Figure 10-21** An adapter for a rotating type of cylinder.

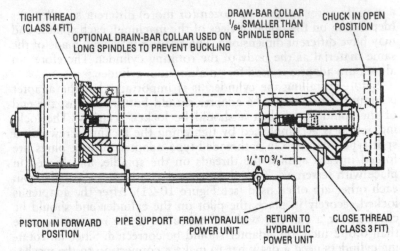

TIGHT THREAD
(CLASS 4 FIT)

OPTIONAL DRAIN-BAR COLLAR USED ON
LONG SPINDLES TO PREVENT BUCKLING

DRAW-BAR COLLAR
$^1/_{64}$ SMALLER THAN
SPINDLE BORE

CHUCK IN OPEN
POSITION

PISTON IN FORWARD
POSITION

PIPE SUPPORT

FROM HYDRAULIC
POWER UNIT

RETURN TO
HYDRAULIC
POWER UNIT

CLOSE THREAD
(CLASS 3 FIT)

$^1/_4$" TO $^3/_8$"

**Figure 10-22** Installation of power-operated chucking equipment on a machine. *(Courtesy Logansport Machine Co., Inc.)*

## Failure

Following are the major causes of failure in rotating cylinders:

- *Lack of lubrication*—Improper lubrication to the bearings causes a hydraulic rotating cylinder to fail. An increase in friction caused by lack of lubrication causes piston and rod packing failure in air cylinders.
- *Dirt*—Dirt inside the cylinder causes packing and bearing failure, scoring of the piston rod, shaft stem, and cylinder body, and the fouling of the passages within the cylinder.
- *Misapplication*—If the rotating cylinders are used at speeds far in excess of their designed speeds, packing and bearing failure usually results.
- *Poor installation*—If a cylinder is not mounted solidly or if it is not mounted concentrically with the spindle, a whipping action on the rear end of the cylinder may soon result in leakage of the shaft packing and in breakdown of the bearings.

## Repair and Maintenance

Following are some important practices and suggestions that can be used in dismantling a rotating cylinder:

- Dismantle the cylinder on a clean workbench. Use proper tools. *Do not use a pipe wrench on the finished surfaces*. If the cylinder is placed in a vise, use soft pads and do not apply too much pressure to the cylinder body. Loosen the rod-packing retainer. Then, remove the cover screws, take off the cover, and remove the piston and piston rod.

- Each part should be cleaned thoroughly as it is removed. Parts with internal passages should be cleaned out with compressed air. A protective coating should be applied to steel or iron parts, and the parts placed in protected storage if the cylinder is to be dismantled for a prolonged length of time.

- Score marks in the cylinder body should be eliminated, or the body should be scrapped if the scores are too deep. In a cast-iron body, the marks can be brazed, and then the interior of the body can be refinished.

- If the piston rod is scored, the marks may be removed with a fine emery cloth, unless they are too deep, in which case it is probably cheaper to replace the rod.

- It is advisable to replace all packing and seals when the cylinder is disassembled. In installing cup packing, apply enough tension so that they do not leak, but do not apply enough tension to turn in the lips. If the piston parts are designed to make metal-to-metal contact, too much cam pressure may damage the cups.

- If the rod bearing in the cylinder body is worn, it should be replaced. If the rod bearing is part of the body, there is stock enough (in the body) to bore out and press in a bronze sleeve-type bearing. The bearing must be tight.

- A worn or scored oil shaft should be replaced, because this is a vital part of the cylinder. Any rough portions on the shaft can be a source of trouble.

- In reassembly, all screws should be tightened securely. Tighten the cover screws evenly. Care should be exercised to prevent cuts in the packing and gaskets as they are placed.

- After the cylinder has been reassembled and before applying pressure, lubricate and rotate the distributor body to ensure that it moves freely and without any binding action.

- Apply fluid pressure to one port of the distributor body, and check piston movement. Place a finger over the pipe port to check for leaks. Then shift the pressure connection to the second port, let the piston move to the end of its travel, and place a finger over the first port to check for leaks. Slight packing leaks that are caused by a slight blow of the shaft can usually be taken care of after the cylinder has been run in, by pulling upward on the shaft packing. If excessive leakage occurs, the cylinder should be dismantled.

## Summary

The hydraulic cylinder receives the fluid (under pressure) from a supply line. The oil in the cylinder acts on a piston to do work in a linear direction. The work that is performed is the product of the fluid pressure and the area of the cylinder bore. The chief types of hydraulic cylinders are nonrotating and rotating.

Three types of nonrotating cylinders are available: double acting, single acting and plunger or ram-type. In the double-acting cylinders, fluid pressure can be applied to either side of the piston. Therefore, work can be performed in either direction. Fluid pressure is applied to only one side of the piston in a single-acting cylinder. The piston is returned to its starting position by action of the spring in a spring-return type of single-acting cylinder. The plunger or ram-type cylinder is another type of single-acting cylinder.

Tremendous forces can be developed in non-rotating cylinders. To determine the force developed when pressure is applied to the blind end (the end opposite the piston rod) of the cylinder, the formula $F = PA$ is used. The formula $F = P(A - A_1)$ is used to determine the force created when fluid pressure is applied to the rod end of the cylinder.

Standard nonrotating hydraulic cylinders are designed for the different pressure ranges, and they should be used for applications within these ranges. These operating ranges are designated (in pounds per square inch of operating pressure) as 0–150, 0–750, 0–1500, 0–2000, 0–3000, and 0–5000.

Installation is an important factor in the performance of a nonrotating cylinder. The different styles of mounting these cylinders are flange-mounted, centerline-mounted, and foot-mounted.

The installation of rotating cylinders is extremely important, because these cylinders often revolve at high speeds. These cylinders are not mounted directly onto the machine spindle. They are mounted on an adapter that has previously been mounted onto the spindle. Therefore, it is important that both the adapter and the cylinder are installed correctly.

## Review Questions

1. What is the chief difference between single-acting and double-acting hydraulic cylinders?

2. What is an application for a nonrotating cylinder?

3. What is the purpose of a cushion collar and nose?

4. List three factors that may cause a nonrotating cylinder to fail.

5. Name at least five types of seals that are used on the pistons.

6. List four factors that may cause a rotating cylinder to fail.

7. Name four types of piston rod seals.

8. How can a nonrotating cylinder be used in conjunction with a rotating cylinder?

9. Name two types of cylinder cover seals.

10. What materials are generally used in cylinder covers?

11. Draw the schematic of a hydraulic cylinder and label the parts.

12. What is the purpose of a cushion adjustment valve?

13. Draw the symbol for a hydraulic cylinder with threaded tube construction.

14. What is the most commonly used method for mounting nonrotating cylinders?

15. How can leakage be reduced to a bare minimum in a hydraulic cylinder?

16. Identify six important parts of a rotating hydraulic cylinder.

17. What are the major causes of rotating cylinder failure?

18. How can dirt cause damage to a rotating cylinder?

19. What are three different styles of mounting cylinders?

20. How are standard non-rotating hydraulic cylinders designated as to pressure ranges?

21. What is an application for a rotating cylinder?

# Chapter 11

## Control Valves

The control valves that are used in hydraulic systems can be divided into three categories:

- Pressure controls
- Flow controls
- Directional controls

*Pressure controls* regulate the pressure intensity in the various portions of the system. *Flow controls* regulate the speed at which the hydraulic fluid is permitted to flow. This, in turn, controls the piston speed in the cylinders, the movement of the valve spools, the rotation speeds of the shafts of fluid motors, and the actuation speeds of other devices. *Directional controls* are used to direct the hydraulic fluid to the various passages in the system. Many types of directional controls are available—from the simple shut-off valves (similar to those used on sill-cocks in the home) to the six- and eight-way control valves that are used to control automatic machinery.

The size of its external openings determines the size of a control valve (or the amount of fluid that can pass through the opening with minimum backpressure). For example, a control valve with external openings or pipe ports threaded for a $1/2$-inch pipe should be capable of passing the same amount of fluid that can be passed normally by a $1/2$-inch pipe. Generally, the flow of hydraulic fluid does not exceed 15 feet per second. However, in some instances, the oil velocity far exceeds this rate, and it may be nearly twice this rate. Excessive velocities of hydraulic fluid create heat in the system, and they contribute to control problems that arise from undesirable pressure drops.

Sub-plate mountings (see Figure 11-1) or manifold-type mountings are often used for control valves. Then the control valve can be replaced without disturbing the piping.

Flange-type connections (see Figure 11-2) are used on many of the larger control valves. These connections are usually used for ports that are larger than 2 inches, and they are used mostly in hydraulics for military or mobile applications (see Figure 11-3).

Materials that are used in construction of a control valve depend largely on the fluid medium, the operating pressure, and the ambient temperature. The bodies of hydraulic valves are made of high-tensile

**Figure 11-1** A bank of control valves with sub-plate mounting.
*(Courtesy Continental Hydraulics)*

cast iron, cast steel, or plate steel, and the alloy steels are used for the interior parts.

Many of the interior parts are specially heat-treated. Bronze or cast-iron alloys are used for the bodies of high-pressure water valves, and heat-treated steels or stainless steels are used for the interior parts. Care should be exercised in selecting valves for high-pressure water systems, because high-velocity water erodes some types of materials. The damage that results is called *wire-drawing effect* or *termite effect*, and it can render a control valve completely inoperable in a relatively short period.

The packing for control valves is made in various configurations and materials. Some of these configurations are: O-rings, quad rings, V packing, U packing, and so on. Some of the materials used are Teflon, Viton, Buna N, and treated leather.

**Figure 11-2**   A larger hydraulic valve with flange connections, used for high-pressure service. *(Courtesy Denison Division, Abex Corporation)*

## Pressure Controls

Hydraulic control valves are designed for pressures of 1000, 2000, 3000, 5000, and even higher. Following are the types of hydraulic pressure control valves:

- Pressure relief
- Sequence
- Pressure reducing
- Counterbalance
- Unloading

## Pressure Relief

The *pressure-relief valve* (see Figure 11-4) safeguards the hydraulic system. The pump and the means of driving the pump are kept from overloading. The other components of the hydraulic system are also protected from excessive pressure. When the preset operating pressure is reached, the operating mechanism in the relief valve causes the oil to spill through the exhaust port, thus relieving the pressure.

Several different types of pressure-relief valves are used:

- *Direct-acting* type, with a spool or piston acting against a heavy spring.

MEETS STANDARDS OF MS 16142 (SHIPS)

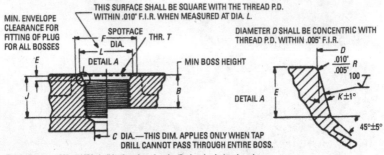

Finish diameters "A" and "D" shall be free from longitudinal and spiral tool marks.
Annular tool marks up to 100 micro-inches will be permissible.

| TUBE OUTSIDE DIAMETER | STRAIGHT THREAD T | | | | | B | C | D | E | F | J | K | L |
| | TH'D SIZE UNF-28 | PITCH DIA. | | MINOR DIA. | | MIN. TH'D DEPTH | MIN. DIA | +.005 -.000 DIA | +.015 -.000 DIA | DIA. | MIN. | ±1° | MIN. DIA. |
| | | MIN. | MAX. | MIN. | MAX. | | | | | | | | |
|---|---|---|---|---|---|---|---|---|---|---|---|---|---|
| 1/8 | 5/16-24 | .2854 | .2902 | .267 | .277 | .390 | .062 | .358 | .074 | .672 | .468 | 12° | .438 |
| 3/16 | 3/8-24 | .3479 | .3528 | .330 | .340 | .390 | .125 | .421 | .074 | .750 | .468 | 12° | .500 |
| 1/4 | 7/16-20 | .4050 | .4104 | .383 | .395 | .454 | .172 | .487 | .093 | .828 | .547 | 12° | .563 |
| 5/16 | 1/2-20 | .4675 | .4731 | .446 | .457 | .454 | .234 | .550 | .093 | .906 | .547 | 12° | .625 |
| 3/8 | 9/16-18 | .5264 | .5323 | .502 | .515 | .500 | .297 | .616 | .097 | .969 | .609 | 12° | .688 |
| 1/2 | 3/4-16 | .7094 | .7159 | .682 | .696 | .562 | .391 | .811 | .100 | 1.188 | .688 | 15° | .875 |
| 5/8 | 7/8-14 | .8286 | .8356 | .798 | .814 | .656 | .484 | .942 | .100 | 1.344 | .781 | 15° | 1.000 |
| 3/4 | 1 1/16-12 | 1.0084 | 1.0158 | .972 | .990 | .750 | .609 | 1.148 | .130 | 1.625 | .906 | 15° | 1.250 |
| 7/8 | 1 3/16-12 | 1.1334 | 1.1409 | 1.097 | 1.115 | .750 | .719 | 1.273 | .130 | 1.765 | .906 | 15° | 1.375 |
| 1 | 1 5/16-12 | 1.2584 | 1.2659 | 1.222 | 1.240 | .750 | .844 | 1.398 | .130 | 1.910 | .906 | 15° | 1.500 |
| 1 1/4 | 1 5/8-12 | 1.5709 | 1.5785 | 1.535 | 1.553 | .750 | 1.078 | 1.713 | .132 | 2.270 | .906 | 15° | 1.875 |
| 1 1/2 | 1 7/8-12 | 1.8209 | 1.8287 | 1.785 | 1.803 | .750 | 1.312 | 1.962 | .132 | 2.560 | .906 | 15° | 2.125 |
| 2 | 2 1/2-12 | 2.4459 | 2.4540 | 2.410 | 2.428 | .750 | 1.781 | 2.587 | .132 | 3.480 | .906 | 15° | 2.750 |

**Figure 11-3** Hydraulic control valves with straight pipe threads are commonly used on military and mobile applications.

*(Courtesy Imperial-Eastern Corp.)*

- *Direct-operated pilot* type, which is pilot-operated, with the piston acting against a small spring.
- *Remote-actuated pilot* type, which is controlled through a remote valve. In the latter valve type, the remote valve may be located at a distance from the relief valve, and connected to the relief valve by piping.

## Sequence

*Hydraulic sequence valves* are used to set up a sequence of operations. In several instances, a second four-way directional control valve can be eliminated by using one or two sequence valves. The sequence valves may be either the direct-acting type or the

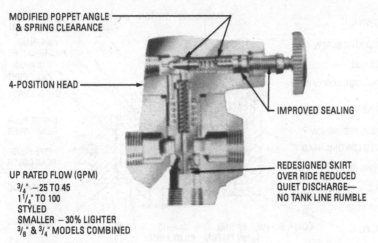

MODIFIED POPPET ANGLE
& SPRING CLEARANCE

4-POSITION HEAD

IMPROVED SEALING

REDESIGNED SKIRT
OVER RIDE REDUCED
QUIET DISCHARGE—
NO TANK LINE RUMBLE

UP RATED FLOW (GPM)
³/₄" – 25 TO 45
1¹/₄" TO 100
STYLED
SMALLER – 30% LIGHTER
³/₈" & ³/₄" MODELS COMBINED

**Figure 11-4** A hydraulic pressure-relief valve.
*(Courtesy Sperry Vickers, Div. of Sperry Rand Corp.)*

direct-operated pilot type of valve. The direct-acting valve (see Figure 11-5) can be used for low-pressure hydraulic service, but the direct-operated pilot-type valve (see Figure 11-6) is used only for higher-pressure hydraulic service. By building the check valve into the sequence valve body to provide free-flow return, piping and fittings can be eliminated at that point.

## Pressure Reducing
*Pressure-reducing valves* are commonly used in hydraulic systems where more complicated system requirements demand more than one operating pressure. Pressure reduction from the upstream side of the valve to the downstream side of the valve can be as much as 10 to 1. If the pressure on the upstream side of the valve is 1000 psi, the pressure on the downstream side of the valve can conceivably be reduced to 100 psi. The hydraulic pressure reducing valves are of two types:

**Figure 11-5** A direct-acting sequence valve used for low-pressure hydraulic service.
*(Courtesy Logansport Machine Co., Inc.)*

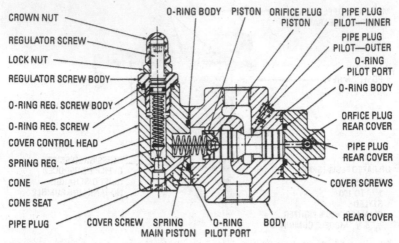

CROWN NUT

REGULATOR SCREW

LOCK NUT

REGULATOR SCREW BODY

O-RING REG. SCREW BODY

O-RING REG. SCREW

COVER CONTROL HEAD

SPRING REG.

CONE

CONE SEAT

PIPE PLUG

O-RING BODY    PISTON    ORIFICE PLUG    PIPE PLUG
                          PISTON          PILOT—INNER

PIPE PLUG
PILOT—OUTER

O-RING
PILOT PORT

O-RING BODY

ORFICE PLUG
REAR COVER

PIPE PLUG
REAR COVER

COVER SCREWS

REAR COVER

COVER SCREW    SPRING       O-RING       BODY
               MAIN PISTON  PILOT PORT

**Figure 11-6** A direct-operated pilot-type sequence valve used for hydraulic service.

- *Direct-acting* valve (see Figure 11-7)
- *Direct-operated pilot-type* valve

## Counterbalance and Unloading
Hydraulic valves, such as the *counterbalance* valve and the *unloading valve*, have fewer applications than the previously mentioned

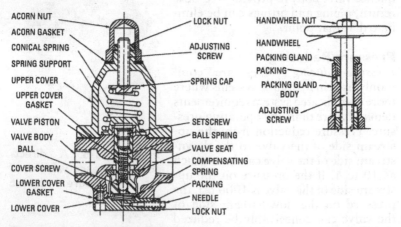

ACORN NUT

ACORN GASKET

CONICAL SPRING

SPRING SUPPORT

UPPER COVER

UPPER COVER
GASKET

VALVE PISTON

VALVE BODY

BALL

COVER SCREW

LOWER COVER
GASKET

LOWER COVER

LOCK NUT

ADJUSTING
SCREW

SPRING CAP

SETSCREW

BALL SPRING

VALVE SEAT

COMPENSATING
SPRING

PACKING

NEEDLE

LOCK NUT

HANDWHEEL NUT

HANDWHEEL

PACKING GLAND

PACKING

PACKING GLAND
BODY

ADJUSTING
SCREW

**Figure 11-7** A direct-acting pressure-reducing valve used for hydraulic service.

valves. The counterbalance valve can be used either to restrict a movement or to balance a load that is being held in position by a cylinder, a motor, or an actuator. The unloading valve is used to unload either a pump or an accumulator; it is actuated from an external signal.

## Operating Signals

The pressure control valves that receive the operating signal from the upstream side are the safety, pressure relief (see Figure 11-8), sequence (see Figure 11-9), and counterbalance valves. The pressure regulating valves and the pressure reducing valves (see Figure 11-10) receive the operating signal from a downstream source. The source of the operating signal is significant primarily in identifying a valve in a circuit diagram or in determining its specific function. This information can also be quite helpful in troubleshooting a malfunctioning system or component.

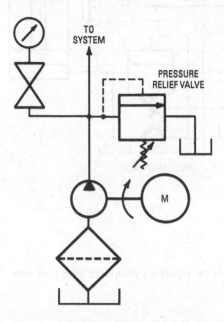

**Figure 11-8** Use of a pressure-relief valve in hydraulic system.

## Flow Controls

The valves that control the amount of flow of fluid in a hydraulic system are called *flow controls*. The *noncompensating* type of control that is designed for pneumatic service can usually be used for low-pressure hydraulic service. Following are the various types of

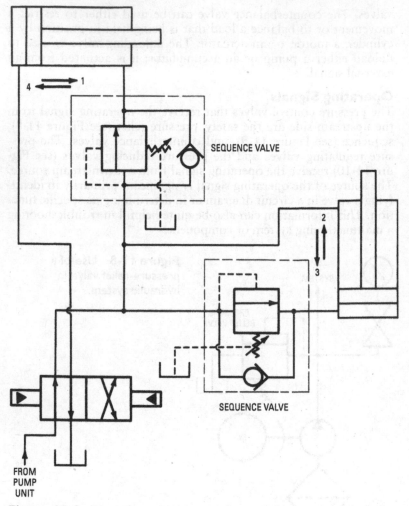

**Figure 11-9** Use of two sequence valves to eliminate one four-way valve.

hydraulic flow controls:

- Needle
- Noncompensating
- Compensating (pressure and pressure temperature)

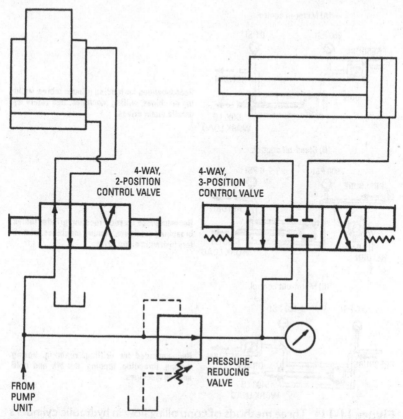

**Figure 11-10**  Pressure-reducing valve used in two pressure systems.

Following are three methods of controlling the flow (see Figure 11-11) from a relatively constant source of fluid:

- Meter-in
- Meter-out
- Bleed-off

In the *meter-in* method, the fluid is throttled before it reaches the device that is to be controlled. In the *meter-out* method, the fluid is throttled after it leaves the device. Here the exhausting fluid is throttled. In the *bleed-off* method, a portion of the hydraulic fluid is bled off before it reaches the device. The devices mentioned may be cylinders, fluid motors, actuators, or large controls.

(A) Meter-in control.

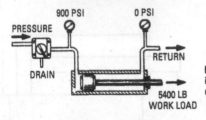

Recommended for feeding grinder tables, welding machines, milling machines, and rotary hydraulic motor drives.

(B) Bleed-off control.

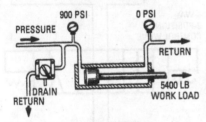

Recommended for reciprocating grinder tables, broaching machines, honing machines, and rotary hydraulic motor drives.

(C) Meter-out control.

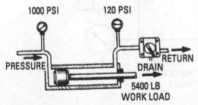

Recommended for drilling, reaming, boring, turning, threading, tapping, cut-off, and cold sawing machines.

**Figure 11-11** Three methods of controlling flow in hydraulic cylinders are: (a) meter-in; (b) bleed-off; and (c) meter-out.

*(Courtesy Sperry Vickers, Div. of Sperry Rand Corp.)*

## Needle

*Needle valves* (see Figure 11-12) are used in both pneumatic and hydraulic systems to meter fluid. The design of the needle is important where fine metering is required. Accuracy problems can be caused by dirty fluid where fine metering must be accomplished.

## Noncompensating

The *noncompensating-type flow controls* are used most commonly because of their low price and their availability. Although the noncompensating-type flow control cannot provide sufficient accuracy for extremely fine machine-tool feeds, it performs satisfactorily in most installations. Figure 11-13 shows a speed-control valve that

**Figure 11-12** High-pressure needle valve for service at 5000 psi.

*(Courtesy Imperial-Eastman Corp.)*

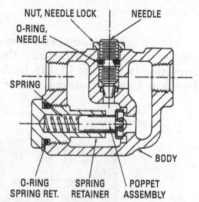

NUT, NEEDLE LOCK    NEEDLE

O-RING, NEEDLE

SPRING

BODY

O-RING SPRING RET.    SPRING RETAINER    POPPET ASSEMBLY

**Figure 11-13** A speed-control valve used for both pneumatic service and low-pressure hydraulic service.

can be used for both pneumatic service and low-pressure oil. Figure 11-14 shows a hydraulic flow-control valve that can be used for high-pressure oil.

## Compensating

*Temperature- and pressure-compensated flow controls* are often found on machine-tool applications where accurate feed rates are essential. A constant feed rate is provided for any temperature setting by the automatic temperature-compensating throttle, even though temperature changes occur in the hydraulic oil. The pressure-compensating device is a built-in pressure hydrostat that automatically compensates for any changes in loads.

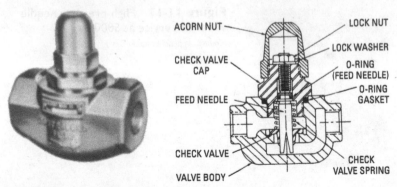

Figure labels: ACORN NUT, CHECK VALVE CAP, FEED NEEDLE, CHECK VALVE, VALVE BODY, LOCK NUT, LOCK WASHER, O-RING (FEED NEEDLE), O-RING GASKET, CHECK VALVE SPRING

**Figure 11-14** A hydraulic flow-control valve.
*(Courtesy Logansport Machine Co., Inc.)*

In each instance, compensation is achieved by automatically varying the size of the orifice to meet the demands of the changing load or condition (see Figure 11-15). In this type of valve, there is a reverse free-flow from the outlet port to the inlet port.

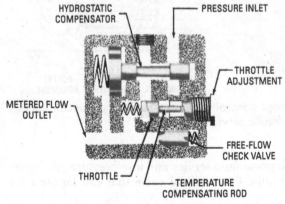

Figure labels: HYDROSTATIC COMPENSATOR, METERED FLOW OUTLET, THROTTLE, PRESSURE INLET, THROTTLE ADJUSTMENT, FREE-FLOW CHECK VALVE, TEMPERATURE COMPENSATING ROD

**Figure 11-15** Temperature- and pressure-compensated flow-control valve with check valve. *(Courtesy Sperry Vickers, Div. of Sperry Rand Corp.)*

Pressure-compensated flow controls are available without the temperature compensator, and they are built with an overload-relief valve (see Figure 11-16). By using the overload-relief valve, the only load that is imposed on the pump is the load that is needed to overcome the work resistance. This reduces the input power and the heat losses in applications where the loads may vary considerably.

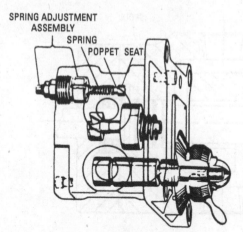

SPRING ADJUSTMENT
ASSEMBLY
SPRING
POPPET SEAT

**Figure 11-16** Flow-control and overload-relief valve.
*(Courtesy Sperry Vickers, Div. of Sperry Rand Corp.)*

By turning the dial on the face of the control to the "zero" setting, the pump can be unloaded completely.

The pressure-compensated flow control ensures accurate flow despite the varying loads. These valves are available as either port-in-body–type or sub-plate mountings, with the latter being used more commonly.

## Directional Controls

*Directional control valves* may be spool-type, piston-type, poppet-type, disk-type, or plug-type valves. The *two-way directional control valve* is one of the most common directional control valves that can be found in a hydraulic system. This valve can be used to close or to open a portion of a system, to close or to open an entire system, or to close or to open the passage to a single component (such as a pressure gage). There are two ports in a two-way valve. In the normal position of the valve actuator, the two ports may be connected, or they may be closed to each other. If the two ports are connected, the valve is called a *normally open* valve. If the two ports are closed to each other, it is called a *normally closed* valve (see Figure 11-17).

Three port connections are found in the *three-way directional* control valves, and they may be "normally open" or "normally closed" when the valve actuator is in the normal or "at-rest" position. In a "normally open" three-way valve, the inlet is connected to the cylinder port, and the exhaust port is blocked. When the

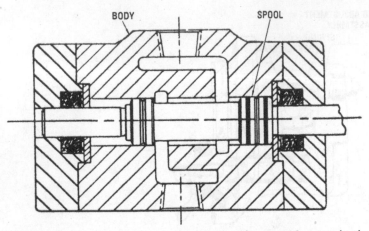

BODY    SPOOL

**Figure 11-17**    A two-way normally-open directional control valve.

actuator is moved to the second position, the inlet port is blocked, and the cylinder port is connected to the exhaust port.

In a *normally closed* valve, the inlet port is blocked, and the cylinder port is connected to the exhaust port when the actuator is in the normal position. When the actuator is moved to the second position, the inlet port is connected to the cylinder port, and the exhaust port is blocked.

In some types of three-way valves, there are three operating positions. This type of valve is called a *three-position three-way directional control valve*. In the center operating position, all three ports can be blocked. This control is often used to actuate a single-acting cylinder (spring-return or gravity-return). The center position is a "hold" position, so that the piston of the cylinder can be positioned and stopped at any point in its range of travel. Three-way valves are employed to actuate single-acting cylinders, large control valves, fluid motors, fluid actuators, and regenerative systems. Two three-way valves can be used to actuate a double-acting cylinder.

**Figure 11-18**    A hydraulic four-way directional control valve.

*Four-way directional control valves* have four port connections: one inlet port, two cylinder ports, and one exhaust port.

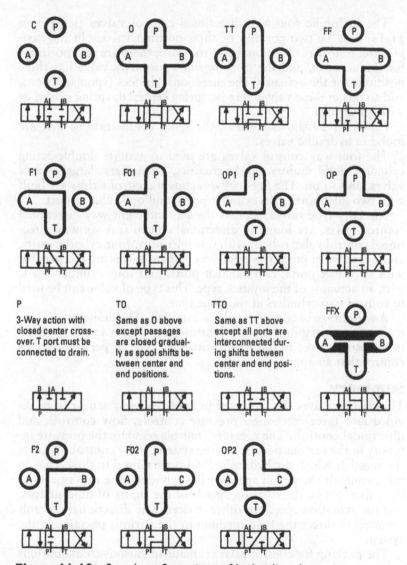

**Figure 11-19** Spool configurations of hydraulic valves.

(Courtesy Double-A Products Co., Sub. of Brown and Sharpe Mfg. Co.)

The hydraulic four-way directional control valves (see Figure 11-18) may be two-position or three-position valves. In the *two-position four-way* directional control valve, there are two positions for the actuator, and the *three-position four-way* valve has three positions for the actuator. The directional devices (spools, pistons, and so on) in these valves may be spring centered, spring offset, or without springs.

Figure 11-19 shows some of the spool configurations that are found in hydraulic valves.

The four-way control valves are used to actuate double-acting cylinders, fluid motors, fluid actuators, intensifiers, large control valves, and so on. The *five-way directional control valves* are built with two inlet ports, two cylinder ports, and one exhaust port.

Specialty-type valves, such as the six- and eight-way directional control valves, are found in directional controls. A six-way directional control valve is built with one inlet port, four cylinder ports, and one exhaust port. The eight-way control valves are constructed with two inlet ports, two exhaust ports, and four cylinder ports, with an actuator of the joystick type. This type of valve can be used to control two cylinders at the same time.

A wide range of controls are now available. The designers and the maintenance personnel can avail themselves of quite a wide selection in choosing valves to fulfill the demanded type of performance or control that an application requires.

## Summary

The control valves that are used in fluid power systems can be divided into three categories: pressure controls, flow controls, and directional controls. The pressure controls regulate the pressure intensity in the various portions of the system. Flow controls regulate the speed at which the hydraulic fluid is permitted to flow. This, in turn, controls the piston speed in the cylinders, the movements of the valve spools, the rotation speeds of the shafts of fluid motors, and the actuation speeds of other devices. The directional controls are used to direct the fluid medium to the various passages in the system.

The packing for control valves is made in various configurations and materials. Some of these configurations are: cups, O-rings, quad rings, V packing, U packing, and so on. Some of the materials used are Teflon, Viton, Buna N, and treated leather.

Hydraulic control valves are designed for pressures of 1000, 2000, 3000, and 5000 psi, and higher.

The hydraulic pressure control valves are pressure relief, pressure reducing, sequence, counterbalance, and unloading.

The valves that control the amount of flow of liquid in a hydraulic system are called flow controls. Three methods of controlling the flow from a relatively constant source of fluid are meter-in, meter-out, and bleed-off.

The directional control valves may be spool-type, piston-type, poppet-type, disk-type, or plug-type valves. The two-way directional control valve is one of the most common valves that can be found in either a pneumatic or a hydraulic system. Many types of directional control valves are available—ranging from the simple shutoff valve to the six- and eight-way control valves that are used to control automatic machinery.

## Review Questions

1. List the three types of control valves that are used in fluid power systems.
2. What configurations are used in making the valve packing?
3. List the hydraulic pressure control valves.
4. List three uses for a two-way directional control valve.
5. Describe the basic operation of a two-way directional control valve.
6. How can a three-way directional control valve be used to operate a single-acting cylinder?
7. List some applications for four-way directional control valves.
8. How would a three-way valve be used to operate a four-way, spring-offset directional control valve?
9. What is the difference between a *normally open* and a *normally closed* three-way directional control valve?
10. What determines the size of a control valve?
11. What is meant by the term *sub-plate mounting*?
12. What determines the material used in the construction of a control valve?
13. What type of pressures is a control valve designed to handle?
14. What are three types of pressure relief valves?
15. How are hydraulic sequence valves used?
16. How is a pressure-reducing valve used in a hydraulic system?

17. What are valves that control the amount of flow of fluid in a hydraulic system called?

18. How are needle valves used in both pneumatic and hydraulic systems?

19. List four types of directional control valves.

20. What valves are most commonly found in either a pneumatic or a hydraulic system?

# Chapter 12

## Hydraulic Control Valve Operators

Chapter 11 discussed the various control valves that are used in hydraulic systems. A means of operating these control valves must be provided if they are to function. In most instances, pressure controls and flow controls utilize valve operators that are different from those used on directional controls. In general, many different types of valve operators are available as standard equipment for directional control valves. It is important to select a suitable valve operator, because a poor choice may lead to problems in creating and maintaining the desired efficiency and controllability in a given fluid system.

### Pressure-Control Valve Operators

Pressure-control operators are made in relatively few general types, including the following:

- Direct-acting screw
- Direct-acting cam roller
- Offset cam roller
- Solenoid
- Pilot types

Nearly all types of hydraulic pressure-control valves use a *screw-type operator* for setting the correct spring tension at which the valve is to function. Figure 12-1 shows a combination of screw and sliding cam operators used on a pressure control. As the screw is advanced, the disk spring is compressed, that provides very accurate control over a wide range of flows.

Figure 12-2 shows a direct-acting spring-operated hydraulic relief valve. As the screw-type operator is advanced, spring tension is placed on the spot to increase the operating pressure of the system.

The *direct* or *offset cam roller type* of operator is often employed on valves for special applications. For example, this type of operator on a relief valve may be advantageous if a considerable increase in pressure must occur at a given point in the travel of the piston in a cylinder. At that point, a cam can be placed on the machine table or other moving member. At the correct position, the cam depresses the

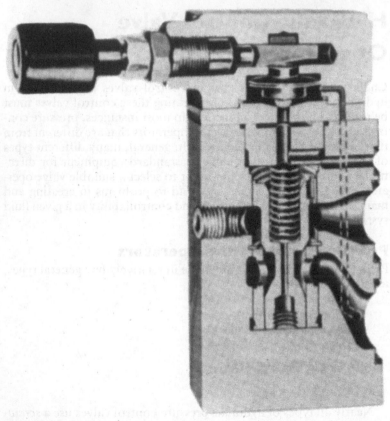

**Figure 12-1** Pressure control that uses a combination of screw and sliding cam operator.

cam-roller–type of operator on the relief valve, causing an increase in operating pressure.

The cam-type operators are also found on hydraulic pressure-relief valves used for testing applications where the cams have an extremely shallow angle. The cam roller is depressed gradually, and pressure readings are recorded. Solenoid operators are sometimes used on relief valves.

## Flow-Control Valve Operators

In nearly all instances, the flow-control valve operator (in hydraulic systems) utilizes a screw-type or threaded mechanism to open and to close the control orifice in the flow control valve. The screw-type

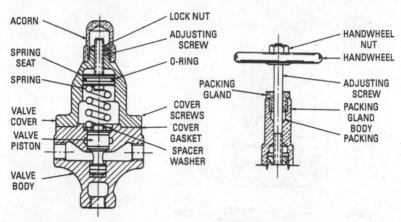

**Figure 12-2** Hydraulic pressure-relief valve with a screw-type operator.

mechanism may be in the form of a needle with slots, as shown in the diagram of the hydraulic valve in Figure 12-3, which is used for pressures up to 3000 pounds per square inch. After the correct flow is achieved, the needle is locked in position by means of a locknut.

A screw-type mechanism in the form of a needle is used in the low-pressure hydraulic flow control valve shown in Figure 12-4. The

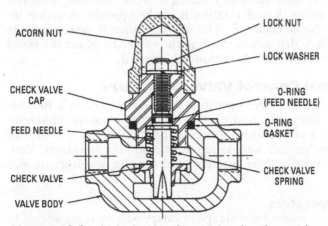

**Figure 12-3** A hydraulic flow-control valve with a screw-type operator.

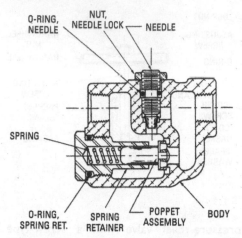

**Figure 12-4** Screw-type operator used in a speed- or flow-control valve.

needle in the valve is installed in such a way that the major diameter cannot be backed out beyond the retaining pin (a safety feature).

Screw or threaded mechanisms may be equipped with a hand wheel, a micrometer-type knob, a locking device, or some other device for controlling it.

In Figure 12-5, the valve operator consists of a throttle mechanism that is so machined that rotating it in one direction decreases the size of orifice $B$ and rotating it in the opposite direction increases the size of orifice $B$. The size of orifice $B$ determines the amount of flow that passes through the valve. To adjust the speed of the actuator, the throttle is manually rotated.

## Directional Control Valve Operators

Directional control valve operators can be classified in a number of different categories, and each category may consist of several different types of operators. Some of these categories are manual, solenoid, mechanical, automatic-return, and pilot operators. Various combinations of these different types of valve operators are also available.

## Manual Operators

In most instances, the manual type of valve operator is considered to be the most positive and least expensive. The manual-type operator is actuated by means of the hand, foot, or some other part of the

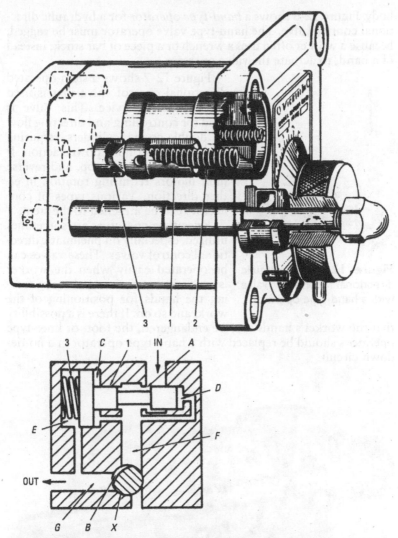

**Figure 12-5**   A flow-control and check valve. This series incorpo-rates an integral check valve consisting of a poppet (1), spring (2), and seat (3). The check valve permits reverse free flow from the valve outlet to its inlet port, bypassing the flow-controlling elements.

*(Courtesy Sperry Vickers, Div. of Sperry Rand Corp.)*

body. Figure 12-6 shows a *hand-type operator* for a hydraulic directional control valve. The hand-type valve operator must be rugged, because a worker often uses a wrench or a piece of bar stock, instead of a hand, to actuate the valve operator.

**Figure 12-6** A hydraulic directional control valve with a hand-type operator.

Figure 12-7 shows a stem-operated directional control valve that is used for hydraulic service. This valve is used for controlling and directing flow to double-acting cylinders requiring push, pull, and hold power action. It is also used to start, stop, and reverse fluid motors requiring rotation in either direction. Various types of connectors can be attached to the stem.

*Knee-type operators* are sometimes utilized, especially on pneumatic directional control valves. These valves can be operated easily when the worker is in a sitting position, thereby freeing the hands for positioning of the work, and so on. If there is a possibility that the worker's hands may be endangered, the foot- or knee-type operators should be replaced with a hand-type operator in a no-tie-down circuit.

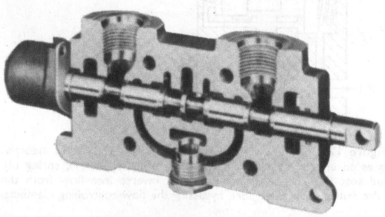

**Figure 12-7** Stem-actuated hydraulic control valve.
*(Courtesy Commercial Shearing, Inc.)*

The hand-type operators may be of the spring-offset type, the spring-centered type, or the detent type.

When the *spring-offset valve operator* (see Figure 12-8) is used, a worker is required to keep a hand on the operator until the direction of flow within the valve is changed. When the operator is released, it shifts automatically to its original position.

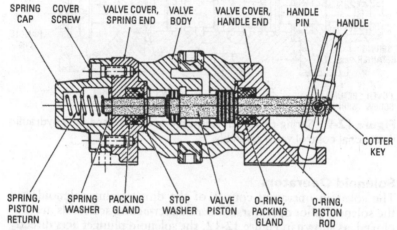

SPRING CAP   COVER SCREW   VALVE COVER, SPRING END   VALVE BODY   VALVE COVER, HANDLE END   HANDLE PIN   HANDLE

COTTER KEY

SPRING, PISTON RETURN   SPRING WASHER   PACKING GLAND   STOP WASHER   VALVE PISTON   O-RING, PACKING GLAND   O-RING, PISTON ROD

**Figure 12-8**  A two-way hydraulic valve with a sliding-spool flow director.

When the *spring-centered valve operator* (see Figure 12-9) is used, a worker must apply force to the handle in either outward position to prevent the handle from returning to the neutral position. When a *detent* on the valve operator is used to locate and to hold the spool position, a worker needs only to move the handle until the detent engages (see Figure 12-10). The handle remains in place until it is moved to another position. Detents are used in two-position and three-position directional control valves.

A directional control valve sometimes requires two valve operators; one of the valve operators is used to shift the valve spool manually in one direction, and the other valve operator is used to reverse the valve spool by means of a solenoid or some mechanical means. With this arrangement, a worker can move some distance from the directional control valve and perform another task while the fluid power equipment is going through its normal operating cycle. Figure 12-11 shows a hydraulic four-way directional control valve in which the spool is reversed by mechanical means.

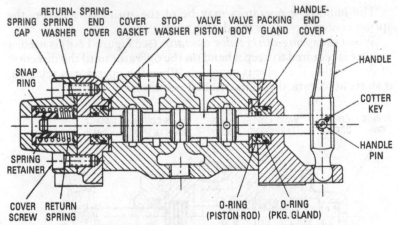

**Figure 12-9** Spring-centered hand-type operator used on a hydraulic directional control valve assembly.

## Solenoid Operators

The solenoid operators consist of the direct-acting solenoid and the solenoid pilot operators. Where direct-acting solenoids are employed, as shown in Figure 12-12, the solenoid plunger acts directly against the end of the valve spool in order to shift it.

The large direct-acting solenoid operators cause considerable noise when they are energized, because of the weight of the solenoid plungers. These large operators are not generally employed on high-cycling operations, because of the harsh and severe impact.

Solenoid operators are available in various voltages (such as 115 volts, 230 volts, and so on) for alternating current (AC). In most instances, they are suitable for continuous duty. This means that the solenoid coil may be energized for an indefinite length of time. Solenoids of this type are also available for direct-current (DC) applications and are usually equipped with a cut-out mechanism, often referred to as a *mouse-trap*. Direct-current solenoids normally are not recommended for continuous service.

The two-position directional control valves that use a direct-acting solenoid operator on one end of the valve and a spring-return type of operator on the other end of the valve are widely used in hydraulic service. This eliminates one solenoid, thus reducing the cost of the control valve. The combination is commonly used in the design of fail-safe circuits. A disadvantage is that the solenoid plunger works against the spring, and it must be kept energized

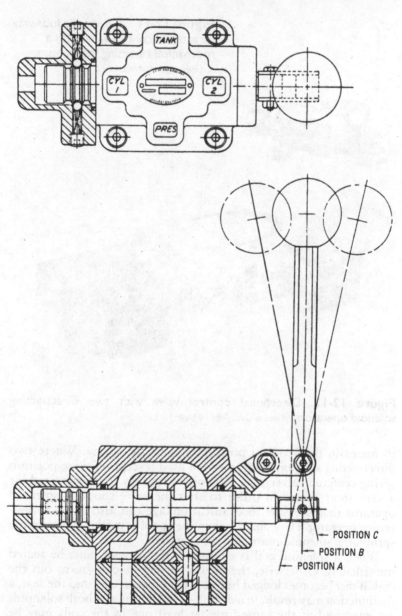

POSITION C

POSITION B

POSITION A

**Figure 12-10**  Cutaway of a hand-type valve operator with detent.

**Figure 12-11** A hydraulic four-way directional control valve with a mechanical reversing mechanism.

*(Courtesy Logansport Machine Co., Inc.)*

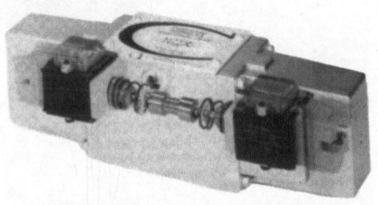

**Figure 12-12** Directional control valve with two direct-acting solenoid operators. *(Courtesy Continental Hydraulics)*

to maintain the spool in position against the spring. Where two direct-acting solenoid operators are used (except on three-position spring-centered valves), the solenoid needs only to be energized for a very short period in order to shift the valve spool. A solenoid operator that uses the mechanical linkage can also use a spring-return operator. Here again, when the spool is working against the spring, the solenoid must be kept energized.

When a solenoid coil is energized, the plunger must be seated immediately. Otherwise, the inrush current quickly burns out the coil. If dirt becomes lodged between the plunger head and the seat, a malfunction may result. In a double-solenoid valve, if both solenoids are energized at the same time, at least one of the coils may be burned out.

Solenoid pilot operators are common, and, in most instances, they are quite inexpensive. They are compact and require only a small space. The inrush and holding currents are quite low (for example, 0.290 ampere inrush current and 0.210 ampere holding current for 115-volt, 60-hertz current). These operators are available for either AC or DC.

Solenoid pilot operators are available as a complete two-way or three-way valve (see Figure 12-13) used to control pilot-operated valves or as a subassembly. Solenoid pilot operators are available with explosion-proof housings for use in hazardous locations.

**Figure 12-13** A solenoid pilot operator for a directional control valve. *(Courtesy Schrader Div., Scovill Mfg. Co.)*

## Mechanical Operators

Mechanical operators on directional control valves play an important role on automatic equipment. The most commonly used mechanical operators are the direct-acting cam-roller type (see Figure 12-14), the toggle lever type (see Figure 12-15), and the pin type (with or without threaded end).

In the *direct-acting cam-roller type of operator*, more force is usually required to depress the roller, since it is acting against a spring. The angle on the cam should not be too abrupt, to avoid placing an excessive side load against the cam roller and the bearing in the valve cover.

The direct-acting cam-roller type of operator is found on some of the smaller directional control valves, but it is more commonly used on the larger, more rugged valves for hydraulic service. This type of operator requires a rather short stroke to complete an actuation. The cam roller should not be over-stroked, although some of these valves provide for a small amount of over-travel. The cam is usually attached either to a machine table or to the piston rod of a cylinder.

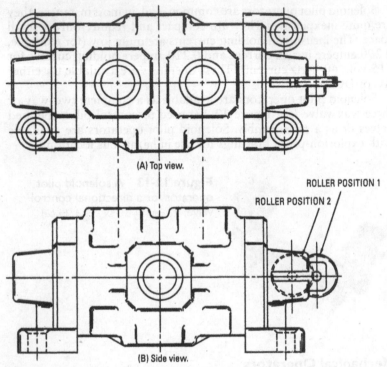

(A) Top view.

ROLLER POSITION 1

ROLLER POSITION 2

(B) Side view.

**Figure 12-14** Cam roller-operated directional control valve.

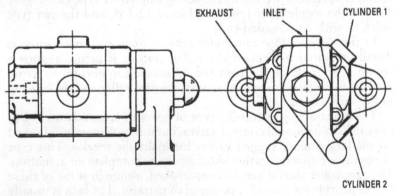

EXHAUST    INLET    CYLINDER 1

CYLINDER 2

**Figure 12-15** Toggle lever-operated hydraulic control valve.

If it is attached to the piston rod of a cylinder, the rod should be supported near the contact point to eliminate deflection of the piston rod. Rotary cams are often used to actuate the cam roller.

The *toggle-lever type of operator* requires only slight effort to actuate. An inexpensive trip mechanism can be employed. This may be attached to a machine slide, to a feed mechanism, to a piston rod of a cylinder, or to some other moving mechanism. A work piece moving down a conveyor can be used to trip the toggle mechanism.

*Return-type operators* for directional controls can be of the spring type. The spring returns the spool or flow director to the neutral position, as in a three-position directional control valve. Or, it may return the spool or flow director to the end opposite the spring, as in a two-position directional control valve. Figure 12-9 shows a valve in which the spring is returned to the neutral position when the spring is released. The spring-return type of operator is used in conjunction with manual, solenoid, mechanical, and pilot types of operators.

## Pilot Operators

Pilot types of operators are of the direct acting, bleed, or differential types, and they are found on valves where high cycling, safety interlocks, and automatic sequencing are required, as shown in Figures 12-16 to 12-18. A direct-acting pilot operator used on a four-way hydraulic control valve is shown in Figure 12-16. The medium that actuates the pilot operator is hydraulic fluid. To provide for better control of the spool as it shifts, chokes are often employed between the pilot operator and the spool. If two direct-acting pilot operators are employed (one at each end of the spool assembly), two chokes (see Figure 12-17) are usually used. Adjustable orifices in these chokes are controlled by a needle, and, in some instances, one choke may be set differently from the other choke, so that the spool shifts faster in one direction of travel.

Although most hydraulic directional control valves that use direct-acting pilot operators make use of hydraulic fluid as the medium for shifting the main valve spool, more applications are now being found for compressed air as the shifting medium. There are several advantages to using compressed air as the medium. The controls that direct the fluid to the direct-acting pilot operators are less expensive, are more compact, and are manufactured in greater variety.

Compressed air provides for faster spool shifting. Interlocks can be set up between a pneumatic system and a hydraulic system, permitting the use of air pressure for clamping and the use of hydraulic

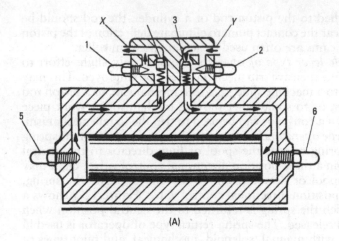

(A)

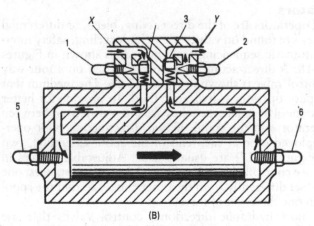

(B)

**Figure 12-16** Two chokes that are adjustable, used on pilot-operated hydraulic valves. *(Courtesy Sperry Vickers, Div. of Sperry Rand Corp.)*

pressure of a heavy work cycle. In addition, the necessary piping between the directional control and the operator is less expensive than when an all-hydraulic system is used.

Air-actuated direct-acting pilot operators are now manufactured so that the diameter of the operator is sufficient for permitting an air pressure of 3 to 5 psi to shift the spool in the directional control

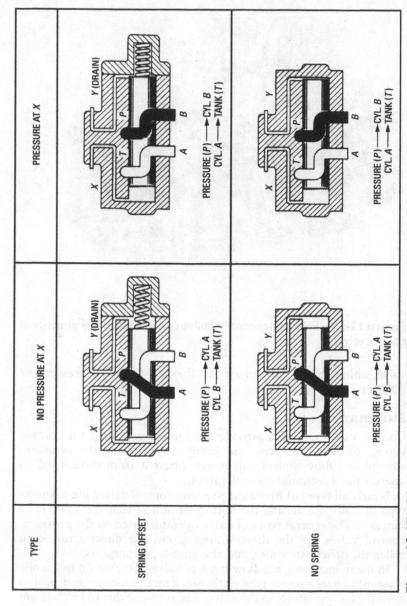

**Figure 12-17** Pilot-operated hydraulic control valve. *(Courtesy Sperry Vickers, Div. of Sperry Rand Corp.)*

**Figure 12-18** Solenoid-controlled pilot-operated four-way directional control valve.

valve, although the valve itself is subjected to an oil pressure of 3000 psi.

## Summary

Control-valve operators provide a means of operating the control valves. In most instances, the valve operators for the pressure-control and flow-control valves are different from those used to operate the directional control valves.

Nearly all types of hydraulic pressure control valves use a screw-type operator for setting the spring tension at that the valve is to function. The general types of valve operators used on the pressure-control valves are the direct-acting screw, the direct-acting cam roller, the offset cam roller, and the pilot-type operators.

In many instances, the flow control valve operator (in hydraulic systems) utilizes a screw-type or threaded mechanism to open and to close the control orifice in the valve. The screw or thread mechanism may be equipped with a handwheel, a micrometer-type knob, a locking device, or some other device for controlling it.

Several different types of operators are used for the directional control valves. They are classified as manual, solenoid, mechanical, automatic-return, and pilot operators. Various combinations of these types of valve operators are also available.

## Review Questions

1. What is the purpose of the valve operator?
2. List three types of pressure control-valve operators.
3. What types of operators are used for flow-control valves?
4. List four types of directional control valves.
5. What applications are most suitable for pilot-operated directional control valves?
6. Name two applications for solenoid-actuated relief valves.
7. Name three applications for cam-operated relief valves.
8. What type of actuator should be used on directional control for high-cycling operations?
9. What are the differences between AC and DC solenoids?
10. What is the most commonly used type of pressure-control valve operator for setting spring tension?
11. What type of valve operator is considered to be the most positive and least expensive?
12. What are three kinds of hand-type operators?
13. Where is the knee-type operator utilized?
14. What type of solenoid is not recommended for continuous service?
15. What is the difference between inrush current and holding current in a solenoid?
16. What are the most commonly used mechanical operators?
17. List two types of pilot operators.
18. What type of operator do nearly all types of hydraulic pressure control valves use?
19. How are operators for directional valves classified?

# Part IV

# Fluids, Lines, and Fittings

# Chapter 13

## Hydraulic Fluids

In hydraulics, hydraulic fluids are usually divided into three categories: petroleum-base fluids, synthetic-base fluids, and water. The first two fluids are used in so-called *packaged-power devices*. Water is generally used as the hydraulic fluid in central hydraulic systems.

The function of a good hydraulic fluid is threefold:

- It is a means of transmission of fluid power.
- It is a means of lubrication of the components of the fluid power system.
- It acts as a sealant.

The selection of the proper hydraulic fluid is important, because it has a direct bearing on the efficiency of the hydraulic system, on the cost of maintenance, and on the service life of the system's components.

### Petroleum-Base Fluids

Three basic types of mineral oils are used:

- Pennsylvania, or paraffin-base oils.
- Gulf coast, or naphthenic- and asphaltic-base oils.
- Mid-Continent, or mixed-base oils.

These contain both naphthenic and paraffin compounds.

To obtain certain characteristics, chemicals are added to oil. These chemicals are called *additives*. Additives cannot make an inferior oil perform as well as good oil, but they can make a good oil perform even better. An additive may be in the form of an antifoam agent, a rust inhibitor, a film-strengthening agent, or an oxidation stabilizer.

You should not attempt to place additives in hydraulic oil. That job is primarily for the oil manufacturer or refiner.

### Synthetic-Base Fluids

Since fire hazards are prevalent around certain types of hydraulically operated machines (especially where open fires are present), much research has been done to develop fire-resistant hydraulic fluids. These fluids are divided into two classifications: synthetic-base mixtures and water-base fluids. Not all synthetic-base fluids are fire-resistant.

473

Synthetic-base fluids include chemical compounds (such as the chlorinated biphenyls, phosphate esters, or mixtures containing each). These hydraulic fluids are fire-resistant because a large percentage of phosphorous and chloride materials are included.

Water-base fluids depend on a high percentage of water to improve the fire-resistant nature of the fluid. In addition to water, these compounds contain antifreeze materials (such as glycol-type thickeners, inhibitor, and additives).

Synthetic-base fluids have both advantages and disadvantages. Following are some of the advantages:

- Many of them are fire-resistant.
- Sludge or petroleum-gum formation is reduced.
- Temperature has little effect on the thickening or thinning or the fluid.

A disadvantage of many synthetic fluids is their deteriorating effect on some materials (such as packing, paints, and some metals used in intake filters).

## Quality Requirements

Certain qualifications are demanded of good hydraulic oil. The oil should not break down and it should give satisfactory service. High-quality hydraulic oils have many of the following properties:

- Prevent rusting of the internal parts of valves, pump, and cylinders.
- Prevent formation of a sludge or gum that can clog small passages in the valves and screens in filters.
- Reduce foaming action that may cause activation in the pump.
- Provide a long service lives.
- Retain their original properties through hard usage (must not deteriorate chemically).
- Resist changing the flow ability or viscosity as the temperature changes.
- Form a protective film that resists wear of working parts.
- Prevent pitting action on the parts of pumps, valves, and cylinders.
- Do not emulsify with the water. Water is often present in the system either from external sources or from condensation.
- Have no deteriorating effect on gaskets and packing.

# Maintenance

Proper maintenance of hydraulic oil is often forgotten. Too often, hydraulic oil is treated in a matter-of-fact manner. Following are a few simple rules regarding maintenance:

- Store oil in clean container. The container should not contain lint or dirt.
- Keep lids or covers tight on the oil containers, so that dirt or dust cannot settle on the surface of the oil. Oil should never be stored in open containers.
- Store oil in a dry place. Do not allow it to be exposed to rain or snow.
- Do not mix different types of hydraulic oils. Oils having different properties may cause trouble when mixed.
- Use a recommended hydraulic fluid for the pump.
- Use clean containers for transporting oil from the storage tank to the reservoir.
- Ensure that the system is clean before changing oil in the power unit. Do not add clean oil to dirty oil.
- Check the oil in the power unit regularly. Have the oil supplier check the sample of the oil from the power unit in the laboratory. Contaminants often cause trouble. These can be detected by frequent tests, which may aid in determining the source of contamination. On machines that use coolants or cutting oil, extreme caution should be exercised to keep these fluids from entering the hydraulic system and contaminating the oil.
- Drain the oil in the system at regular intervals. It is difficult to set a hard-and-fast rule as to the length of the interval. In some instances, it may be necessary to drain the oil only every two years. However, once each month may be necessary for other operating conditions. This depends on the conditions of operation and on the original quality of the hydraulic oil. Thus, several factors should be considered in determining the length of the interval.

Before placing new oil in the hydraulic system, it is often recommended that the system be cleaned with a hydraulic system cleaner. The cleaner is placed in the system after the oil has been removed. The hydraulic system cleaner should be used while the hydraulic system is in operation, and usually requires 50 to 100 hours to clean the system. Then, the cleaner should be drained; the filters, strainers, and oil reservoir cleaned; and the system filled with good hydraulic oil.

If the hydraulic oil is spilled on the floor in either changing or adding oil to the system, it should be cleaned up at once. Good housekeeping procedure is important in reducing fire and other safety hazards.

## Change of Fluids in a Hydraulic System

If the fluid in a hydraulic system is to be changed from a petroleum-base fluid to a fire-resistant fluid or vice versa, the system should be drained and cleaned completely.

Keep in mind the following when changing from a petroleum-base fluid to a water-base fluid:

- Drain out all the oil (or at least as much as possible) and clean the system.
- Either remove lines that form pockets, or force the oil out with a blast of clean, dry air.
- Strainers should be cleaned thoroughly. Filters should be cleaned thoroughly and the filter element replaced.
- Check the internal paint in all components. It is likely that the paint should be removed.
- Check the gaskets and packing. Those that contain either cork or asbestos may cause trouble.
- Flush out the system. Either a water-base fluid or a good flushing solution is recommended.
- Since hydraulic fluids are expensive, the system should be free of external leaks.

Keep in mind the following when changing from a water-base fluid to a petroleum-base fluid:

- Remove all of the water-base fluid. This step is very important. A small quantity of water-base fluid left in the system can cause considerable trouble with the new petroleum-base fluid.
- The reservoir should be scrubbed and cleaned thoroughly. If the interior of the reservoir is not painted, it should be coated with a good sealer that is not affected by hydraulic oil.
- The components should be dismantled and cleaned thoroughly. Cleaning with steam is effective.
- Flush the system with hydraulic oil and then drain.
- Fill the system with good hydraulic oil.

A similar procedure should be used in changing from synthetic-base fluids to petroleum-base fluids—and vice versa. If phosphate-base fluids are used, the packing should be changed. If satisfactory performance from a hydraulic fluid and a hydraulic system is expected, use a good grade of hydraulic fluid, keep it clean, change at regular intervals, do not allow it to become overheated, and keep contaminants out of the system.

## Selection of a Hydraulic Fluid

The main functions of the hydraulic fluid are to transmit a force applied to one point in the fluid system to some other point in the system, and to reproduce quickly any variation in the applied force. Thus, the fluid should flow readily, and it should be relatively incompressible. The choice of the most satisfactory hydraulic fluid for an industrial application involves two distinct considerations:

- Fluid for each system should have certain essential physical properties and characteristics of flow and performance.
- The fluid should have desirable performance characteristics over a period. The oil may be suitable when initially installed. However, its characteristics or properties may change, resulting in an adverse effect on the performance of the hydraulic system.

The hydraulic fluid should provide a suitable seal or film between moving parts to reduce friction. It is desirable that the fluid should not produce adverse physical or chemical changes while in the hydraulic system. The fluid should not promote rusting or corrosion in the system, and it should act as a suitable lubricant to provide film strength for separating the moving parts, to minimize wear between them.

Certain terms are required to evaluate the performance and suitability of a hydraulic fluid. Important items are discussed in the following sections.

## Specific Weight

The term *specific weight* of a liquid indicates the weight per unit of volume. For example, water at 60°F weighs 62.4 pounds per cubic foot. The specific gravity of a given liquid is defined as *the ratio of the specific weight of the given liquid divided by the specific weight of water*. For example, if the specific gravity of the oil is 0.93, the specific weight of the oil is approximately 58 pounds per cubic foot (0.93 × 62.4). For commercially available hydraulic fluids, the specific gravity may range from 0.80 to 1.45.

## Viscosity

*Viscosity* is a frequently used term. In many instances, the term is used in a general, vague, and loose sense. To be definite and specific, the term viscosity should be used with a qualifying term.

The term *absolute* or *dynamic viscosity* is a definite, specific term. As indicated in Figure 13-1, the hydraulic fluid between two parallel plates adheres to the surface of each plate, which permits one plate to slide with respect to the other plate (like playing cards in a deck). This results in a shearing action in which the fluid layers slide with respect to each other. A shear force acts to shear the fluid layers at a certain velocity (or rate of relative motion) to provide the shearing action between the layers of fluid. The term *absolute* or *dynamic viscosity* denotes a physical property of the hydraulic fluid that indicates the ratio of the shear force and the rate or velocity at which the fluid is being sheared.

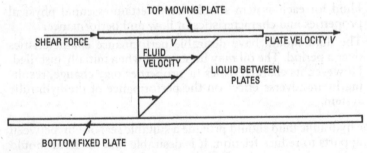

**Figure 13-1**   Shearing action of a liquid.

To simplify, a very *viscous* fluid or a fluid having a high dynamic viscosity is a fluid that does not flow freely. Fluid having a low dynamic viscosity flows freely. The term *fluidity* is the reciprocal of *dynamic viscosity*. A fluid having a high dynamic viscosity has a low fluidity, and fluid having a low dynamic viscosity has a high fluidity. Generally speaking, the dynamic viscosity of a liquid decreases as temperature increases. Therefore, as oil is heated, it flows more freely. Because of pressure effects, it is difficult to draw general, firm conclusions for all oils. It is possible for an increase in fluid pressure to increase the viscosity of oil.

## Saybolt Universal Viscosimeter

The term *dynamic viscosity* is sometimes confused with the reading taken from the Saybolt Universal Viscosimeter. In actual industrial practice, this instrument has been standardized arbitrarily for testing of petroleum products. Despite the fact that it is called a viscosimeter, the Saybolt instrument does not measure dynamic viscosity. Figure 13-2 illustrates the principle of the Saybolt viscosimeter.

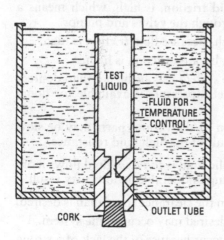

**Figure 13-2** Basic operating principle of the Saybolt Viscosimeter.

In operating the instrument, the liquid to be tested is placed in the central cylinder, which is a short, small-bore tube having a cork at its lower end. Surrounding the central cylinder, a liquid bath is used to maintain the temperature of the liquid that is being tested. After the test temperature has been reached, the cork is pulled, and the time in seconds that is required for 69 milliliters of the test fluid to flow out of the cylinder is measured with a stopwatch. This measured time (in seconds) is called the *Saybolt Universal Reading*.

The Society of Automotive Engineers (SAE) has established standardized numbers for labeling of the oils. For oils tested at 130°F in a standard Saybolt Universal instrument, Table 13-1 indicates *SAE* numbers for the corresponding ranges of Saybolt Universal readings. For example, if the oil is labeled "SAE 10," the Saybolt Universal reading at 130°F is in the range from 90 to less than 120 seconds.

**Table 13-1    Range of Saybolt Readings, in Seconds**

| SAE Numbers | Minimum | Maximum |
|---|---|---|
| 10 | 90 | less than 120 |
| 20 | 120 | less than 185 |
| 30 | 185 | less than 255 |

## Viscosity Problems

If the viscosity of the hydraulic fluid is *too high* (fluid does not flow as freely as desired), the following undesirable actions may result:

- Internal resistance, or fluid friction, is high, which means a high resistance to flow through the valves and pumps.
- Power consumption is high, because fluid friction is high.
- Fluid temperature is high, because friction is high.
- Pressure drop through the system may be higher than desired, which means that less useful pressure is available for doing useful work.
- The motion and operation of the various parts may be slow and sluggish. This is a result of the high fluid resistance.

If the viscosity of the hydraulic fluid is *too low* (fluid flows more freely than desired), the following undesirable actions may result:

- More leakage may occur in the clearance space than is desired.
- A lower pressure than is desired may occur in the system.
- An increase in wear may occur because of the lack of a strong fluid film between mechanical parts that move in relation to each other.
- Pump leakage may increase, resulting in reduced pump delivery and efficiency.
- A loss of control may occur because fluid film strength is reduced.

With respect to Saybolt readings, the viscosimeter readings of oils in service should not exceed 4000 seconds, and they should not read less than 45 seconds.

## Viscosity Index

Ideally, the dynamic viscosity of any oil should change only slightly as the temperature changes. In the automobile engine, the oil in the crankcase is operated over a wide range of temperatures. On a very cold winter morning, until the car has been operated for some length of time, the temperature of the oil may be very low, and the dynamic viscosity of the oil may be very high. If the dynamic viscosity of the oil is excessively high, large forces and large amounts of power may be required to shear the oil films. Also, after the engine has been operated for a period of time on a hot summer day, the temperature of the oil may be very high, and the dynamic viscosity of the oil may be too low. Therefore, the oil may not form a suitable lubricating film between the sliding surfaces. A breakdown of the oil film may result in excessive wear of the metal surfaces and a loss of power in the engine.

The term *viscosity index* is an arbitrarily defined ratio. It indicates the relative change in Saybolt Universal reading, with respect to temperature. The most desirable oils are those that have a high viscosity index (that is, the change in Saybolt reading is relatively small as the temperature changes). Oils having a small viscosity index register a relatively large change in Saybolt reading as the temperature changes.

## Lubricating Value

The terms *oiliness* and *lubricity* are used to refer to the lubricating value of any oil. These terms are most often used when the moving surfaces are relatively close and may make metal-to-metal contact. At the same pressure and temperature, oil *A* may be a better lubricant than another oil *B*. Therefore, oil *A* possesses more oiliness or lubricity than oil *B*. The lubricating value of a fluid depends on its chemical structure and its reaction with various metal surfaces when the metal surfaces are relatively close to each other. Thus, oiliness and lubricity are extremely important in the performance of the oil.

## Pour Point

The *pour point* of a fluid is defined as the lowest temperature at which the fluid flows when it is chilled under given conditions. The pour point is important when the hydraulic system is exposed to low temperatures. As a general rule, the most desirable pour point should be approximately 20°F below the lowest temperature to that the fluid will be exposed.

## Oxidation and Contamination

*Oxidation* is a chemical reaction in which oxygen combines with another element. Because air contains oxygen, the oxygen that is involved in fluid oxidation comes from exposing or mixing the fluid with air. The oxidation reaction increases with increased exposure of the oil to the air.

Undesirable quantities of air in hydraulic systems can be caused by mechanical factors, such as air leakage into the oil suction line, low fluid level in the oil reservoir, and leakage around the packing. Air leakage may result in the erratic motion of mechanical parts, and it may cause the fluid to oxidize more rapidly. All oils contain some air in solution that may not cause any trouble. If the air is not in solution, a foaming action may result. If trapped in a cylinder, air that is not in solution is highly compressible. However, oil is not as highly compressible as air. Irregular action of a cylinder, for example, may result if a significant quantity of air becomes undissolved.

Ferrous metals are destroyed by rust. Rust can develop in a hydraulic system if moisture is present. This moisture may be the result of condensation from air that enters through leaks on the intake (low pressure) side of a pump.

The *oxidation stability* of any oil refers to the inherent ability of oil to resist oxidation. Oxidation increases with increases in temperature, pressure, and agitation. Oxidation also increases as the oil becomes contaminated with such substances as grease, dirt, moisture, paint, and joint compound. Various metals also promote oil oxidation, and the various fluids have different oxidation characteristics. Table 13-2 lists the essential properties of the commercially available hydraulic fluids.

**Table 13-2   Properties of Available Hydraulic Fluids**

| | |
|---|---|
| **Petroleum-Base Fluids** | |
| Viscosity range, Saybolt Universal reading, in seconds, at 100°F | 40 to 5000 |
| Operating temperature, in °F | −75 to 500 |
| Minimum viscosity index | 76 to 225 |
| **Fire-Resistant Fluids (water-oil emulsions, water-glycol, phosphate-ester, chlorinated hydrocarbon, silicate ester, silicon)** | |
| Viscosity range, Saybolt Universal reading, in seconds, at 100°F | 20 to 5000 |
| Operating temperature, in °F | −100 to 600 |

## Hydraulic Filters

Hydraulic filters are needed to aid in eliminating many of the potential causes of hydraulic system failures. Proper filters and proper filter maintenance are important in obtaining satisfactory results in a hydraulic system.

Although most hydraulic systems are considered closed, they are not free from contaminants. Following are four sources of contaminants common in hydraulic systems:

- *Wear*—As the sliding members of components move, small particles of metal and seals enter the fluid. A typical example of this action is the movement of a cast-iron piston within a steel cylinder tube. Wear begins as soon as the cylinder is placed in operation, although it may not be visible to the eye for a long period.
- *Formation of sludge and acids due to fluid breakdown*—When extreme heat and pressure are encountered, chemical reaction

within the fluid causes the formation of sludge and acids that are harmful to the precision parts of the components. For example, resinous coatings may cause a valve spool to freeze within the valve body by forming on moving parts, or small orifices may become clogged. Acids cause pitting and corrosive conditions.

- *Built-in contaminants in the manufacture of components*—In castings with intricate cored passages, core sand is difficult to remove, and small quantities of sand can enter the system as the fluid flows through the cored passages under high pressure. Lint and small metal chips are also encountered.

- *Contaminants from outside the system*—Lint may enter the system if the filter cap on the oil reservoir is not replaced. Dirt that clings to the piston rod of a cylinder or to the stem of a valve may enter the system. Water or coolant may enter a system.

Factors that should be considered in selecting a hydraulic filter are flow rate, pressure drop, degree of filtration, capacity, ease of servicing, compatibility with the fluid in the system, and pressure to that the filter is subjected. The filter in a hydraulic system can be located in a number of places, including the following:

- *In the sump or oil reservoir*—This is a sump-type filter (see Figure 13-3). The filter should have a capacity twice that of the hydraulic pump in keep pressure drop at a minimum and to eliminate the possibility of cavitation in the pump.

**Figure 13-3** A sump-type filter equipped with magnetic rods for collecting minute particles of steel and iron. *(Courtesy Marvel Engineering Co.)*

- *In the discharge line from the relief valve*—Since, in most systems, a considerable quantity of fluid passes through the discharge of the relief valve, a low-pressure filter with fine filtration is recommended. The filter should be equipped with a low-pressure bypass valve to avoid filter or system failure. The capacity of the filter should be large enough to handle the full flow of the pump without imposing a backpressure on the relief valve. Backpressure on the exhaust of the relief valve can cause malfunctions with the system.

- *In the bypass line from the pump*—In bypass line filtration, a small percentage (approximately 10 percent) of the flow from the pump passes through the bypass filter, and returns to the reservoir as clean oil. A pressure-compensated flow control should be installed between the pump and the filter to maintain constant flow at minimum pressure through the filter. An internal bypass valve in the filter is recommended to avoid filter failure if the filter becomes clogged completely.

- *In the pressure line between the pump and the directional control valve*—To protect the components of the system that are located beyond the pump, a high-pressure filter that can handle full pump pressure and flow is often employed. Although a 25-micron filtration is an often-used standard in industry, a 5-micron, or less, filter is sometimes desirable if close-fitting parts are to be protected. The flow capacity through the filter should be as high as possible (four to five times pump capacity). A built-in bypass relief valve should be incorporated within the filter, because the filter may become overloaded with contaminants. Figure 13-4 shows a high-pressure filter equipped with a warning switch that provides a signal when the filter requires cleaning.

**Figure 13-4** A high-pressure filter equipped with a warning switch that provides a signal when the filter requires cleaning.
*(Courtesy Marvel Engineering Co.)*

- *In the intake line between the sump and the pump*—The intake line filter is similar to the sump-type filter, except that it is encased and is mounted outside the reservoir. A 100-mesh filter is commonly used for hydraulic oils, and a 60-mesh filter is usually specified for aqueous-base fluids. The filter elements can be replaced without disturbing the piping (see Figure 13-5).

- *In the exhaust line between the four-way directional control valve and the reservoir*—The exhaust line filter is helpful when the bulk of the fluid is returned to the reservoir through the four-way directional control valve. However, if most

**Figure 13-5** Intake-line filter placed between the sump and the pump. *(Courtesy Marvel Engineering Co.)*

of the fluid is returned through the relief valve or bypass valve, this type of filter is of little value. When the return-line filter is used, it should have more capacity than the maximum flow of the return line, to reduce backpressure to a minimum on the exhaust of the control valve. Sudden surges and shock can have a detrimental effect on the element in this type of filter.

## Mobile-type Hydraulic Filter Units

Figure 13-6 shows a mobile-type filter unit. Such a unit has many uses in industry, including the following:

- Cleaning old hydraulic systems.
- Filtering make-up oil (ideal for adding oil to servo-controlled systems and other systems where clean oil is absolutely essential).
- Cleaning new, in-plant systems before startup and systems on OEM machines and equipment before shipment to customer.
- Filtering out hydraulic system contamination resulting from failure of a component in the system.
- Recirculating and filtering power unit oil without having to shut down the power unit.

The unit in Figure 13-6 is equipped with a gear-type hydraulic pump driven by an electric motor; a 10-micron nominal paper element final filter; two hoses, a suction hose and a pressure hose; a utility box for storing spare element and small tools; and a drip tray that catches oil spills from filters and hoses. Figure 13-7 is similar to Figure 13-6, except that the hydraulic pump is driven by an air motor that receives its air through a filter, regulator, and lubricator unit. This unit is also equipped with a stored barrel caddy. In operation, the barrel caddy wheels rest on the floor and the oil drum is placed on the caddy.

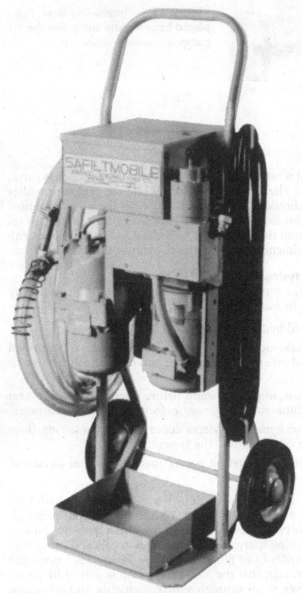

**Figure 13-6**   Mobile-type hydraulic filter unit. *(Courtesy Marvel Engineering Co.)*

**Figure 13-7**   Mobile-type hydraulic filter unit operated by pneumatic motor. *(Courtesy Marvel Engineering Co.)*

## Summary

Hydraulic fluids are usually divided into three categories: petroleum-base fluids, synthetic-base fluids, and water. The function of a good hydraulic fluid is threefold: it is a means of transmitting fluid power; it is a means of lubricating the components of a fluid power system; and it acts as a sealant.

To obtain certain desired characteristics, chemicals called additives are added to any oil. An additive may be in the form of an antifoam agent, a rust inhibitor, a film-strengthening agent, or an oxidation stabilizer.

The main functions of a hydraulic fluid are to transmit a force applied at one point in a system to some other point in the system, and to reproduce quickly any variation in the applied force. Thus, the fluid should flow readily, and it should be relatively incompressible. The choice of the most satisfactory hydraulic fluid for an industrial application involves two distinct considerations: the fluid should have certain essential physical properties and characteristics of flow and performance; and the fluid should have desirable performance characteristics over a period.

A very viscous fluid or a fluid having a high dynamic viscosity is a fluid that does not flow freely; a fluid having a low dynamic viscosity flows freely. The term fluidity is the reciprocal of dynamic viscosity. A fluid having a high dynamic viscosity has a low fluidity, and a fluid having a low dynamic viscosity has a high fluidity.

The terms oiliness and lubricity refer to the lubricating value of any oil. The lubricating value of a fluid depends on its chemical structure and its reaction with various metal surfaces when the surfaces are relatively close to each other.

Hydraulic filters are needed to aid in eliminating many of the potential causes of failure in hydraulic systems. Proper filters and proper filter maintenance are important factors in obtaining satisfactory results in a hydraulic system. In selecting a hydraulic filter, factors that should be considered are flow rate, pressure drop, degree of filtration, capacity, ease of servicing, compatibility with the fluid in the system, and the pressure to that the filter is subjected.

## Review Questions

1. List three disadvantages of using water in a fluid power system.
2. What may be the cause of hydraulic oil becoming overheated in a hydraulic system?
3. In what ways can air enter a hydraulic system?

4. In what ways can dirt get into a hydraulic system, despite the fact that a suitable filter is used?

5. What precautions should be taken in changing the oil in a hydraulic system?

6. In storing hydraulic fluids, what precautions should be exercised?

7. What hazards are presented when hydraulic oil remains on the floor?

8. List three types of commonly used hydraulic fluids.

9. What is the effect of some fire-resistant fluids on packing, gaskets, and filters?

10. What are three categories in that hydraulic fluids may be divided?

11. What is the function of a good hydraulic fluid?

12. List three types of mineral oils.

13. What are some of the requirements of good hydraulic oil?

14. What are the steps that need to be followed in changing from a water-base fluid to a petroleum-base fluid?

15. What is the main function of a hydraulic fluid?

16. What does water weigh at 60°F?

17. What is the range in specific gravities of hydraulic fluids?

18. Define *viscosity*.

19. What does *absolute* or *dynamic viscosity* mean?

20. What does the Saybolt meter measure?

21. What does SAE stand for?

22. What are the minimum and maximum Saybolt readings for SAE 20 oil?

23. What is the *viscosity index*?

24. What is the *pour point*?

25. What is the purpose of the filter in a hydraulic system?

# Chapter 14

# Fluid Lines and Fittings

The efficiency of a fluid power system is often limited by the lines (fluid carriers) that carry the fluid operating medium from one fluid power component to another. The purpose of these carriers is to provide leakproof passages at whatever operating pressure may be required in a system. A poorly planned system of fluid carriers for the system often results in component malfunctions caused by the following:

- Restrictions (creates a backpressure in the components)
- Loss of speeds (reduces efficiency)
- Broken carriers, especially in high-pressure systems (creates fire hazards and other problems)

Selection of the proper carrier is as important as the proper selection of the fluid components.

Fluid may be directed through either lines or manifolds. Fluid lines or piping fall into three categories:

- Rigid
- Semi-rigid, or tubing
- Flexible, or hose

In many instances, all three categories are employed in a single fluid system. The pressures involved and the fluid medium used will determine the type of carrier and the connectors and fittings.

## Rigid Pipe

Steel pipe is the original type of carrier used in fluid power systems, and it is available in the following four different weights:

- *Standard (STD)*, or *Schedule 40*—This pipe (seamless) is designed for test pressures of 700 psi in the $\frac{1}{8}$-inch size to 1100 psi in the 2-inch size.
- *Extra Strong (XS)*, or *Schedule 80*—This weight of pipe is used in the medium pressure range of hydraulic systems. This pipe (seamless) is designed for test pressures of 850 psi for $\frac{1}{8}$-inch size to 1600 psi in the 2-inch size (Grade B).
- *Schedule 160*—This pipe is designed for test pressures up to 2500 psi.

- *Double Extra Heavy* (XXS)—This pipe is also used for test pressures up to 2500 psi, even though the wall thickness is somewhat heavier.

Sizes of pipe are listed by the nominal inside diameter (ID), which is actually a misnomer. The sizes are nearly equivalent to Schedule 40, but there is a difference. For example, the ID of a ¼-inch Schedule 40 pipe is 0.364 inch, and the ID of a ½-inch Schedule 40 pipe is 0.622 inch.

As the schedule number of the pipe increases, the wall thickness also increases. This means that the ID of the pipe for each nominal size is smaller, but the outside diameter (OD) of the pipe for each nominal size remains constant (see Table 14-1).

Some of the fittings that are used with steel pipe are tees, crosses, elbows, unions, street elbows, and so on. The pipe fittings are listed in nominal pipe sizes.

Steel pipe is one of the least-expensive fluid carriers for hydraulic fluid as far as material costs are concerned, but installation costs often consume considerable time in comparison to installation costs of other types of fluid carriers. Pipe is suited for handling large fluid volumes and for running long lines of fluid carriers. Pipe is commonly used on suction lines to pumps and for short connections between two components. It is also useful on piping assemblies that are seldom disassembled. Pipe provides rigidity for holding various components in position, such as valves that are designed for in-line mounting with no other method of support. Pipe should be cleaned thoroughly before it is installed in a fluid power system.

## Semi-Rigid (Tubing)

Two types of steel tubing are utilized in hydraulic systems, as recommended by the United States of America Standards Institute (USASI) hydraulic standards. These types of tubing are *seamless* and *electric-welded*. Tubing sizes are measured on the OD of the tubing. Seamless steel tubing is manufactured from highly ductile, dead-soft, annealed low-carbon steel, and an elongation in 2 inches of 35 percent minimum. In tubes with an OD of ³⁄₈ inch and/or a wall thickness of 0.035 inch, a minimum elongation of 30 percent is permitted. The diameter of the tubing does not vary from that specified by more than the limits shown in Table 14-2.

The process used for making steel tubing is the cold drawing of pierced or hot-extruded billets. Table 14-3 shows the nominal sizes of seamless steel tubing that are readily available for hydraulic systems.

# Table 14-1 Sizes of Steel Pipe

| Nominal Pipe Size (in.) | Pipe OD (in.) | Schedule 40 (Standard) | | Schedule 80 (Extra Heavy) | | Schedule 160 | | Double Extra Heavy | |
|---|---|---|---|---|---|---|---|---|---|
| | | Pipe ID (in.) | Burst Press (psi) | Pipe ID (in.) | Burst Press (psi) | Pipe ID (in.) | Burst Press (psi) | Pipe ID (in.) | Burst Press (psi) |
| 1/8 | 0.405 | — | — | — | — | — | — | — | — |
| 1/4 | 0.540 | 0.364 | 16,000 | 0.302 | 22,000 | — | — | — | — |
| 3/8 | 0.675 | 0.493 | 13,500 | 0.423 | 19,000 | — | — | — | — |
| 1/2 | 0.840 | 0.622 | 13,200 | 0.546 | 17,500 | 0.466 | 21,000 | 0.252 | 35,000 |
| 3/4 | 1.050 | 0.824 | 11,000 | 0.742 | 15,000 | 0.614 | 21,000 | 0.434 | 30,000 |
| 1 | 1.315 | 1.049 | 10,000 | 0.957 | 13,600 | 0.815 | 19,000 | 0.599 | 27,000 |
| 1 1/4 | 1.660 | 1.380 | 8400 | 1.278 | 11,500 | 1.160 | 15,000 | 0.896 | 23,000 |
| 1 1/2 | 1.900 | 1.610 | 7600 | 1.500 | 10,500 | 1.338 | 14,800 | 1.100 | 21,000 |
| 2 | 2.375 | 2.067 | 6500 | 1.939 | 9100 | 1.689 | 14,500 | 1.503 | 19,000 |
| 2 1/2 | 2.875 | 2.469 | 7000 | 2.323 | 9600 | 2.125 | 13,000 | 1.771 | 18,000 |
| 3 | 3.500 | 3.068 | 6100 | 2.900 | 8500 | 2.624 | 12,500 | — | — |

### Table 14-2 Specifications for Tubing Diameters

| OD, Nominal, | OD (inches) | ID (inches) |
|---|---|---|
| $\frac{1}{4}$ to $\frac{1}{2}$ inch, incl. | ± 0.003 | — |
| Above $\frac{1}{2}$ to $1\frac{1}{2}$ inch, incl. | ± 0.005 | ± 0.005 |
| Above $1\frac{1}{2}$ to $3\frac{1}{2}$ inch, incl. | ± 0.010 | ± 0.010 |

### Table 14-3 Steel Tubing Sizes and Safety Factors

| Tube OD | Fitting Size | $4/_1$ Safety Factor Working Pressure in psi | | | | $5/_1$ Safety Factor Working Pressure in psi | | |
|---|---|---|---|---|---|---|---|---|
| | | 1000 | 2000 | 3000 | 5000 | 1000 | 2000 | 3000 |
| $\frac{1}{8}$ | 2 | 0.020 | 0.020 | 0.020 | 0.025 | 0.020 | 0.020 | 0.020 |
| $\frac{3}{16}$ | 3 | 0.020 | 0.020 | 0.020 | 0.035 | 0.020 | 0.020 | 0.028 |
| $\frac{1}{4}$ | 4 | 0.020 | 0.020 | 0.028 | 0.049 | 0.020 | 0.022 | 0.035 |
| $\frac{5}{16}$ | 5 | 0.020 | 0.025 | 0.035 | 0.056 | 0.020 | 0.028 | 0.042 |
| $\frac{3}{8}$ | 6 | 0.020 | 0.028 | 0.042 | 0.072 | 0.020 | 0.035 | 0.049 |
| $\frac{1}{2}$ | 8 | 0.020 | 0.042 | 0.056 | 0.095 | 0.025 | 0.04 | 0.065 |
| $\frac{5}{8}$ | 10 | 0.025 | 0.042 | 0.072 | 0.120 | 0.032 | 0.058 | 0.083 |
| $\frac{3}{4}$ | 12 | 0.028 | 0.058 | 0.083 | 0.134 | 0.035 | 0.072 | 0.109 |
| $\frac{7}{8}$ | 14 | 0.035 | 0.072 | 0.095 | 0.165 | 0.042 | 0.083 | 0.120 |
| 1 | 16 | 0.042 | 0.083 | 0.109 | 0.180 | 0.049 | 0.095 | 0.134 |
| $1\frac{1}{4}$ | 20 | 0.049 | 0.095 | 0.134 | 0.238 | 0.058 | 0.120 | 0.165 |
| $1\frac{1}{2}$ | 24 | 0.058 | 0.120 | 0.165 | 0.284 | 0.072 | 0.134 | 0.203 |
| 2 | 32 | 0.072 | 0.148 | 0.220 | 0.375 | 0.095 | 0.180 | 0.259 |

| Tube OD | Fitting Size | $7.5/_1$ Safety Factor Working Pressure in psi | | | | $10/_1$ Safety Factor Working Pressure in psi | | |
|---|---|---|---|---|---|---|---|---|
| $\frac{1}{6}$ | 2 | 0.020 | 0.020 | 0.025 | 0.042 | 0.020 | 0.025 | 0.035 |
| $\frac{3}{16}$ | 3 | 0.020 | 0.025 | 0.042 | 0.065 | 0.020 | 0.035 | 0.056 |
| $\frac{1}{4}$ | 4 | 0.020 | 0.035 | 0.058 | 0.095 | 0.022 | 0.049 | 0.072 |
| $\frac{5}{16}$ | 5 | 0.022 | 0.042 | 0.065 | 0.109 | 0.028 | 0.056 | 0.083 |
| $\frac{3}{8}$ | 6 | 0.025 | 0.058 | 0.083 | 0.134 | 0.035 | 0.072 | 0.109 |
| $\frac{1}{2}$ | 8 | 0.035 | 0.072 | 0.109 | 0.220 | 0.049 | 0.095 | 0.134 |
| $\frac{5}{8}$ | 10 | 0.042 | 0.095 | 0.134 | 0.220 | 0.058 | 0.120 | 0.180 |
| $\frac{3}{4}$ | 12 | 0.058 | 0.109 | 0.148 | 0.259 | 0.072 | 0.134 | 0.203 |

(*continued*)

**Table 14-3 (continued)**

| Tube OD | Fitting Size | 7.5/1 Safety Factor | | | | 10/1 Safety Factor | | |
| --- | --- | --- | --- | --- | --- | --- | --- | --- |
| | | Working Pressure in psi | | | | Working Pressure in psi | | |
| $^7/_8$ | 14 | 0.065 | 0.120 | 0.180 | 0.320 | 0.083 | 0.165 | 0.238 |
| 1 | 16 | 0.072 | 0.134 | 0.203 | 0.350 | 0.095 | 0.180 | 0.259 |
| $1^1/_4$ | 20 | 0.095 | 0.180 | 0.259 | 0.450 | 0.120 | 0.238 | 0.350 |
| $1^1/_2$ | 24 | 0.109 | 0.203 | 0.320 | 0.500 | 0.134 | 0.284 | 0.450 |
| 2 | 32 | 0.134 | 0.284 | 0.400 | — | 0.180 | 0.375 | — |

## Manufacturing Process

Electric-welded steel tubing is manufactured by shaping a cold-rolled strip of steel into a tube and then performing a welding and drawing operation. The chemical and physical properties of electric-welded steel tubing are similar to those of seamless steel tubing.

To use steel tubing (or any other type of tubing) in a fluid power system, it is necessary to attach some type of fitting to each end of the tubing. Numerous methods are employed to accomplish this. However, in the final analysis, the fitting holds the tubing securely, providing a leakproof assembly that can withstand the pressures for which it was designed. In some instances, the fitting is welded to the tubing. In other applications (air systems), friction between the tubing and the fitting is sufficient to hold the pressure.

Figure 14-1 shows an assembly in which a sleeve is brazed to the tubing, and the nut is then screwed onto the fitting. Figure 14-2 shows a sleeve that digs into the tubing wall when the nut is tightened on the fitting. Figure 14-3 shows some of the fittings that are most commonly used (such as tees, elbows, crosses, and straights).

**Figure 14-1** An assembly in which a sleeve is brazed to the tubing, and the nut is then screwed onto the fitting.

*(Courtesy Imperial-Eastman Corporation)*

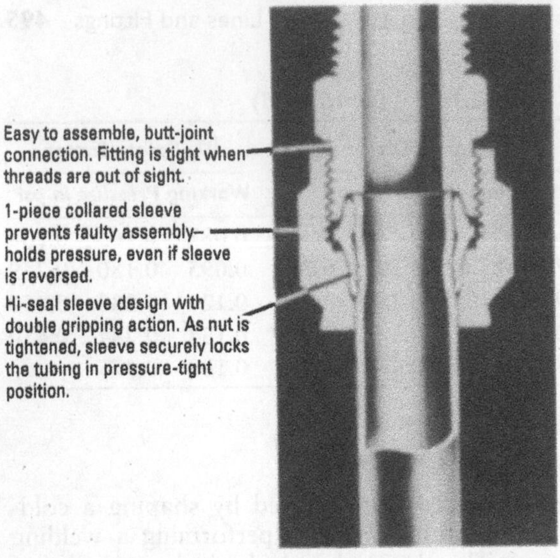

Easy to assemble, butt-joint connection. Fitting is tight when threads are out of sight.

1-piece collared sleeve prevents faulty assembly— holds pressure, even if sleeve is reversed.

Hi-seal sleeve design with double gripping action. As nut is tightened, sleeve securely locks the tubing in pressure-tight position.

**Figure 14-2** A fitting for high-pressure tubing in which the sleeve of the fitting grips the tubing. *(Courtesy Imperial-Eastman Corporation)*

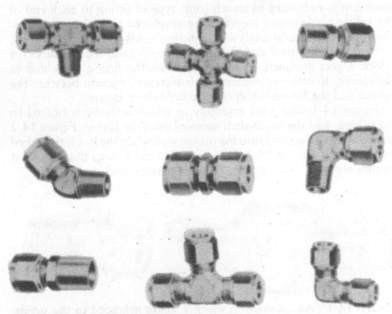

**Figure 14-3** Tubing fittings used in fluid-power systems.
*(Courtesy Imperial-Eastman Corporation)*

## Other Types

Following are other types of tubing employed in fluid power systems:

- *Copper tubing*—This type of tubing is often found on air circuits that are not subject to USASI standards. Because of the work hardening when flared, and since it is an oil-oxidation catalyst, USASI standards restrict the use of copper tubing for hydraulic service. Copper tubing can be worked easily to make bends that reduce fitting requirements.

- *Aluminum tubing*—Seamless aluminum tubing is approved for low-pressure systems. This tubing has fine flaring and bending characteristics.

- *Plastic tubing*—Plastic tubing for fluid power lines is made from several basic materials. Among these materials are nylon, polyvinyl, polyethylene, and polypropylene.

  - *Nylon tubing*—This tubing is used on fluid power applications in the low-pressure range, up to 250 psi. It is suitable for a temperature range of minus 100°F to plus 225°F. This tubing possesses good impact and abrasion resistance. It can be stored without deteriorating or becoming brittle, and it is not affected by hydraulic fluids. One of the newer developments is the self-storing type of nylon tubing that looks like a coil spring; it is very popular for use on pneumatic tools.

  - *Polyvinyl chloride tubing*—For air lines with pressures up to 125 psi, this type of tubing may be used. Temperatures should not exceed 100°F continuously. It may be used intermittently for temperatures up to 160°F.

  - *Polyethylene tubing*—This is ideal tubing for pneumatic service, and it is also used for other fluids at low pressures. It possesses great dimensional stability and resists most chemicals and solvents. Polyethylene tubing is manufactured in several different colors that lend it readily to color-coding. Tubing sizes are usually available up to and including $1/2$-inch OD.

  - *Polypropylene tubing*—This type of tubing is suitable for operating conditions with temperatures of minus 20°F to plus 280°F, and it can be sterilized repeatedly with steam. It possesses surface hardness and elasticity that provide good abrasion resistance. It is usually available in sizes up to and including $1/2$-inch OD, in natural or black colors.

## Installation of Tubing

Following are some precautions that should be observed when installing tubing (see Figure 14-4):

- Avoid straight-line connections wherever possible, especially in short runs.

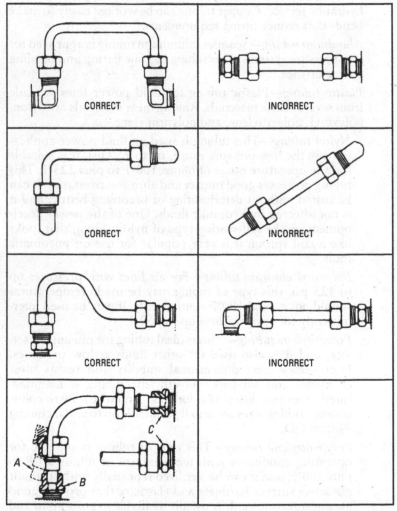

**Figure 14-4**   Correct method (a) and incorrect method (b) of installing tubing in a system. *(Courtesy of Weatherhead Co.)*

- Design the piping systems symmetrically. They are easier to install and present a neater appearance.
- Care should be taken to eliminate stress from the tubing lines. Long tubing should be supported by brackets or clips. All parts installed on tubing lines (such as heavy fittings, valves, and so on), should be bolted down to eliminate tubing fatigue.
- Before installing tubing, inspect the tube to ensure that it conforms to the required specifications, that it is of the correct diameter, and that the wall thickness is not out of round.
- Cut the ends of tubes reasonably square. Ream the inside of the tube and remove the burrs from the outside edge. Excessive chamfer on the outside edge destroys the bearing of the end of the tube on the seat of the fitting.
- To avoid difficulty in assembly and disconnection, a sufficient straight length of tube must be allowed from the end of the tube to the start of the bend. Allow twice the length of the nut as a minimum.
- Tubes should be formed to assemble with true alignment to the centerline of the fittings, without distortion or tension. A tube that has to be sprung (position *A* in Figure 14-4) to be inserted into the fitting has not been properly fabricated, and when so installed and connected, places the tubing under stress.
- When assembling the tubing, insert the longer leg to the fitting as at *C* (see Figure 14-4), and then insert the other end into fitting *D*. Do not screw the nut into the fitting at *C*. This holds the tubing tightly, and restricts any movement during the assembly operation. With the nut free, the short leg of the tubing can be moved easily, brought to position properly, and inserted into the seat in the fitting *D*. The nuts can then be tightened as required.

## Flexible Piping (Hose)

Hose is employed in a fluid power system in which the movement of one component of the system is related to another component. An example of this utilization is a pivot-mounted cylinder that moves through an arc while the valve to that the cylinder is connected with fluid lines remains in a stationary position. Hose may be used in either a pneumatic or a hydraulic system.

Many different types of hose are used in fluid power systems, and nearly all of these types of hose have three things in common

**Figure 14-5** Section of hose assembly showing the layers of material, including the inner liner, reinforcement, and outer cover.
*(Courtesy Imperial-Eastman Corporation)*

(see Figure 14-5):

- *A tube or inner liner that resists penetration by the fluid being used*—This tube should be smooth to reduce friction. Some of the materials used for the tube are neoprene, synthetic rubber, and so on.

- *A reinforcement that may be in the form of rayon braid, wire braid, or spiral-bound wire*—The strength of the hose is determined by the number of thickness and type of reinforcement. If more than one thickness of reinforcement is used, a synthetic type of separator is utilized. A three-wire braid type of hose is made of three layers of wire braid.

- *An outer cover to protect the inner portion of the hose and to enable the hose to resist heat, weather, abrasion, and so on*—This cover can be made of synthetic rubber, neoprene, woven metal, or fabric.

The nominal size of hose is specified by ID (such as $3/16$-inch, $1/4$-inch, $3/8$-inch, $1/2$-inch, and so on). The outside diameter of hose depends on the number of layers of wire braid, and so on. Fluid power applications may require hose with working pressure ratings ranging from approximately 300 psi to 12,000 psi. Table 14-4 shows specifications for a typical two-wire braid hose. The hose and the couplings that are attached at each end of the hose make up the hose assembly for fluid power applications. Several methods are used to attach these couplings to the hose, including the following:

- *Pressed* on by mechanical crimping action. Production machines can be used to make large quantities of the assemblies.

- *Screwed* on by removing the outer cover of the hose for a required distance that is marked on the shell of the fitting. The shell is then threaded onto the braid, and the male body can be screwed into the hose and shell assembly. In some types of

**Table 14-4  Specifications for a Typical Two-Wire Braid Hose**

| Size | | Pressure | | Bend Radius |
|---|---|---|---|---|
| Hose ID (in.) | Hose OD (in.) | Recommended Maximum Working Pressure (psi) | Minimum Burst Pressure (psi) | Minimum Bending Radius (in.) |
| $3/16$ | $5/8$ | 5000 | 20,000 | 4 |
| $1/4$ | $11/16$ | 5000 | 20,000 | 4 |
| $5/16$ | $3/4$ | 4250 | 17,000 | 5 |
| $3/8$ | $27/32$ | 4000 | 16,000 | 5 |
| $1/2$ | $31/32$ | 3500 | 14,000 | 7 |
| $5/8$ | $1 3/32$ | 2750 | 11,000 | 8 |
| $3/4$ | $1 1/4$ | 2250 | 9000 | $9 1/2$ |
| $7/8$ | $1 3/8$ | 2000 | 8000 | 11 |
| 1 | $1 9/16$ | 2000 | 8000 | 12 |
| $1 1/4$ | 2 | 1625 | 6500 | $16 1/2$ |
| $1 1/2$ | $2 1/4$ | 1250 | 5000 | 20 |
| 2 | $2 3/4$ | 1125 | 4500 | 25 |

couplings, it is unnecessary to remove the outer cover. Screwed-on couplings can be disassembled and used again.

- *Clamped* on by screwing the hose onto the coupling stem until it bottoms against the collar on the stem. Then, the hose clamp is attached with bolts. Some couplings require two bolts, and others require four bolts for tightening the clamp on the hose.

- *Pushed* on by pushing the hose onto the coupling. This type of coupling is used in low-pressure applications up to 250 psi. No tools except a knife to cut the hose to length are needed to make this type of assembly. This type of coupling is reusable.

Figure 14-6 shows various types of ends that are used on hose couplings. This permits the user a wide choice.

Hose assemblies are measured with respect to their overall length from the extreme end of one coupling to the extreme end of the other coupling (see Figure 14-7). In applications, using an elbow coupling, the length is measured from the centerline of the sealing surface of the elbow end to the centerline of the coupling on the opposite end.

The length of a hose assembly that is to be looped can be determined from Figure 14-8. In addition, the diameter of hose required

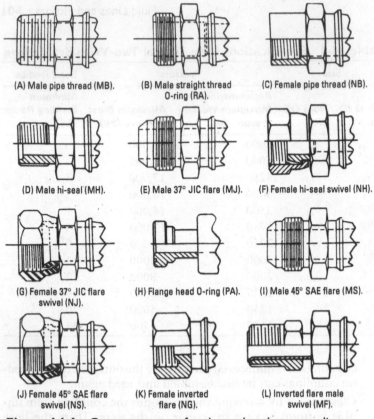

(A) Male pipe thread (MB).

(B) Male straight thread O-ring (RA).

(C) Female pipe thread (NB).

(D) Male hi-seal (MH).

(E) Male 37° JIC flare (MJ).

(F) Female hi-seal swivel (NH).

(G) Female 37° JIC flare swivel (NJ).

(H) Flange head O-ring (PA).

(I) Male 45° SAE flare (MS).

(J) Female 45° SAE flare swivel (NS).

(K) Female inverted flare (NG).

(L) Inverted flare male swivel (MF).

**Figure 14-6** Common types of ends used on hose couplings.

*(Courtesy Imperial-Eastman Corporation)*

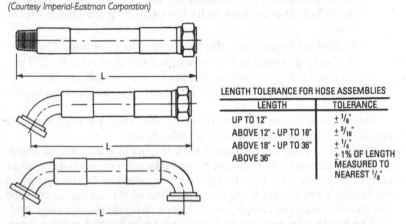

| LENGTH TOLERANCE FOR HOSE ASSEMBLIES | |
|---|---|
| LENGTH | TOLERANCE |
| UP TO 12" | $\pm \frac{1}{8}$" |
| ABOVE 12" - UP TO 18" | $\pm \frac{3}{16}$" |
| ABOVE 18" - UP TO 36" | $\pm \frac{1}{4}$" |
| ABOVE 36" | + 1% OF LENGTH MEASURED TO NEAREST $\frac{1}{8}$" |

**Figure 14-7** Method of measuring the length of a hose assembly.

*(Courtesy Imperial-Eastman Corporation)*

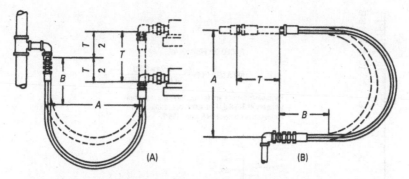

Typical Dimensions for One- and Two- Wire Braid Hose

If bending diameters other than those below are used,  apply the following formulas:

(A) OVERALL LENGTH $= B + 1.57A + \frac{1}{2}T$

(B) OVERALL LENGTH $= B + 1.57A + T$

| I.D. OF HOSE | B CONSTANT FOR STRAIGHT PORTION INCLUDING COUPLIG | MIN. A | MINIMUM OVERALL LENGTH | |
|---|---|---|---|---|
| | | | (A) | (B) |
| $\frac{3}{16}$" | 10" | 8" | $23'' + \frac{1}{2}T$ | $23'' + T$ |
| $\frac{1}{4}$" | 10" | 8" | $23'' + \frac{1}{2}T$ | $23'' + T$ |
| $\frac{3}{8}$" | 10" | 10" | $26'' + \frac{1}{2}T$ | $26'' + T$ |
| $\frac{1}{2}$" | 12" | 14" | $34'' + \frac{1}{2}T$ | $34'' + T$ |
| $\frac{3}{4}$" | 14" | 19" | $44'' + \frac{1}{2}T$ | $44'' + T$ |
| 1" | 16" | 22" | $51'' + \frac{1}{2}T$ | $51'' + T$ |
| $1\frac{1}{4}$" | 18" | 32" | $68'' + \frac{1}{2}T$ | $68'' + T$ |
| $1\frac{1}{2}$" | 20" | 44" | $87'' + \frac{1}{2}T$ | $87'' + T$ |
| 2" | 20" | 48" | $95'' + \frac{1}{2}T$ | $95'' + T$ |

**Figure 14-8**   Typical dimensions for one-wire and two-wire braid hose in determining length.

to ensure proper performance of hose for hydraulic service can be determined in Figure 14-9.

In applications in which it is desirable to disconnect one end of a hose assembly repeatedly, quick-disconnect couplings are recommended. The quick-disconnect couplings save considerable time in making or breaking the connection. When properly chosen, quick-disconnect couplings provide positive shutoff, so that the fluid is not lost. Figure 14-10 shows a quick-disconnect coupling with double shutoff; when the coupling is disconnected, the valved-nipple and the valved-coupler prevent the escape of fluid. Quick-disconnect couplings are also furnished in various other combinations, such as

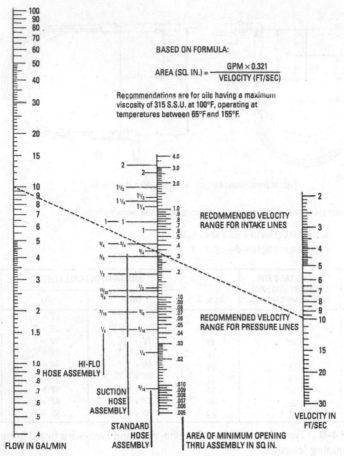

**Figure 14-9** Method of determining the correct size of hose.

*(Courtesy Imperial-Eastman Corporation)*

a single shutoff coupling with a plain nipple and a valved-coupler, and a no-shutoff coupler with a plain nipple and a plain coupler. Quick-disconnect couplings are available in a wide range of sizes from ¼-inch to 4-inch pipe size. Larger sizes are usually available as "specials." Various metals (such as brass, aluminum, stainless steel, and alloy steel) are used. Seals in these couplings depend on the type of service involved. The fitting connections on the ends of the couplings are available in several forms, such as female (NPT), male (NPT), hose shank, female (SAE), male flare, bulkhead, and so on.

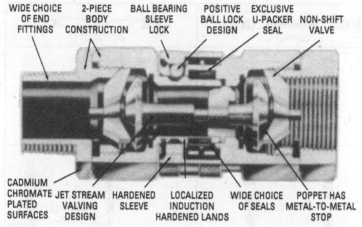

WIDE CHOICE OF END FITTINGS   2-PIECE BODY CONSTRUCTION   BALL BEARING SLEEVE LOCK   POSITIVE BALL LOCK DESIGN   EXCLUSIVE U-PACKER SEAL   NON-SHIFT VALVE

CADMIUM CHROMATE PLATED SURFACES   JET STREAM VALVING DESIGN   HARDENED SLEEVE   LOCALIZED INDUCTION HARDENED LANDS   WIDE CHOICE OF SEALS   POPPET HAS METAL-TO-METAL STOP

**Figure 14-10**   Cutaway of quick-disconnect double shutoff.
(Courtesy Snap-Tite, Inc.)

Figure 14-11 shows correct and incorrect installation methods for hose assemblies.

## Manifolds

Manifolds are designed to eliminate piping, to reduce joints (which are often a source of leakage), to conserve space, and to help streamline equipment. Manifolds are usually one of the following types:

- Sandwich
- Cast
- Drilled
- Fabricated-tube

The *sandwich* type of manifold is made of flat plates in which the center plate or plates are machined for the passages, and the porting is drilled in the outer plates. The passages are then bonded together to make a leakproof assembly. A *cast*-type manifold is designed with cast passages and the drilled ports. The casting may be steel, iron, bronze, or aluminum, depending on the fluid medium to be used. In the *drilled* type of manifold, all the porting and passages are drilled in a block of metal. The *fabricated-tube* type of manifold is made of tubing to that the various sections have been welded. This makes an assembly that may contain welded flange connections, valve sub plates, male or female pipe connectors, and so on. These assemblies are usually produced in large quantities for use on the hydraulic systems of mobile equipment. The assemblies

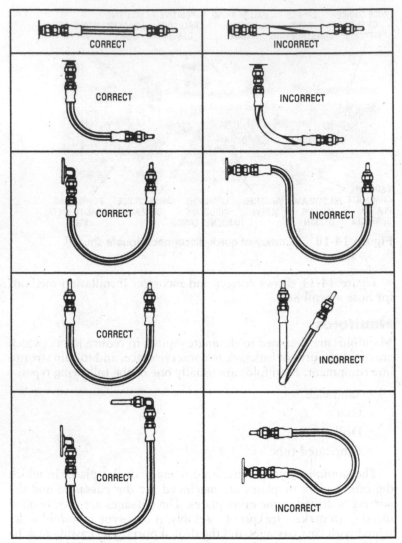

**Figure 14-11** Installation of hose assembly. *(Courtesy The Weatherhead Co.)*

are held to close tolerances, because they are manufactured in fixtures. Although manifolds are used mostly on hydraulic systems, the demand for them in pneumatic systems is increasing.

## Summary

Fluid lines or piping is found in three categories: rigid, semi-rigid, or tubing (flexible or hose). In many instances, all three categories of piping are employed in a single fluid system.

The size of rigid pipe is indicated by the nominal inside diameter (ID), that is actually a misnomer. For example, the inside diameter of a $^1/_4$-inch Schedule 40 pipe is 0.364 inch, and the inside diameter of a $^1/_2$-inch Schedule 40 pipe is 0.622 inch.

Two types of steel tubing are utilized in hydraulic systems, as recommended by USASI hydraulic standards: seamless and electric-welded. Tubing size is indicated by the nominal outside diameter (OD) of the tubing.

Flexible pipe (hose) is utilized in a fluid power system in which the movement of one component of the system is related to another component. The size of a hose is indicated by the nominal ID (such as $^3/_{16}$-inch, $^1/_4$-inch, and so on).

Manifolds are designed to eliminate piping, to reduce joints that are often a source of leakage, to conserve space, and to help streamline modern-day equipment. Manifolds are usually one of the following types: sandwich, cast, drilled, and fabricated-tube.

## Review Questions

1. List the three types of fluid lines or piping.
2. What are the advantages of steel pipe as a fluid carrier?
3. What are three types of tubing commonly used in fluid power systems?
4. What is the chief advantage of hose in fluid power systems?
5. What is the chief advantage of manifolds in fluid power systems?
6. How may the efficiency of a fluid power system become limited?
7. What does Schedule 40 pipe mean as compared to Schedule 80?
8. Describe where XXS pipe is used.
9. What is one of the least expensive fluid carriers for hydraulic fluids?

10. What is the outside diameter of a ¾-inch pipe?

11. What is the ID of a 1-inch Schedule 80 pipe?

12. Why is copper tubing restricted for hydraulic service?

13. For what type of systems is seamless aluminum tubing used?

14. What are the types of materials used in making plastic tubing for fluid power lines?

15. What is the recommended maximum working pressure for a 1-inch ID two-wire braid hose?

16. What does NPT mean?

17. List four types of manifolds.

18. What are the two types of steel tubing used in hydraulic lines?

# Appendix A

## Pump Resources

The Hydraulic Institute (HI) is one of the 140,000 associations that exist today. It has been serving the pump industry for the past 84 years. The Institute's mission is "serving member companies and pump users by providing product standards and a forum for the exchange of industry information." The technical and managerial leadership that the Institute enjoys is world-class. Currently, companies that manufacture pumps in North America, or suppliers to the industry in six product areas, are eligible to join. Today, more than 200 individuals employed by member companies are actively involved on HI committees that are developing new standards, statistics and meetings for the industry. Nearly 200 pump users, contractors, and other interested parties are actively involved in reviewing and approving HI standards.

Launched in April 2000, the HI Web site was a significant milestone in the history of the association. It provides free downloadable executive summaries of the *LCC Guide*, the *Master Index to HI Standards*, more than 100 pump diagrams and software useful for evaluating pumping systems and selecting motors, and a *Supplier Finder* with seven ways to identify suppliers (including by product, market served, and trade name). The site also has a new Energy Section offering useful information and supporting HI's role as an Allied Partner of the U.S. Department of Energy.

Table A-1 lists some valuable links for those interested in contacting related associations and organizations on the national, international, and industry levels.

### Table A-1  Key Contacts

| Organization | Internet Web Site |
| --- | --- |
| American National Standards Institute (ANSI) | www.ansi.org |
| Hydraulic Institute | www.pumps.org/public/news/ pumps_associations_print.htm |
| **National Associations** | |
| Air-Conditioning and Refrigeration Institute | www.ari.org |
| Electrical Apparatus Service Association, Inc. (EASA) | www.easa.com |
| Fluid Sealing Association | www.fluidsealing.com |

*(continued)*

## Table A-1   (continued)

| Organization | Internet Web Site |
| --- | --- |
| **International Associations** | |
| Asso Pompe | www.assopompe.it |
| British Pump Manufacturers Association | www.bpma.org.uk |
| Europump | www.europump.org |
| French Pump and Valve Association (AFCP) | www.afcp-afir.org |
| Swedish Pump Supplier's Association | www.vibab.se/swepump |
| **Industry Publications** | |
| National Manufacturing Week | http://manufacturingweek.com |
| WaterInfoCenter | www.waterinfocenter.com |

# Appendix B

## Oils and Fluids

A number of synthetic oils and fluids are available for use with high-speed compressor pumps and are recommended for use in single-stage and multi-stage rotary screw, vane, and reciprocating compressor crankcases and cylinders, vacuum pumps, and other compressor applications. Most of these are available in five viscosity grades. These synthetic oils have a number of additives that make them capable of inhibiting carbon build-up and that make them suitable for operating equipment at higher temperatures. Most are formulated to keep water from mixing with the oil and deteriorating its lubricating ability. Manufacturers also claim it is easier for water removal with no need for specialized water-removal processes.

For example, the synthetic compressor oil is formulated with synthetic ester technology that produces a longer life fluid that effectively prevents wear, oxidation, foam, and rust, while its inherent lubricity and thermal conductivity act to reduce heat and energy consumption. This increases the operating efficiency and reduces maintenance costs.

Performance features are maintained across the spectrum with a wide operating temperature range. It has a low pour-point, high-viscosity index and the lack of paraffin (wax) makes synthetic oils an excellent all-weather lubricant.

Hydraulic oils are formulated for long life. They typically reduce maintenance costs. This is done by extending drain intervals. AMSOIL's synthetic AW series of anti-wear hydraulic oils, for example, are recommended for high- and low-pressure gear, vane, and piston stationary and mobile hydraulic systems. They are available in many viscosity grades. The equipment manufacturer usually recommends the viscosity for the device. The oil will operate efficiently in temperatures of –20°F or lower. These oils are shear-stable, long-life lubricants based on high-quality synthetic oil technology. These oils are formulated with an additive system that inhibits oxidation and prevents acid and viscosity increase, inhibits rust, and inhibits foam formation, as well as preventing spongy hydraulics.

There is also thermally stable biodegradable hydraulic oil that is designed to biodegrade to its natural state when subjected to sunlight, water, and microbial activity. It has a very low toxicity level because the oil is formulated with ashless additives that do not contain heavy metals.

Traditional oil-lubricated pumps use oil for lubricating, cooling, and sealing. Oil reduces the friction of pistons or vanes that must slide against the cylinder walls. Older-style rotary vane pumps used drip oilers. The oil merely reduced friction, but newer oil-flooded designs use copious amounts of oil and actually cause the vanes to hydroplane. This not only reduces the friction, but since the vanes do not actually touch the cylinder wall, the vane life is greatly extended. It is common to obtain 30,000 to 40,000 hours of vane life.

Oil removes the heat of friction. When oil returns to the oil sump (or reservoir), the heat is radiated away from the pump. A fan usually blows across cooling fins on the housing. This exchange of heat is important. If oil runs too hot, it carbonizes. Carbonization can cause the pump to overheat and eventually cause its failure.

The amount of oil being fed to the pump affects sealing. Drip-type oilers on rotary vane pumps need sufficient oil to seal the clearances around the vanes. Oil is used primarily for lubrication. The oil seal has a significant internal leakage of air to the pump. This limits the vacuum level the pump can produce. A typical rotary-vane pump of this type can generally produce a maximum vacuum level of about 26 to 27 inches of mercury. The more recent oil-flooded rotary-vane designs use a greater quantity of oil (sometimes as much as 2 gallons per minute) to seal these clearances. This increases the efficiency of the pump and allows maximum vacuum levels of as much as 29.83 inches of mercury. The maximum vacuum achievable at sea level is 29.92 inches of mercury. Rotary-screw pumps are oil-flooded so they can achieve higher vacuum levels with 29.7 inches of mercury being typical. In addition to cooling, oil is used to keep the rotors from touching each other. Most rotary-screw pumps do not use timing gears to drive the rotors. To prevent contact, one rotor drives the other.

# Appendix C

## Latest Pumps Available

One of the manufacturers of pumps is Gould Pumps, a division of ITT Industries. Figures C-1 to C-13 provide specifications sheets for a selected number of their latest models.

**APPLICATIONS**

Specifically designed to remove water from:
• Drainage ditches
• Trenches
• Basements
• Manholes
• Excavating drainage in the building trades

**SPECIFICATIONS**

Pump:
• Discharge size: 2" threaded-hose coupling design.
• Capacities: up to 80 GPM.
• Total heads: up to 52 ft.
• Max. solids: any particles passing through strainer.
• Mechanical seals: outer seal—silicon carbide, inner seal—carbon ceramic.
• Temperature limit: 95°F (35°C) maximum.
• Depth of immersion: 16.5 ft (5 m) maximum.

Motor:
• Single phase: 3500 RPM, 1/2 HP and 1HP, 115 and 230 V, 60 Hz
• Built-in starter with full overload and temperature protection.
• Class F insulation.
• Air filled design.
• Upper and lower heavy-duty ball bearing construction.
• Power cord: 50 ft.

**FEATURES**

■ Impeller: Polyurethane for wear and corrosion resistance.
■ Diffuser: Polyurethane for wear and corrosion resistance.
■ Mechanical Seal: Dual seals for double leakage protection, outer seal—silicon carbide.
■ Rubber Liner: Protects against wear around impeller.
■ Bottom Strainer: Made of impact absorbing EPDM rubber—suction holes allow for low pump down.

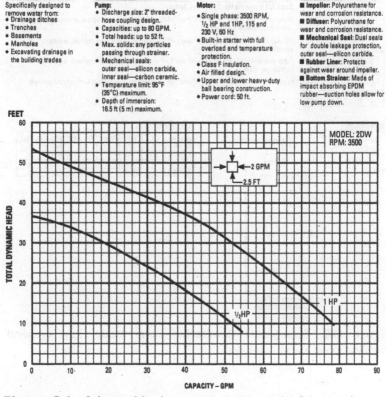

**Figure C-1** Submersible dewatering pump model 2dw (application specifications). *(Courtesy Gould Pumps)*

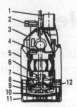

## COMPONENTS

| Item No. | Description |
|----------|-------------|
| 1 | Discharge connection |
| 2 | Cable entry |
| 3 | Handle/cover |
| 4 | Support bearing |
| 5 | Pump casing |
| 6 | Main bearing |
| 7 | Inner mechanical seal |
| 8 | Outer mechanical seal |
| 9 | Impeller |
| 10 | Suction cover |
| 11 | Strainer |
| 12 | Oil plug |

## REPLACEMENT KITS

**Each kit contains the following parts:**

Impeller kit (15K11 for $1/2$ HP, 15K12 for 1 HP)—Impeller, impeller screw, protective plug, washer, assembly instruction

Diffuser kit (15K13 for both $1/2$ HP and 1 HP)—Diffuser, barrel nuts, screws, washers, assembly instruction, sticker

Outer seal kit (15K14 for both $1/2$ HP and 1 HP)—Mechanical face-seal unit, assembly instruction, sticker

Fastener kit (15K15 for both $1/2$ HP and 1HP)—Barrel nuts, washers, socket head screws

## DIMENSIONS

$1/2$ HP unit:
$W = 7^5/_{16}"$
$H = 16^3/_{18}"$

1 HP unit:
$W = 7^5/_{16}"$
$H = 17^3/_4"$

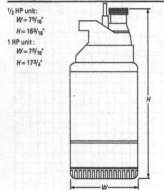

## ELECTRICAL DATA

| Order No. | HP | Volts | Phase | Max. Amp | RPM | Wt (lbs) |
|-----------|-----|-------|-------|----------|------|----------|
| 2DW0511 | $1/2$ | 115 | 1 | 5.5 | 3500 | 20 |
| 2DW0512 | $1/2$ | 230 | 1 | 2.9 | 3500 | 20 |
| 2DW1011 | 1 | 115 | 1 | 9.8 | 3500 | 25 |
| 2DW1012 | 1 | 230 | 1 | 4.9 | 3500 | 25 |

**Figure C-2** Submersible dewatering pump model 2dw (general specifications). *(Courtesy Gould Pumps)*

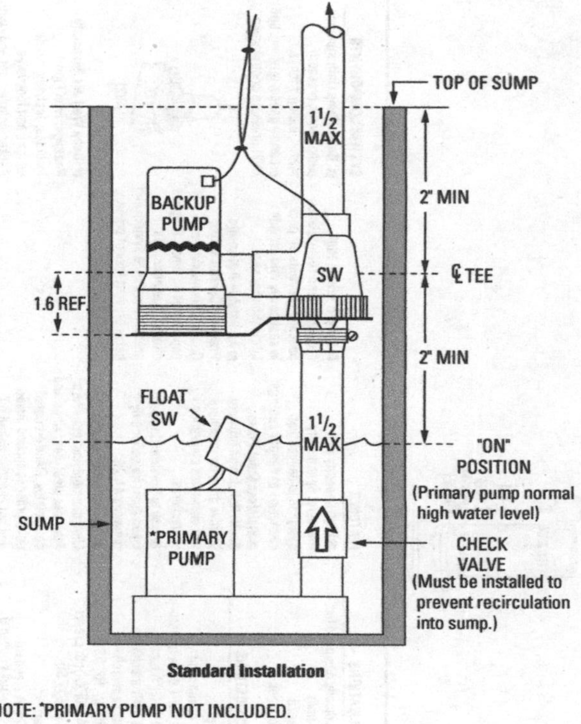

**Figure C-3** Battery backup sump pump model SPBB. *(Courtesy Gould Pumps)*

**APPLICATIONS**

Designed to provide emergency backup service for primary pump in the event of a power outage. Will also operate if main pump can't keep up with inflow.

**SPECIFICATIONS**

- 12-volt pump
- Float switch
- 10-amp battery charger
- Check valve
- Pipe fittings
- Battery box
- System requires minimum 105-amp deep-cycle marine battery (not included).

**FEATURES**

■ Automatic start-up
■ Self-charging
■ Indicator lights on charger
■ Alarm sounds when pumping and battery won't charge.

**PERFORMANCE CHART**

| Discharge Height | GPH | Battery Life |
|---|---|---|
| 5' | 1,380 | 9 hrs |
| 10' | 900 | 9 hrs |
| 13' | 480 | 11 hrs |

TOP OF SUMP

2' MIN

₵ TEE

2' MIN

"ON" POSITION
(Primary pump normal high water level)

CHECK VALVE
(Must be installed to prevent recirculation into sump.)

1 1/2 MAX

SW

1 1/2 MAX

BACKUP PUMP

FLOAT SW

*PRIMARY PUMP

1.6 REF

SUMP

**Standard Installation**

NOTE: *PRIMARY PUMP NOT INCLUDED.

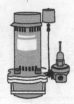

## APPLICATIONS

Specifically designed for:
- Homes
- Farms
- Cottages

## SPECIFICATIONS

**Pump:**
Pipe connection —
 1" NPT drive (pressure)
 1¼" NPT suction
 1" NPT discharge
Pressure switch —
 ¾–1½ HP, AS4FX; 2 HP,
 AS10FX; both preset
 (30–50 PSI)

**Motor:**
- NEMA standard
- ¾–1½ HP, 115/230 V,
 60 Hz, capacitor start
- 2HP, 230 V only, 60 Hz
- Single phase
- 3,500 RPM
- Built-in overload with
 automatic reset
- Stainless steel shaft
- Rotation: clockwise when
 viewed from motor end.

**Maximum temperature:** 140°F

## FEATURES

■ **High Capacity and Pressure:** Specifically designed to deliver high capacities at deeper settings.

■ **Stainless Steel Pump Shaft:** Hex design provides positive drive for impellers and eliminates clearance adjustments.

■ **Easy to Service:** Can be taken apart for service by removing 4 bolts.

■ **Corrosion Resistant:** Glass filled thermoplastic impellers. Stainless steel wear rings and coverplates. Electro-coated paint process applied inside and out and then baked on.

■ **Powered for Continuous Operation:** Pump is designed for continuous operation. All ratings are within the motor manufacturer's recommended working limits.

■ **Easy to Prime:** High discharge and recirculation passage prevents air pocket in seal cavity and provides lubrication for seal.

■ **Adjustable Automatic Pressure Control Valve:** Guarantees maximum capacity at all times. By-pass check valve equalizes pressure in entire system and helps prevent loss of prime.

## SYSTEM COMPONENTS

■ **Basic Pump Unit:** Includes pump, motor, pressure switch, tubing, fittings, pressure gauge, bushing, and AV21 pressure control valve.

AV21

■ **Deep Well Jet Assembly Package:** Twin Pipe includes: Jet body, nozzle, venturi, and foot valve. Packer includes: Jet body with built-in check valve, nozzle, and venturi.

■ **Additional Accessories Required:** For Packer: well casing adapter AWD2 (2") or AWD3 (3").

**Figure C-4** Vertical jet pumps deep well model SJ (application specifications). *(Courtesy Gould Pumps)*

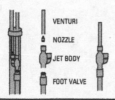

VENTURI
NOZZLE
JET BODY
FOOT VALVE

## PERFORMANCE RATINGS — TWIN PIPE SYSTEMS

| HP/Model | ¾ HP–SJ07 | | | 1 HP–SJ10 | | | 1½ HP–SJ15 | | | 2 HP–SJ20 | | |
|---|---|---|---|---|---|---|---|---|---|---|---|---|
| Well Casing Min. I.D. in. | 4 | | | | | | | | | | | |
| Pressure — Pipe Size (in.) | 1 | | | | | | | | | | | |
| Suction — Pipe Size (in.) | 1¼ | | | | | | | | | | | |
| Jet Assembly Package (1) | FT4-48 | FT4-47 | FT4-06 | FT4-47 | FT4-24 | FT4-08 | FT4-45 | FT4-30 | FT4-29 | FT4-45 | FT4-30 | FT4-29 |
| Jet Body (Only) | AT4 | | | | | | | | | | | |
| Nozzle | AN012 | AN013 | AN014 | AN013 | AN015 | AN015 | AN013 | AN013 | AN013 | AN013 | AN013 | AN013 |
| Venturi | AD727 | AD724 | AD720 | AD724 | AD724 | AD720 | AD726 | AD722 | AD719 | AD726 | AD722 | AD719 |
| Control Valve Setting (PSI) | 37 | 40 | 40 | 43 | 43 | 43 | 62 | 65 | 67 | 76 | 78 | 80 |

**Depth to Jet Assembly (Based on Submergence of 5 feet)** — Gallons per Minute (GPM) 30–50 Pressure Switch Setting

| Feet | | | | | | | | | | | | |
|---|---|---|---|---|---|---|---|---|---|---|---|---|
| 30 | 15.2 | | | 16.0 | | | 13.0 | | | 13.0 | | |
| 40 | 13.5 | | | 15.2 | | | 13.0 | | | 13.0 | | |
| 50 | 11.8 | | | 14.7 | | | 13.0 | | | 13.0 | | |
| 60 | 10.7 | | | 13.7 | | | 12.8 | | | 12.8 | | |
| 70 | 8.5 | 9.0 | | 11.7 | 11.7 | | 12.2 | | | 12.8 | | |
| 80 | | 7.5 | | | 10.5 | | 10.6 | | | 12.0 | | |
| 90 | | 6.2 | 6.2 | | 9.0 | | 9.2 | 9.3 | | 11.3 | | |
| 100 | | | 5.8 | | 7.5 | | | 9.0 | | 9.5 | 9.5 | |
| 110 | | | 5.0 | | 6.3 | 6.0 | | 8.0 | | | 9.4 | |
| 120 | | | 4.6 | | | 5.0 | | 7.0 | | | 8.5 | |
| 130 | | | 4.0 | | | 4.3 | | 6.2 | 5.8 | | 7.7 | |
| 140 | | | | | | 4.0 | | | 5.6 | | 6.8 | |
| 150 | | | | | | 3.5 | | | 5.2 | | 6.0 | 5.7 |
| 160 | | | | | | | | | 4.7 | | | 5.5 |
| 170 | | | | | | | | | 4.2 | | | 5.3 |
| 180 | | | | | | | | | 3.7 | | | 5.0 |
| 190 | | | | | | | | | 3.2 | | | 4.7 |
| 200 | | | | | | | | | | | | 3.8 |

(1) Jet assembly package includes jet body, venturi, nozzle, foot valve.

NOTE: An offset of 50 feet will result in a decrease of about 25% from ratings as shown.

**Figure C-5**   Vertical jet pumps deep well model SJ (performance ratings for twin pipe systems). *(Courtesy Gould Pumps)*

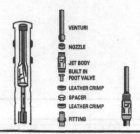

VENTURI
NOZZLE
JET BODY BUILT IN FOOT VALVE
LEATHER CRIMP
SPACER
LEATHER CRIMP
FITTING

## PERFORMANCE RATINGS — PACKER SYSTEM

| HP/Model | 3/4 HP SJ07 | | | | | 1 HP SJ10 | | | | | 1 1/2 HP SJ15 | | | | | | 2 HP SJ20 | | | | | |
|---|---|---|---|---|---|---|---|---|---|---|---|---|---|---|---|---|---|---|---|---|---|---|
| Well Casing in. | 2 | | | 3 | | 2 | | | 3 | | 2 | | | 3 | | | 2 | | | 3 | | |
| Suction Pipe Size (in.) | 1 1/4 | | | 1 1/2 | | 1 1/4 | | | 1 1/2 | | 1 1/4 | | | 1 1/2 | | | 1 1/4 | | | 1 1/2 | | |
| Jet Assembly Package (1) | FP2-51 | FP2-50 | FP2-06 | FP3-40 | FP3-42 | FP2-50 | FP2-07 | FP2-06 | FP3-40 | FP3-49 | FP2-49 | FP2-30 | FP2-29 | FP3-47 | FP3-46 | FP3-34 | FP2-49 | FP2-30 | FP2-29 | FP3-47 | FP3-46 | FP3-34 |
| Jet Body (Only) | AP2 | | | AP3 | | AP2 | | | AP3 | | AP2 | | | AP3 | | | AP2 | | | AP3 | | |
| Nozzle | AN012 | AN013 | AN014 | AN013 | AN014 | AN013 | AN014 | AN015 | AN013 | AN015 | AN013 | | | | | | | | | | | |
| Venturi | AD727 | AD724 | AD720 | AD724 | AD722 | AD724 | | | AD720 | AD724 | AD723 | AD726 | AD722 | AD719 | AD727 | AD722 | AD719 | AD726 | AD722 | AD719 | AD727 | AD722 AD719 |
| Pressure Control Valve Setting (PSI) | 40 | | | 36 | 40 | 45 | | | 40 | 43 | 62 | 65 | 67 | 62 | 64 | 66 | 76 | 78 | 80 | 75 | 78 | 80 |

Gallons per Minute 30–50 Pressure Switch Setting

Depth to Jet Assembly (based on submergence of 5 feet)

| Feet | FP2-51 | FP2-50 | FP2-06 | FP3-40 | FP3-42 | FP2-50 | FP2-07 | FP2-06 | FP3-40 | FP3-49 | FP2-49 | FP2-30 | FP2-29 | FP3-47 | FP3-46 | FP3-34 | FP2-49 | FP2-30 | FP2-29 | FP3-47 | FP3-46 | FP3-34 |
|---|---|---|---|---|---|---|---|---|---|---|---|---|---|---|---|---|---|---|---|---|---|---|
| 30 | 14.9 | | | 15.3 | | 15.8 | | | 16.3 | | 13.0 | | | 14.7 | | | 13.0 | | | 14.7 | | |
| 40 | 13.1 | | | 13.5 | | 14.0 | | | 16.0 | | 13.0 | | | 14.5 | | | 13.0 | | | 14.7 | | |
| 50 | 11.8 | | | 12.5 | | 12.5 | | | 15.5 | | 12.9 | | | 14.3 | | | 13.0 | | | 14.7 | | |
| 60 | 10.5 | | | 12.2 | | 11.2 | 11.3 | | 14.2 | | 12.7 | | | 14.3 | | | 13.0 | | | 14.5 | | |
| 70 | 8.4 | 8.8 | | 11.8 | | 10.0 | | | 12.5 | | 11.8 | | | 14.2 | | | 12.8 | | | 14.5 | | |
| 80 | | 7.8 | | 10.2 | 8.5 | 8.7 | | | 10.0 | | 10.2 | | | 12.2 | | | 11.8 | | | 13.5 | | |
| 90 | | 6.1 | 6.2 | 7.8 | | 7.3 | | | 8.3 | 9.0 | 8.3 | 9.1 | | 10.0 | | | 10.5 | | | 12.7 | | |
| 100 | | 5.7 | 6.9 | 5.5 | | 6.2 | | | 8.3 | | 7.8 | 9.1 | | | | | 8.5 | 9.1 | | 10.5 | | |
| 110 | | 4.3 | 6.0 | | | 4.2 | 4.8 | | 7.3 | | 7.3 | | | 8.8 | | | 6.9 | | | 8.3 | 9.1 | |
| 120 | | 3.8 | 5.3 | | | 4.0 | | | 6.7 | | 6.2 | | | 7.8 | | | 7.9 | | | 9.1 | | |
| 130 | | 2.2 | 4.5 | | | 3.5 | | | 5.8 | | 5.4 | 5.8 | | 6.9 | | | 6.9 | | | 8.6 | | |
| 140 | | | | | | 3.0 | | | 4.8 | | 5.3 | | | 6.2 | | | 6.0 | | | 8.2 | | |
| 150 | | | | | | 2.4 | | | 4.1 | | 4.8 | | | 5.4 | 5.6 | | 5.2 | 5.6 | | 7.3 | | |
| 160 | | | | | | | | | | | 4.3 | | | 5.5 | | | 5.4 | | | 8.5 | | |
| 170 | | | | | | | | | | | | | | 5.0 | | | 4.9 | | | 5.7 | 5.7 | |
| 180 | | | | | | | | | | | | | | 4.7 | | | 4.3 | | | 5.6 | | |
| 190 | | | | | | | | | | | | | | 4.3 | | | 3.8 | | | 5.5 | | |
| 200 | | | | | | | | | | | | | | 3.8 | | | 3.3 | | | 5.3 | | |
| 210 | | | | | | | | | | | | | | 3.3 | | | | | | 4.8 | | |
| 220 | | | | | | | | | | | | | | | | | | | | 4.5 | | |
| 230 | | | | | | | | | | | | | | | | | | | | 4.0 | | |

(1) Jet assembly package includes: jet body, venturi, nozzle, built-in check valve, crimps, spacer and fittings.
NOTE: An offset of 50 feet will result in a decrease of about 25% from ratings as shown.

**Figure C-6** Vertical jet pumps deep well model SJ (performance ratings for packer systems). *(Courtesy Gould Pumps)*

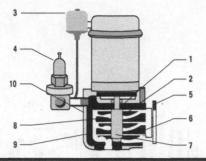

## COMPONENTS

| Item No. | Description |
|---|---|
| 1 | Mechanical seal |
| 2 | Corrosion resistant impellers |
| 3 | Pressure switch |
| 4 | Pressure control valve |
| 5 | Rugged cast iron construction |
| 6 | Electro-coat paint |
| 7 | Stainless steel hex shaft |
| 8 | Stainless steel wear rings |
| 9 | Stainless steel cover plates |
| 10 | O-ring seals |

## DIMENSIONS AND WEIGHTS

| HP/Model | ¾-SJ07 | 1-SJ10 | 1½-SJ15 | 2-SJ20 |
|---|---|---|---|---|
| Stages | 2 | 2 | 3 | 3 |
| Weight | 58 | 63 | 77 | 88 |
| Length | 14 | 14 | 14 | 14 |
| Width | 8 | 8 | 8 | 8 |
| Height | 15½ | 16 | 20 | 20½ |

Pressure Control Valve included in dimensions.
(All dimensions are in inches and weight in lbs. Do not use for construction purposes.)

## MODEL SJ

For twin pipe or packer systems.

**Figure C-7**  Vertical jet pumps deep well model SJ (general specifications). *(Courtesy Gould Pumps)*

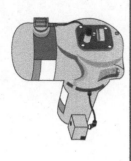

## APPLICATIONS

Specifically designed for:
- Small homes
- Camps
- Mobile homes

## PUMP SPECIFICATIONS
- Pipe connections:
  $3/4''$ suction,
  $3/4''$ discharge
- Pressure switch: AS3FX

## MOTOR SPECIFICATIONS
- NEMA standard
- 115/230 V, 60 Hz
- Single phase
- 3500 RPM
- Built-in overload with automatic reset.
- Capacitor start type
- Stainless steel shaft
- $1/2$ HP—see back ①

## FEATURES
- **Self-Adjusting Capacity:** Automatically adjusts to the water demand within the pump's capacity. When one faucet is in use, there's no change in the water flow when a second outlet is opened.
- **Compact:** Can be easily installed in locations where other larger pumps are difficult to install.
- **Tankless:** Self-contained water system, no separate tank required. Fresh water is always delivered directly from the well to the faucet.
- **Exclusive Air Handling Ability:** When pump is primed, it never needs priming again even if the water level drops below the end of the suction pipe.

It resumes pumping as soon as the water level rises again. Gas or air in the water will not air-bind the pump.
- **Protected Mechanical Seal:** Water is retained in the casing so the seal can never run dry.
- **Easy to Service:** Pump has nozzle clean-out plug. Provides easy access to nozzle and diffuser for cleaning. Back pull-out design—pipes do not have to be disturbed.
- **Powered for Continuous Operation:** Pump ratings are within the motor manufacturer's recommended working limits, can be operated continuously without damage.
- **Corrosion Resistant:** Electro-coated paint process is applied inside and out, then is baked on.

## SYSTEM COMPONENTS
- **Basic Pump Unit:** Includes pump, motor, pressure switch, tubing and fittings, air volume control valve, suction check valve, nozzle and diffuser.

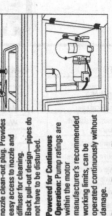

**Figure C-8** Shallow well jet pump model BFO3S (application specifications). *(Courtesy Gould Pumps)*

## FEATURES

1. Mechanical Seal
2. Rugged Cast-Iron Construction
3. Corrosion-Resistant Impeller and Guide Vane
4. Drain Plug
5. Two-Compartment Motor
6. Stainless Steel Shaft
7. Built-In Air Volume Control
8. Built-In Check Valve
9. Nozzle Clean-Out Plug
10. Priming Plug

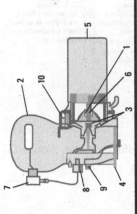

## PERFORMANCE RATINGS

| HP/Model | ① ½ HP - BFO3S | | | Maximum Shut-Off (PSI) |
|---|---|---|---|---|
| Shallow Well Pkg. | None Required | | | |
| Adapter | Built-In | | | |
| Nozzle | AN015 | | | |
| Venturi (Diffuser) | 4K5 | | | |
| Total Section Lift (Feet) | Discharge Pressure (PSI) | | | |
| | 20 | 30 | | |
| | Gallons Per Hour | | | |
| 5 | 525 | 380 | | 51 |
| 10 | 480 | 330 | | 49 |
| 15 | 410 | 290 | | 46 |
| 20 | 350 | 255 | | 44 |
| 25 | 265 | 220 | | 42 |

① Pump now uses a ¹/₂ HP motor due to ¹/₃ HP motors being obsolete.

## DIMENSIONS AND WEIGHTS

(All dimensions in inches and weight in lbs.) (Do not use for construction purposes.)

| HP/Model | ① ½ HP — BFO3S |
|---|---|
| Weight | 69 |
| Length | 20 |
| Width | 10½ |
| Height | 15 |

**Figure C-9   Shallow well jet pump model BFO3S (performance specifications).** *(Courtesy Gould Pumps)*

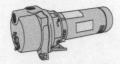

## APPLICATIONS

Specifically designed for the following uses:
- Lawn sprinkling
- Irrigation
- Air-conditioning systems
- Heat pumps
- Water transfer
- Dewatering

## SPECIFICATIONS

**Pump:**
- Pipe connections: $1\frac{1}{2}$" NPT suction $1\frac{1}{2}$" NPT discharge
- Capacities: to 110 GPM at 5-foot suction lift.
- Heads: to 128 feet.
- Reprime capabilities: to 25 feet suction lift.
- Maximum working pressure: 125 PSIG.
- Maximum water temperature: 140°F (60°C).
- Rotation: clockwise when viewed from motor end.

**Motor:**
- NEMA standard open drip proof.
- 60 Hz, 3,500 RPM.
- Stainless steel shaft.
- Single phase: $\frac{3}{4}$—$1\frac{1}{2}$ HP, 115/230 V; 2 and 3 HP, 230 V only. Built-in overload with automatic reset.
- Three phase: 230/460 V. Overload protection must be provided in starter unit. Starter and heaters (3) must be ordered separately.
- Optional TEFC motors are available. See price book for order numbers.

## FEATURES

■ **Self-Priming Design:** Once pump is primed it never needs priming again even if water level drops below the end of the suction pipe. Pumping resumes once the water level rises above the end of the suction pipe.

■ **Serviceable:**
- Back pullout design allows disassembly of pump for service without disturbing piping.
- Two compartment motor for easy access to motor wiring and replaceable components.

■ **Diffuser (Guidevane):** Bolt down diffuser provides positive alignment with impeller. Diffuser also has stainless wear ring for extended performance in abrasive conditions. FDA compliant, injection molded, food grade, glass filled Lexan® for durability and abrasion resistance.

■ **Impeller:** FDA compliant, glass filled Noryl®. Corrosion and abrasion resistant.

■ **Corrosion Resistant:** Electro-coated paint process is applied inside and out, then baked on.

■ **Casing:** Cast-iron construction. 4-bolt, back pull-out design. Tapped openings provided for vacuum gauge and casing drain.

■ **Powered for Continuous Operation:** Pump ratings are within the motor manufacturer's recommended working limits. Can be operated continuously without damage.

■ **Mechanical Seal:** Carbon/ceramic faces, BUNA elastomers. 300 series stainless steel metal parts. Exclusive design prevents the seal from running dry.

## STANDARD ODP MODELS

| Model | HP | Phase |
|-------|-----|-------|
| GT07 | $\frac{3}{4}$ | |
| GT10 | 1 | |
| GT15 | $1\frac{1}{2}$ | 1 |
| GT20 | 2 | |
| GT30 | 3 | |
| GT073 | $\frac{3}{4}$ | |
| GT103 | 1 | |
| GT153 | $1\frac{1}{2}$ | 3 |
| GT203 | 2 | |
| GT303 | 3 | |

## AGENCY LISTINGS

 Canadian Standards Association

 Underwriters Laboratories

Goulds Pumps is ISO 9001 Registered.

## SELF-PRIMING

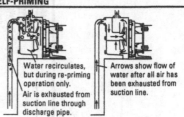

Water recirculates, but during re-priming operation only.

Air is exhausted from suction line through discharge pipe.

Arrows show flow of water after all air has been exhausted from suction line.

**Figure C-10** Irri-Gator Self-Priming Centrifugal Pump model GT (application specifications). *(Courtesy Gould Pumps)*

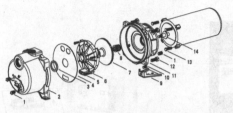

## COMPONENTS

| Item No. | Description |
|---|---|
| 1 | Plug – $1/4$ NPT |
| 2 | Casing |
| 3 | Seal ring—diffuser |
| 4 | Diaphragm |
| 5 | Machine screw |
| 6 | Diffuser |
| 7 | Impeller |
| 8 | Mechanical seal |
| 9 | Foot |
| 10 | Bolt—foot to adapter |
| 11 | Motor adapter |
| 12 | Bolt—casing to adapter |
| 13 | Bolt—adapter to motor |
| 14 | Deflector |

## DIMENSIONS AND WEIGHTS

| Model | GT07 | GT10 | GT15 | GT20 | GT30 | GT073 | GT103 | GT153 | GT203 | GT303 |
|---|---|---|---|---|---|---|---|---|---|---|
| HP | $3/4$ | 1 | $1 1/2$ | 2 | 3 | $3/4$ | 1 | $1 1/2$ | 2 | 3 |
| Length L | $19^{9}/16$ | $19^{7}/8$ | $21^{3}/16$ | $20^{9}/16$ | $21^{11}/32$ | 19 | $19^{1}/4$ | $20^{1}/16$ | $20^{13}/16$ | $21^{3}/16$ |
| Width | | | | | $8 1/4$ | | | | | |
| Height | | | | | $9 1/4$ | | | | | |
| Wt (lbs) | 48 | 52 | 60 | 65 | 78 | 49 | 52 | 55 | 68 | 71 |
| Phase | | | Single | | | | | Three | | |

(All dimensions are in inches and weights in lbs. Do not use for construction purposes.)

## PERFORMANCE RATINGS

| Model | PSI Disch. Pressure | Suction Lift in Feet | | | | |
|---|---|---|---|---|---|---|
| | | 5 | 10 | 15 | 20 | 25 |
| GT07/ GT073 | 20 | 44 | 41 | 36 | 31 | 24 |
| | 30 | 34 | 31 | 26 | 22 | 14 |
| | 40 | 10 | 4 | 0 | 0 | 0 |
| GT10/ GT103 | 20 | 53 | 51 | 49 | 46 | 41 |
| | 30 | 43 | 41 | 38 | 36 | 32 |
| | 40 | 29 | 22 | 16 | 8 | 0 |
| GT15/ GT153 | 20 | 63 | 59 | 54 | 49 | 39 |
| | 30 | 60 | 55 | 51 | 46 | 37 |
| | 40 | 45 | 38 | 33 | 20 | 14 |
| GT20/ GT203 | 20 | 86 | 77 | 70 | 59 | 46 |
| | 30 | 80 | 72 | 67 | 57 | 44 |
| | 40 | 65 | 60 | 57 | 50 | 43 |
| GT30/ GT303 | 20 | 105 | 100 | 88 | 76 | 60 |
| | 30 | 92 | 90 | 84 | 75 | 57 |
| | 40 | 73 | 67 | 62 | 55 | 50 |

Performance ratings are in GPM.

## PERFORMANCE CURVE

**Figure C-11**  Irri-Gator Self-Priming Centrifugal Pump model GT (performance specifications). *(Courtesy Gould Pumps)*

**Motor Minder**

**Motor Check**

## SPECIFICATIONS

- Power Consumption: 3 watts (max.)
- Control voltage (+/–10%): 115/230 VAC
- Response times: dry well—3 seconds (max.) air lock—30 seconds rapid cycle (2 seconds start/stop)—6 cycles
- Pilot duty rating: 960 VA
- Term block rating: 230 V, 28 amp
- Output contact rating: 2.5 HP at 230 VAC
- Horsepower range: (see model listing on other side)
- Operating temperature: –20°C to 85°C
- Weight: Panel mount: 1.5 lbs. With 3R enclosure: 3.5 lbs.
- Enclosure dimensions: 8.5" × 4.5" × 3"

## FEATURES

■ **Full Adjustability:** Adjust the load setting and trip point threshold to meet your precise sensitivity requirements. Automatic restart timer adjusts from 9 minutes to 4 hours.

■ **Light Bar Display:** Used for calibration and monitoring load status. As the current drops, the lights descend toward your pre-set threshold. If the system is about to go off, you can adjust consumption accordingly.

■ **Easy to Install:** Installs in minutes on new or existing systems. No plumbing required. Heavy-duty lug-type terminal. Rain-tight housing.

■ **Linear Toroid Sensing:** INTEGRA uses an exclusive linear toroid that senses current flow magnetically. This method is more reliable under locked rotor or other high-current conditions.

■ **Choose from 6 Models:** There is a Motor Minder™ to fit virtually any pump system—from single phase to three phase, from 115 V to 460 V, from 1/20th HP to 50 HP. Available in code-approved enclosure or panel-mount.

■ **Built-in Transient Protector:** Protects the control from damage caused by lightning or line-voltage surges.

■ **One-Year Warranty:** Solid state construction, state-of-the-art engineering and quality craftsmanship make Motor Minder™ an affordable device you can count on for many years.

## SPECIFICATIONS

**Model 401B**
- Operating voltage: 115 VAC (+/–10%)
- Horsepower rating: 1/3 to 1.0 HP

**Model 401A**
- Operating voltage: 230 VAC (+/– 10%)
- Horsepower rating: 1/3 to 2.0 HP
- Motor type 1 phase: indication
- Power consumption: 1 watt (max.)
- No load response time: 2 to 4 seconds
- Output contact rating: 2 HP 230 VAC
- Sensitivity calibration: field adjustable
- Restart timer range: 5 to 75 minutes or manual reset
- Indicator: trip point LED
- Enclosure: UL listed weather proof bell box and cover
- Outside dimensions: 2³/₄" × 4¹/₂" × 3" deep
- Color: aluminum gray
- Weight: 1 lb, 4 oz.

## FEATURES

■ **Pump Motor:** Works on any 2- or 3-wire motor (up to 2 HP, single phase).

■ **Calibration:** Calibrate to your pumping system conditions.

■ **Installation Friendly:** Comes complete—nothing more to buy. Standard weatherproof enclosure and mounting hardware are included. Compact single-gang enclosure fits anywhere. Mounts anywhere—before or after the pressure switch, on the control box or at the pump (nonsubmersible).

■ **Bypass Setting:** No need to wire around device to test other components.

■ **Restart:** Automatic or manual restart. Adjustable from 5 to 75 minutes.

■ **Reliable Performance:** High-quality sophisticated solid-state technology meets the demands of the most grueling conditions.

■ **Warranty:** We are totally committed to ensuring your total satisfaction. One-year warranty.

**Figure C-12** Pump monitoring controls (application specifications).
*(Courtesy Gould Pumps)*

**INTEGRA MOTOR CHECK**

| Order No. | VAC | Phase | Horsepower |
|---|---|---|---|
| AW401A | 208/230 | Single | $1/2$ to 2 |
| AW401B | 115 | Single | $1/3$ to 1 |

**FEATURES**

- Protects pumps from damage caused by running dry—low water, clogged suction, frozen lines or closed discharge.
- Load adjustable settings with automatic or manual restart times from 5 to 75 minutes.

**SELECTION REFERENCE**

Three phase motors used with magnetic starters or contactor (not included):

| HP | Load Voltage | | | | Control or Pilot Voltage | |
|---|---|---|---|---|---|---|
| | | | | | 115 VAC | 208/230 VAC |
| $1/2$ | 208 | 230 | 460 | 575 | | |
| $3/4$ | 208 | 230 | 460 | 575 | | |
| 1 | 208 | 230 | 460 | 575 | | |
| $1^1/2$ | 208 | 230 | 460 | 575 | | |
| 2 | 208 | 230 | 460 | 575 | | |
| 3 | 208 | 230 | 460 | 575 | Model AW301B-CT1** | Model AW301A-CT1** |
| 5 | 208 | 230 | 460 | 575 | | |
| $7^1/2$ | 208 | 230 | 460 | 575 | | |
| 10 | – | – | 460 | 575 | | |
| 15 | – | – | 460 | 575 | | |
| 20 | – | – | – | 575 | | |
| 10 | 208 | 230 | – | – | | |
| 15 | 208 | 230 | – | – | | |
| 20 | 208 | 230 | 460 | – | | |
| 25 | 208 | 230 | 460 | 575 | Model AW301B-CT2** | Model AW301A-CT2** |
| 30 | 208 | 230 | 460 | 575 | | |
| 40 | – | – | 460 | 575 | | |
| 50 | – | – | 460 | 575 | | |

**INTEGRA MOTOR MINDER**

| Order No. | (pilot) or Control Voltage | Load Voltage | Phase | Horsepower |
|---|---|---|---|---|
| AW101A | 208/230 | 208/230 | Single | Up to 1.5 |
| AW101B | 115 | 115 | Single | Up to 1.0 |
| AW201A* | 208/230 | 208/230 | Single | 2 to 5 |
| | | | Three | 2 to 7.5 |
| AW201B* | 115 | 208/230 | Single | 2 to 5 |
| | | | Three | 2 to 7.5 |
| AW201A1015* | 208/230 | 208/230 | Single | 10 to 15 |
| AW275C-A*** | 208/230 | 208/230 | Single | 2 to 7.5 |
| AW275C-B*** | 115 | | | |
| AW301A-CT1* | 208/230 | See selection reference chart below. | Three | See selection reference chart below. |
| AW301A-CT2* | | | | |
| AW301B-CT1* | 115 | | Three | |
| AW301B-CT2* | | | | |

**FEATURES**

- Protects pumps from burnouts caused by underload or rapid-cycle conditions, plugged suctions, and low water conditions.
- Fully adjustable settings with automatic or manual restart. Light bar for calibration and monitoring status.

**MOTOR MINDER "ADD-ON OPTIONS"**

| Option | Description |
|---|---|
| 1 | Auxiliary Contact |
| 2 | Trip Counter |
| 3 | Auto Restart Timer |
| 4 | Remote Reset |
| 5 | Restart Probe |

Options are non-stock.

Please consult our customer service department for estimated lead time.

To order motor minder with option(s), add option numbers as suffix to the order number.

EXAMPLE: AW101A with an auxiliary contact (1) and an auto restart timer (3) would be AW101A13.

\* Motor minder models AW201 and AW301 require a magnetic contactor.

\*\* Includes an external CT (current transformer) with 12" leads.

\*\*\* Model AW275C includes built-in contactor.

AW401 motor checks and AW101 motor minders carry full load current. A contactor is not required.

Contact factory for 50 Hz application.

**Figure C-13** Pump monitoring controls (performance specifications).

*(Courtesy Gould Pumps)*

Buffalo Pumps, Inc., has been making centrifugal pumps for more than a century. Commercial use dates back to the company's founding in 1887, and marine applications predate World War I. While the defense market has been a significant segment of their business, commercial pumps have been the mainstay of their operation. They furnish Original Equipment Manufacturer (OEM) for refrigeration, paper, and lube oil markets for many of the major companies. They are capable of repairing and producing parts for pumps that were made 40 to 50 years ago.

Figures C-14 to C-21 show a few of their models, along with specifications provided by the manufacturer. For more information on these and other models see their Web site at www.buffalopumps. com.

**Figure C-14** Buffalo Pumps Can-o-Matic, model HCR.

Seal-Less/Leakproof Pumps
Total Heads to 320 Feet
Capacities to 1000 U.S. GPM
Request Bulletin 929

*(Courtesy Buffalo Pumps)*

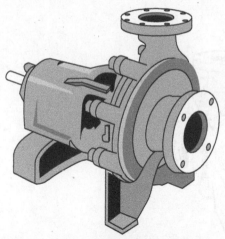

**Figure C-15**  Buffalo Pumps model CRE/CRO.

Frame Mounted ANSI Pumps
Total Heads to 690 Feet
Capacities to 5000 U.S. GPM
Request Bulletin 903

*(Courtesy Buffalo Pumps)*

**Figure C-16**  Buffalo Pumps model CR-SP.

Self-Priming Pumps
Total Heads to 550 Feet
Capacities to 800 U.S. GPM
Request Bulletin 918

*(Courtesy Buffalo Pumps)*

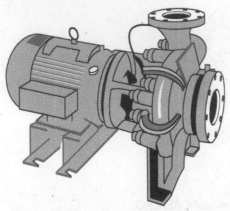

**Figure C-17**   Buffalo Pumps model CCRE.

Close-Coupled ANSI Pumps
Total Heads to 600 Feet
Capacities to 1900 U.S. GPM
Request Bulletin 904

*(Courtesy Buffalo Pumps)*

**Figure C-18**   Buffalo Pumps model RR.

Two-Stage and Four-Stage Pumps
Total Heads to 875 Feet
Capacities to 1000 U.S. GPM
Request Bulletin 980

*(Courtesy Buffalo Pumps)*

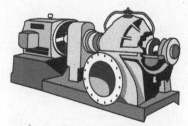

**Figure C-19**  Buffalo Pumps model HSM.

Double Suction Pumps
Total Heads to 400 Feet
Capacities to 8000 U.S. GPM
Request Bulletin 900

*(Courtesy Buffalo Pumps)*

**Figure C-20**  Buffalo Pumps model VCRE.

Lube Oil Pumps
Total Heads to 690 Feet
Capacities to 5000 U.S. GPM
No Bulletin Available

*(Courtesy Buffalo Pumps)*

**Figure C-21**   Buffalo Pumps model VCRO.

Vertical Submerged Pumps
Total Heads to 250 Feet
Capacities to 4800 U.S. GPM
Request Bulletin 905

*(Courtesy Buffalo Pumps)*

# Index

lubricating values, 481
lubrication
  for corrosion-resisting
      centrifugal pumps, 157
  imperfect, 35–36
  perfect, 35
  of rotating cylinders, 430
  for tubing pumps, 301

**M**
machine tools, 398, 399
magma pumps, 276, 313
maintenance
  of centrifugal pumps, 149–152
  of corrosion-resisting
      centrifugal pumps, 157
  of hydraulic oil, 475–476
  of nonrotating cylinders,
      423–424
  of rotating cylinders, 430–432
manifolds, 505, 507
manifold-type mountings, 435,
  436
manual-type operators, 456,
  458–460, 461–462
matter, 3–4
mechanical balancing, 125–127
mechanical efficiency, 255–256
mechanical operators, 463–465
mechanical powers, 38, 40–46
mechanical seal, 198, 200
medical pumps, 273–274
mercury, 12, 21
metallic scraper, 410
metals
  heat conductivity of, 13
  linear expansion of, 15
  specific heat of, 12
  used for gear-type pumps,
      191–196
meter-in method, 443–444
meter-out method, 443–444
mill-type cylinders, 408, 409
mineral oils, 473

mixed-flow impellers,
  121–123
mobile-type filter units,
  485–487
moisture, effect on air
  compressors, 61
moment, 75
momentum, 33–34
motion, 61–62
  acceleration, 30
  apparent versus actual,
      29, 30
  constant, 32
  defined, 28
  and inertia, 34
  and momentum, 33–34
  Newton's laws of, 29–30
  oscillating, 31, 32
  reciprocating, 31–32
  resistance to (see friction)
  tangential, 30–31
  types of, 30
  variable, 32, 33
  velocity, 30
motor drives, 137
motors
  constant-displacement, 385,
      386, 388
  direction of rotation in, 211
  in Dynapower transmission
      system, 365
  fixed-displacement, 360–362
  hydraulic (see hydraulic
      motors)
  internal combustion engines,
      209
  pressure-compensated control
      of, 365
  for rotary pumps, 208–209
  selecting, 209
  for submersible pump, 294
  for sump pumps, 277, 280
  symptoms of problems with,
      159